# Geospatial Technologies for Effective Land Governance

Moha El-Ayachi
*Institut Agronomique et Vétérinaire Hassan II, Morocco*

Loubna El Mansouri
*Institut Agronomique et Vétérinaire Hassan II, Morocco*

A volume in the Advances in Geospatial Technologies (AGT) Book Series

IGI Global
DISSEMINATOR OF KNOWLEDGE

Published in the United States of America by
IGI Global
Engineering Science Reference (an imprint of IGI Global)
701 E. Chocolate Avenue
Hershey PA, USA 17033
Tel: 717-533-8845
Fax: 717-533-8661
E-mail: cust@igi-global.com
Web site: http://www.igi-global.com



Library of Congress Cataloging-in-Publication Data

Names: El-Ayachi, Moha, 1970- editor. | El Mansouri, Loubna, 1978- editor.
Title: Geospatial technologies for effective land governance / Moha El-Ayachi and Loubna El Mansouri, editors.
Description: Hershey, PA : Engineering Science Reference, 2018. | Includes bibliographical references.
Identifiers: LCCN 2017060575| ISBN 9781522559399 (hardcover) | ISBN 9781522559405 (ebook)
Subjects: LCSH: Land use--Morocco--Geographic information systems. | Land use--Morocco--Remote sensing. | Geospatial data--Morocco.
Classification: LCC HD972.Z63 G46 2018 | DDC 333.1/80285--dc23 LC record available at https://lccn.loc.gov/2017060575

This book is published in the IGI Global book series Advances in Geospatial Technologies (AGT) (ISSN: 2327-5715; eISSN: 2327-5723)

British Cataloguing in Publication Data
A Cataloguing in Publication record for this book is available from the British Library.

# Advances in Geospatial Technologies (AGT) Book Series

Nilanjan Dey
Techno India College of Technology, India

ISSN:2327-5715
EISSN:2327-5723

Mission

The geospatial technology field is a fast-paced, high growth industry that is involved in a variety of fields, including military planning, public health, land use, environmental protection, and Google Earth mapping. With such a diverse body of applications, the research in geospatial technologies is always evolving and new theories, methodologies, tools, and applications are being developed. **Advances in Geospatial Technologies (AGT) Book Series** is a reference source and outlet for research that discusses all aspects of geographic information, including areas such as geomatics, geodesy, GIS, cartography, remote sensing, and other areas. Because geospatial technologies are so pervasive in such a variety of areas, AGT also includes books that address interdisciplinary applications of the technologies.

Coverage

- Remote Systems
- Surveying Technologies
- Cartography
- Spatial Reference Systems
- Disaster Informatics
- Land Surveying
- Public Sector Use of Geospatial Technologies
- Digital Terrain Modeling
- Environmental Monitoring
- Geovisualization

IGI Global is currently accepting manuscripts for publication within this series. To submit a proposal for a volume in this series, please contact our Acquisition Editors at Acquisitions@igi-global.com or visit: http://www.igi-global.com/publish/.

The Advances in Geospatial Technologies (AGT) Book Series (ISSN 2327-5715) is published by IGI Global, 701 E. Chocolate Avenue, Hershey, PA 17033-1240, USA, www.igi-global.com. This series is composed of titles available for purchase individually; each title is edited to be contextually exclusive from any other title within the series. For pricing and ordering information please visit http://www.igi-global.com/book-series/advances-geospatial-technologies/73686. Postmaster: Send all address changes to above address. 

## Titles in this Series

*For a list of additional titles in this series, please visit: www.igi-global.com/book-series*

*Emerging Trends in Open Source Geographic Information Systems*
Naveenchandra N. Srivastava (Bhaskaracharya Institute for Space Applications and Geo-Informatics, India)
Engineering Science Reference • copyright 2018 • 302pp • H/C (ISBN: 9781522550396) • US $195.00 (our price)

*Geospatial Technologies in Urban System Development Emerging Research and Opportunities*
Alok Bhushan Mukherjee (Birla Institute of Technology, India) Akhouri Pramod Krishna (Birla Institute of Technology, India) and Nilanchal Patel (Birla Institute of Technology, India)
Engineering Science Reference • copyright 2018 • 179pp • H/C (ISBN: 9781522536833) • US $145.00 (our price)

*Spatial Analysis Techniques Using MyGeoffice®*
João Garrott Marques Negreiros (University of Saint Joseph, China & Macau Polytechnic Institute, China)
Engineering Science Reference • copyright 2018 • 336pp • H/C (ISBN: 9781522532705) • US $225.00 (our price)

*Handbook of Research on Geospatial Science and Technologies*
Joyce Gosata Maphanyane (University of Botswana, Botswana) Read Brown Mthanganyika Mapeo (University of Botswana, Botswana) and Modupe O. Akinola (University of Lagos, Nigeria)
Engineering Science Reference • copyright 2018 • 457pp • H/C (ISBN: 9781522534402) • US $345.00 (our price)

*Emerging GIS Applications for Emergency and Disaster Management*
Shruti Mantri (Kirti College, India) and Seema Purohit (Kirti College, India)
Information Science Reference • copyright 2017 • 117pp • H/C (ISBN: 9781683180098) • US $165.00 (our price)

*Volunteered Geographic Information and the Future of Geospatial Data*
Cláudio Elízio Calazans Campelo (Federal University of Campina Grande, Brazil) Michela Bertolotto (University College Dublin, Ireland) and Padraig Corcoran (Cardiff University, UK)
Information Science Reference • copyright 2017 • 357pp • H/C (ISBN: 9781522524465) • US $205.00 (our price)

*Remote Sensing Techniques and GIS Applications in Earth and Environmental Studies*
Abhisek Santra (Haldia Institute of Technology, India) and Shreyashi Santra Mitra (Haldia Institute of Technology, India)
Information Science Reference • copyright 2017 • 335pp • H/C (ISBN: 9781522518143) • US $185.00 (our price)

*Handbook of Research on Geographic Information Systems Applications and Advancements*
Sami Faiz (LTSIRS Laboratory, University of Tunis El Manar, Tunisia) and Khaoula Mahmoudi (LTSIRS Laboratory, University of Tunis El Manar, Tunisia)
Information Science Reference • copyright 2017 • 650pp • H/C (ISBN: 9781522509370) • US $315.00 (our price)

IGI Global
DISSEMINATOR OF KNOWLEDGE

701 East Chocolate Avenue, Hershey, PA 17033, USA
Tel: 717-533-8845 x100 • Fax: 717-533-8661
E-Mail: cust@igi-global.com • www.igi-global.com

# Table of Contents

**Section 2**
**The Geospatial Technology for Effective Natural Resources' Management**

# Detailed Table of Contents



**Section 1**
**Land Governance Concepts and Tools**



Unlike urban development, rural development is not getting adequate attention in Africa. This is true in the aspects of development policy and as a research interest. It is still not a priority in the development agenda of many African countries. The consequence is that the governance of rural development in many African countries is still treated as a linear affair. The consequence being a negation of the complexities that impede the improvement of living conditions in rural communities in Africa. This chapter is based upon an explorative study that challenges existing notions of rural development. It uses the structured literature review and focused group discussion conducted with African doctoral researchers in Germany to identify how land management can be used as an intervention for rural development. As a way forward, it presents a land managed-based approach to rural development that can be adaptable or applicable in African countries.



In a globalized world evolving into knowledge societies and economies, geographic information plays a very important role, as illustrated by the spatially enabled society (SES) concept. In this context, the role of a modern cadastre in the good governance of land for sustainable development is worth noting. The international and regional agendas and various initiatives related to these themes are examined to deduce the emerging global trends and the possible transformation of the national statistical. One of the prerequisites for an integrated information system is the availability of harmonized data and the interoperability between the various dedicated systems. This depends on the effectiveness of the national statistical system and the establishment of a national spatial infrastructure. The institutional and organizational arrangements for better land governance are discussed. The resulting technological infrastructure to be implemented is then addressed.



Moroccan land and real estate are characterized by a large number of properties and an advanced and varied typology. Among the related shortcomings is how to identify various land statuses and the stakeholders in the sector. These characteristics of land are multidimensional, including urban, suburban, and rural; land use plan; lands under formal and informal registration systems and multiple regulatory frameworks; the importance for the state, the public, and private individuals; regarding its heritage, urban planning, and taxation; the development projects in each sector; etc. The complex character and distinct fields of land administration in Morocco necessitates the creation of a unified and standardized land data model for use by all public and private actors who have a relationship to land. The present study contributes to the perception of an applied taxonomy and aims to define the basic concepts of the land domain ontology. Such expertise is a prerequisite for land governance and land policy that integrate land and real estate into national development processes.



For a decade, many countries have re-engineered their cadastral systems focusing closely on organizational and technological aspects in order to meet requirements of the new digital era. Appropriate configurations of the cadastral systems have been designed. The cadastral system, as a public institution, has a large relationship with all kinds of customers and includes a number of structural, operational, and functional processes that must be improved. This chapter has the aim to illustrate a new conceptual framework by describing the requirements and the technical perspectives of the Moroccan multipurpose cadastre. Basic concepts are discussed in accordance with various assumptions to provide a deeper understanding of the vision and the new structure. The development of an integrated system for administering land and the establishment of the cadastral information systems are carried out to highlight the fundamental aspects of the multipurpose cadastre.



A spatial data infrastructure (SDI) is a platform for coordinating the exchange and sharing of spatial data between several producers or users of spatially referenced data. In Morocco, there is a massive production of spatial data and several generally public administrations are starting to feel the need for geographic information governance through a mechanism of exchange and management of data to optimize their efforts and avoid a redundant production. The purpose of this chapter is to draw up an inventory of the state of the art of geo-spatial data, systems, and tools existing in the central administrations in Morocco in relation with the collection, management, storage, and dissemination of geographical information. Through this study, it was found that the problem is more a question of global governance, and that the current context has assets for the establishment of a spatial data infrastructure in Morocco.



This chapter highlights time series image processing for accurate agriculture characterization through two Moroccan experiences. The first case aims at crop mapping. A new classification approach based on multiple classifiers combination (MCC) was developed and applied to multi-temporal enhanced vegetation index (EVI) bands. The whole process is performed in three stages: (1) Landsat data preparation and multi-temporal staked EVI image extraction, (2) MCC construction from six advanced and supervised classifiers, and (3) stacked EVI image classification using the build-up MCC. Some post-classification contextual rules were also added in order to optimize the crops classification and the final parcel shape. In the second case, a post-classification change detection process was implemented to detect changes in forest area. Many classification schemes with different vegetation and texture indices were investigated. The two experiences are cost-effective, reproducible, and transferable. Consequently, they can regularly be used to produce up-to-date land use maps.



Recent geospatial technologies offer an infinite number of opportunities in the domain of land management and governance. This is due principally to the fact that new satellites allow the collection and monitoring of important information about soil, crops, weather, and climate. In this context, this chapter presents a review of studies conducted in three semi-arid plains in Morocco during the last two decades on the combined use of geospatial technologies, modelling tools, meteorological, and ground data to manage irrigated lands in semi-arid zones. The three studied regions are located in three different sites well distributed over the country. This chapter presents a simple and clear summary of what has been performed using geospatial technologies by Moroccan researchers to improve agricultural management in irrigated areas in Morocco. Such work could help policymakers in developing efficient land governance policies by taking into account the latest scientific results.



Irrigated agriculture is an important strategic sector for Morocco, contributing to food security and employment. Nowadays, irrigation scheme managers shall ensure that water is optimally used. The main objective was to support the irrigation monitoring and management of wheat in the irrigated perimeter using optical remote sensing and crop modeling. The potential of spectral indices derived from SPOT-5 images was explored for quantifying and mapping surface water content changes at large scale. Indices were computed using the reflectance in red, near infrared, and shortwave infrared bands. A field crop model (AquaCrop) was adjusted and tested to simulate the grain yield and the temporal evolution of soil moisture status. This research aimed at providing a scientific and technical approach to assist policymakers and stakeholders to improve monitoring irrigation and mitigating wheat water stress at field and irrigation perimeter levels in semi-arid areas. The approach could lead to operational management tools for an efficient irrigation at field and regional levels.



Land use land cover dynamics is a widespread phenomenon in many parts of Ethiopia and in Ribb watershed. The main objective of the research was assessing land use land cover dynamics and its implication to the sustainability of Ribb Dam in 1973, 1986, 2001, and 2016 by using Landsat image and household questioner. During the last 44 years, cultivated and settlement land and forest cover showed an increment from 26.29% to 54.89% and 9.45% to 12.86%, respectively. The bush land, grazing land, water body, and wetland, however, showed a relative decrement from 29.48% to 17.09%, 21.45% to 12.70%, 4.64% to 2.39%, and 8.70% to 0.08%, respectively. Population pressure, poverty, weak policy and institutional enforcement, and tenure insecurity revealed as a major cause of the change in land use land cover. Soil erosion, lack of fuel wood, and impact on livelihood are major consequences of land use land cover change.

Within the agricultural strategy called Green Morocco Plan launched by the Moroccan government, the funds are allocated by the agricultural development fund to encourage investments. In this context, to manage the grant requests, the agricultural department uses an information system, which processes the files from their submission until the payment of the farmers. However, this system does not take into account the spatial data. This research aims to develop a geospatial model to study the distribution and dynamics of the allocated fund. Through this approach, policymakers will have a strategic key to survey the behavior of the spatial component, to manage funding applications, and to better manage allowed funds. The methodological approach adopted consists of carrying out a mapping of the real world identifying the needs of the future system and modeling the process of the geo-spatial tool. The result of the model will be used to develop and implement a spatial tool for policymakers to improve land information management related to the funded agricultural project.



Forest resources management requires a variety of information related to social systems and to land and its supported resources and their dynamics (land cover, forest stocking, and growth). Such information is, by nature, spatio-temporal and scale dependent and its quality relay on costs for obtaining it. Geosciences and forest geomatics offer valuable methods for ensuring a good compromise between the quality of required information and its costs. This chapter will review and discuss the contribution of geoscience to forest and land inventory. After presentation of information needed and their acquisition methods, through traditional forest inventory, the chapter will focus on technologies aiming at forest resources characterization and assessment such as aerial photogrammetry, satellite imagery, LiDAR data.



Ourika basin in Morocco has very steep slopes with impermeable ground favoring water flows and flooding. History has shown deadly flood events. Floods are becoming recurrent and exacerbated not only by human activities that degrade soil and vegetation cover, accelerating erosion and quick water flows, but also by climate change. In fact, the basin has experienced a very strong dynamic of its vegetation cover, during the last 30 years, and has been subject to climate change impacts. This study is devoted to evaluating the impact of land cover change, mainly vegetation cover, on hydrological functioning of the basin. The HEC-HMS model was used to simulate basin hydrological response, according to two scenarios of land cover change. The first scenario simulates deforestation and urbanization impacts on peak flows, showing an increase of the peak flow by 31.68%. The second evaluates the impact of both reforestation actions and proscription of forest harvesting in the region. The simulated results showed a decline of 17.25% in peak flows, except for heavy precipitation events.

## Section 3
## Land Policy Assessment



The cadastral system as a civilian register of a property has a fundamental role in every society because it ensures the relationship between the land and the humankind. This relationship is evident in the form of property rights and established in many different manners from the full control, through communal forms of tenure, to the individual property rights. This chapter has the aim to deal with the nature and categories of the existing cadastral systems. It will highlight the new approaches and visions to establish cadastral systems that will meet the requirements of re-engineering the framework of land development projects. An overview of the state of the art of the modern concepts of cadastral systems will be provided such as Cadastre 2014, 3D Cadastre, and Marine Cadastre. The goal behind this approach is to enable decision makers to have an idea on the existing paradigms of land administration and to master the relationships between humankind and land, infrastructures and business systems, and between the human being and technological issues.



The land heritage held by the public authorities represents an important part of the Moroccan land system. The management of this land heritage obeys various logics arising from its functional, legal, and institutional complexity. The objective of this study is to identify the numerous logics that drive the management of land assets by various actors and to describe the impact of these logics on the land management. The adopted methodology is to analyze the theoretical aspects and functions of the land tenure, study the regulatory framework of the land and the territorial land, and deal with the institutional framework of the land heritage management. The achieved consequences of this study have enabled the authors to distinguish three logics: the conservative logic, the regulatory logic, and the promoting logic. The goal behind the chapter is to provide the stakeholders with a functional paradigm to take into account for creating the conciliation tools reinforcing the coordination to carry out the urban development of the country.

At independence, the leaders of the member states, in the search for ways of development of their young nations, have encountered the thorny problem of land management which was under the control of the customary regimes. West African land legislation is not sufficiently clarified and suffers from a legal dualism that hinders development and can lead to widespread conflicts. The increasingly visible development of a single regional market and the desire to create conditions for lasting social peace in the region make it imperative to secure the means of subsistence and to harmonize their governance. This work is an attempt to develop a benchmark study approach applicable to West African land. It represents a support for land reforms taking into account present and future land issues in the region. It is a methodology that develops both a set of indicators to evaluate and compare the performance of land practices as an approach to analyzing the results and proposing reforms.

# Foreword

Many stakeholders have been approaching for decades, in parceled way land administration, land planning, and land governance at the national, regional and international levels. Research and education tackled all these issues. Researchers could tackle and answer technical questions related to different aspects of Geospatial Sciences.

Hence, the present book entitled the Geospatial Technologies for Effective Land Governance is the outcome of a wide joint research of many experts and researchers. It is also the outcome of long years of professional experience of the authors, who are able to review, analyze and synthesize a large and dynamic field of expertise and innovations.

This is an issue of paramount importance as it, firstly, will specifically responds to governments, which demonstrate a great need to implement functional land administration and to develop harmonized and standardized practices. Specialized in the Geo-Spatial Sciences, I was able to steer his work towards Information technology tools and solutions in order to feed my field of expertise. The objective is to adopt new thoughtful perspectives that holistically approach land management, planning, administration and governance issues and make them respond to strategic sustainable development issues.

Secondly, the choice of the topic is very interesting and up to date in the sense that it digs into and responds to the "how" questions: how we can enhance the land and strategic planning, how we can approach and implement strategic land governance while using information technology tools and solutions. The aim of this work is to demonstrate the importance and efficiency of Land Information System (LIS) and Geo-Spatial Technologies (GST) use to achieve an effective land administration (land registration, cadastral systems and land planning) and governance at the national, regional and international levels.

Interestingly enough, the authors go over many case studies that review varieties of modern tools and solutions that enable land data collection, processing, visualization, and dissemination. They particularly highlight the use of LIS and GST to meet land administration and governance needs and challenges.

Furthermore, it is to underline that the publication intends to present the current contributions as related to this issue. It also considers the potential areas for developing capacity-building programs as related to specific and innovative issues. Many areas have been explored and examined such as spatial information managements, policy formulation and land development, national spatial data infrastructures for LIS/LIM initiatives, land data modelling, land data sharing and visualization, land management and climate change, spatio-temporal land information management, land information search engines, real time and iterative land information analysis, LIS/GIS applications and land information web based technology.

In addition, the authors asserts that effective land administration can be ensured using: remote sensing for land use assessment, land mapping strategies, land consolidation, Unmanned Aerial Vehicle in cadastral mapping, Mobile LIS, and Land Assessment and valuation. Thus, the book brings major contributions to this field of study by making applied research and science concretely serve development issues such as agriculture and food production, water distribution and access to resources.

Finally, the contributors intend to develop a network of educational experts, scientists who will be in charge of disseminating these concepts and putting them into practice using the already available national regional and international frameworks and research institutes and centers. The ultimate goal is to improve land administration at the national, regional and strategic levels, foster land tenure security, and develop more opportunities for investments contributing by that to sustainable development.

*Nabila Zouhiri*
*University of Ain Chock, Morocco*

**Nabila Zouhiri** *has a Bachelor in Social Sciences majoring in Global Interdependence from Al Akhawayn University in Ifrane, Morocco, a Master in Environmental Sciences from Lund University in Sweden and is involved as researcher in PhD studies in Public law and political sciences at the University of Ain Chock in Casablanca, Morocco. She also worked as program officer with many international institutions and organizations such UNDP (2002-2004) and the embassy of Switzerland (2008-2009). She worked as international consultant with Trelleborg Company in Sweden (1999-2000), the embassy of Sweden in Morocco (2013), the Netherlands Red Cross Hague (2005-2006), Marko-Fonds in Amsterdam (2006-2007). She worked with the top ranking representatives of civil society in Morocco (OMDH – 2010-2012). She participated in many research studies and fieldwork as related to environmental, human rights and social issues (1999-2018). She is specialized in strategic planning and in design, monitoring and evaluation of development projects. She organized many trainings in M&E to the Moroccan civil society. She worked as programs officer, communication officer and studies inspector at the Institute of Agriculture and Veterinary Sciences in Rabat-Morocco (2011-2017). She got UN certificate of English and French and the ENLIGHT International Computer Driving License certificate (UNDP-2002). At last and not least, she got the African prize of the best expert in development (2017) and participated in the Young Arab leaders legislative fellowship program that was organized by the US State Department (2012). She currently works as a regional GIZ Cooperation advisor on SLGA Land Governance Programme since 2018.*

# Preface

The land and its resources must be put to good effect. In order to meet the requirements of the new era, every state has developed and implemented a series of emerging policies according to its national specificities and to the international regulations and trends.

The formalization of land rights has begun early and allowed significantly increasing the participation of individuals and groups in the formal land and labor markets. Landowners with the lowest incomes can borrow their lands to others without fearing to lose their ownerships.

What are the challenges that constrain allowing the existing systems to integrate rights of individuals and groups in formal and informal actions to reduce transactions' costs made by investors?

Do the effective land governance implemented in some countries contribute in reducing the danger of wasting natural resources caused by deforestation or ecological destruction?

What are the offered opportunities by the regulations on land governance to enable authorities transferring land to communities and individual farmers and defining the restrictions on using forest and abandoned areas?

Land as a fundamental element in humane life provides major opportunities for farming, housing, urban planning, and financing. While Land governance has the aim to permit more security for investors, to enhance the supply of credit, to provide the social stability, and to ensure sustainable development.

However, the diversity of land data, data sources, and data users, is affecting the functionalities of administrating and managing land, which are becoming increasingly restrictive.

The use of the geospatial technologies such as GNSS, remote sensing, and GIS will enable producing, manipulating, and displaying land information in easier manner. It can provide communities and decision makers with modern and enhanced tools to interact with their environment and enable them to better evaluate and distribute the natural resources.

Land administration, encompassing both the cadastral systems and land registration, plays an important role in society. Land registration is the process of officially recording rights in land through deeds or title. It provides an official record of the rights in land registers, responsibilities and restrictions on land, and the relationship of the people to the land. It may be established for fiscal purposes (valuation and taxation), legal purposes (conveyancing), or to assist in the management of land and land-use planning and administration(multipurpose).

It is necessary to develop a land governance strategy to establish a methodically arranged public inventory of data related to land resources within a certain country or district, based on a survey of their locations and boundaries. Many developing and developed countries have known a phenomenal growth in population, which has increased demand on land and land-based conflicts in the urban and rural communities.

Thus, it is pressing to develop a strategy-based land governance and formalize land rights to sustain continuous use of land resources. Indeed, it is required to implement a functional land governance system that improves land tenure security and promote greater incentives to investments.

Therefore, there is an astonishing need to promote economy growth by intensifying, modernizing, and diversifying designed solutions.

The adopted tools to reach the goal of land resources management and monitoring are fundamentally based on the geospatial technologies. These tools enable sustainable development and environmental improvement based on secure land, which is a key condition to achieve the promotion of food production and fight land conflicts.

As part of the land governance enhancement, it is necessary to integrate the geospatial technology to implement nationwide land information systems-based GIS that ensures the consistent and accurate collection of all spatial data.

The book will focus on the impact of land transaction practices. The impact-practice analysis will enable us to identify effectively the best opportunities for investors and for the government to facilitate sharing experiences in insuring the achievement of various development projects. It is obviously important to analyze and understand the ongoing policies linked to land governance.

The book is intended to encompass both land governance concepts and the geospatial technologies applications through a series of chapters dealing with case studies and providing concise results.

It has the aim to outline the diversity of concepts and analyze the technology capabilities used in land governing systems and their role as key factors leading to achieving social equity and economical outcomes. The book will highlight features of the adopted and implemented instruments to guarantee efficient land governance.

Regarding the domain of the geospatial technology (GST), a bunch of tools and solutions have been developed enabling land data collection, processing, visualisation, and dissemination. the GST has evolving rapidly and brought modern solutions in land governance. The book will present the current releases conducted by researchers throughout the world.

The researches and studies on the assessment of experiences integrating the geospatial technology for an effective land governance will have a potential impact on the new societies of the digital era.

The assessment covers the identification of the potential uses of land information, the forms of managing and maintaining land data, the needed personal resources, and the possible prospective advantages through an introduction of the geospatial technologies.

The book will affect mainly the domain of research in two ways. First, it will produce a general overview of several experiences that enrich the knowledge on land governance. This will enable students and researchers taking advantage of the pool of gathered knowledge. Secondly, the book will be the unique reference dealing both with the geospatial technology and land governance. This will provide scientists and professionals with a large chain of studies reflecting the last and modern achievements in using the geospatial technologies for effective land governance.

## THE TARGET AUDIENCE OF THE BOOK

The book is focusing to reach various readers and researchers interested by land information and land governance matters. It serves scientists welling to study the impact of integrating geospatial science for effective land governance.

It will be a reference for the academic institutions providing curricula on geomatics, surveying, mapping, and land policy issues related to the use of the geospatial technology. The professionals such as surveyors, chartered surveyors, quantity surveyors, and land assessors will take benefits of its content.

The specialists and scientists that implement solutions using the new geospatial technology will find this book as a tool enhancing land information mining, collecting, processing, and visualization.

For the land policy decision makers, this book will play a major sensitive role for them to improve land management by adopting modern solutions and tools based on the geospatial science.

The book has the goal to build a pool of educational experts and scientists who will disseminate the concepts of integrating geospatial technologies in land governance. There is a need develop harmonized and normalized practices in order to encourage integrating modern tools and solutions in the improvement of land administration. In order to reach this goal, three major areas are identified.

The first area is dealing with the land governance concepts and tools, which encompasses five (5) chapters. The second area is related to the geospatial technology for effective natural resources' management developed around seven (7) chapters. While the last part is dealing with an area linked to the Land policy assessment. This encompasses three (3) chapters.

## Land Governance Concepts and Tools

Chapter 1 dealing with the concept and Approach of Land Management interventions for rural development in Africa highlights the interest of conducting studies in the rural development. This is true in the aspects of development policy and as a research interest. It is still not a priority in the development agenda of many African countries. The consequence is that the governance of rural development in many African countries is still treated as a linear affair. The consequence being a negation of the complexities that impede the improvement of living conditions in rural communities in Africa. This chapter is based upon an explorative study that challenges existing notions of rural development. It uses the structured literature review and focused group discussion conducted with African doctoral researchers in Germany to identify how land management can be used as an intervention for rural development.

Chapter 2 about the Land Governance and The Information Society, the author presents and illustrates the concept of the Spatially Enabled Society (SES). In this context, the role of a modern cadastre in the good governance of land for sustainable development is worth noting. The international and regional agendas and various initiatives related to these themes are examined to deduce the emerging global trends and the possible transformation of the national statistical. One of the prerequisites for an integrated information system is the availability of harmonized data and the interoperability between the various dedicated systems. This depends on the effectiveness of the national statistical system, and the establishment of a National Spatial Infrastructure.

Chapter 3 describes the Moroccan Land Administration Domain - Ontology Model and Land Governance. The characteristics of land are multidimensional, including: urban, suburban and rural; land use plan; lands under formal and informal registration systems and multiple regulatory frameworks; the importance for the state, the public, and private individuals; regarding its heritage, urban planning and taxation; the development projects in each sector; etc. The complex character and distinct fields of land administration in Morocco necessitates the creation of a unified and standardized land data model for use by all public and private actors who have a relationship to land. The present chapter contributes to the perception of a taxonomy applied to the MLAD and aims to define the basic concepts of the land domain ontology. Such expertise is a prerequisite for land governance and land policy that integrate land and real estate into national development processes.

Chapter 4 presents the conceptual framework for implementing a multipurpose cadastral system. This chapter has the aim to illustrate a new conceptual framework by describing the requirements and the technical perspectives of the multipurpose cadastre. Basic concepts are discussed in accordance with various assumptions to provide a deeper understanding of the vision and the new structure. The development of an integrated system for administering land and the establishment of the cadastral information systems are carried out to highlight the fundamental aspects of the multipurpose cadastre.

Chapter 5 describes and analyzes the Spatial Data Infrastructure (SDI) as a platform for coordinating the exchange and sharing of spatial data between several producers or users of spatially referenced data. The purpose behind this chapter is to draw up an inventory of the state of the art of geo-spatial data, systems and existing tools regarding the central administrations in Morocco in relation with collecting, managing, storage and dissemination of geographical information.

## The Geospatial Technology for Effective Natural Resources' Management

The Chapter 6 highlights time series image processing for accurate agriculture characterization through two case studies. The first case aims crop mapping where a new classification approach based on multiple classifiers combination (MCC) was developed and applied to multi-temporal enhanced vegetation index (EVI) bands. The whole process is performed in three stages: (1) Landsat data preparation and multi-temporal staked EVI image extraction, (2) MCC construction from six advanced and supervised classifiers and (3) stacked EVI image classification using the build-up MCC. Some post-classification contextual rules were also added in order to optimize the crops classification and the final parcels shape. In the second case, a post classification change detection process was implemented to detect changes in forest area. Many classification schemes with different vegetation and texture indices were investigated. The two experiences are cost effective, reproducible and transferable. Consequently, they can regularly be used to produce up-to-date land use maps.

Chapter 7 treats an integrated approach based on the geospatial technics, modelling, meteorological, and ground data to improve crop management in semi-arid zones. Recent new satellites allow collecting and monitoring various information on soil, crops, weather and climate. In this context, the chapter presents a review of conducted studies in three semi-arid plains during the last two decades. The three studied regions are located in three different sites well distributed. The chapter presents a simple and clear summary of what has been performed using geospatial technologies by researchers to improve agricultural management in irrigated areas in Morocco. Such work could help policy makers in developing efficient land governance policies by taking into account the latest scientific results.

Chapter 8 has the objective to highlight the role of the optical remote sensing and crop modeling to support irrigation monitoring and management of wheat in the irrigated perimeter using the potential of spectral indices derived from SPOT-5 images that was explored for quantifying and mapping surface water content changes at large scale. Indices were computed using the reflectance in red, near infrared and shortwave infrared bands. A field crop-model (AquaCrop) was adjusted and tested to simulate the grain yield and the temporal evolution of soil moisture status. The chapter provides a scientific and technical approach to assist policy makers and stakeholders to improve monitoring irrigation and mitigating wheat water stress at field and irrigation perimeter levels in semi-arid areas.

Chapter 9 presents a research conducted in Ribb dam in South Gondar of Ethiopia. The main objective of the research is to assess land use land cover dynamics and its implication to the sustainability of Ribb dam in 1973, 1986, 2001, and 2016 by using Landsat image and household questioner. During the last 44 years, cultivated and settlement land and forest, cover showed an increment. The bush-land, grazing land, water body and wetland, however, showed a relative decrement. Population pressure, poverty, weak policy and institutional enforcement, and tenure insecurity revealed as a major cause of the change in land use land cover. Soil erosion, lack of fuel wood and impact on livelihood are a major consequence of land use land cover change.

Chapter 10 presents a Geo-Decisional Tool to manage the agricultural development Fund under the Green Morocco Plan. The chapter aims to develop a geospatial model that enables analyzing the spatial distribution and dynamics of the allocated fund. Through this approach, policymakers will have a strategic key to survey the behavior of the spatial component, to manage funding applications, and to better manage allowed funds. The methodological approach adopted consists, first, of carrying out a realty check study to identify the needs of a modern system. The result of the model will be used to develop and implement a spatial tool policy maker to improve land information management related to the funded agricultural project.

Chapter 11 reviews and discusses the contribution of geoscience to forest and land inventory in order to improve assessing forest resources to sustain their management. Forest resources management requires a variety of information related to social systems and to land and its supported resources and their dynamics (land cover, forest stocking and growth). Such information is, by nature, spatio-temporal and scale dependent and its quality relay on costs for obtaining it. Geosciences and forest geomatics offer valuable methods for ensuring a good compromise between the quality of required information and its costs. the chapter will focus on technologies aiming at forest resources characterization and assessment such as aerial photogrammetry, satellite imagery, LiDAR Data.

Chapter 12 is devoted to evaluate impact of land cover change, mainly vegetation cover, on hydrological functioning of the basin. The HEC-HMS model was used to simulate basin hydrological response, according to two scenarios of land cover change. The first scenario simulates deforestation and urbanization impacts on peak flows, showing an increase of the peak flow by 31.68%. The second evaluates the impact of both reforestation actions and proscription of forest harvesting in the region. The simulated results showed a decline of 17.25% in peak flows, except for heavy precipitation events.

## Land Policy Assessment

Chapter 13 has the aim to deal with the nature and categories of the existing cadastral systems. It highlights the new approaches and visions to establish cadastral systems that will meet the requirements of re-engineering the framework of land development projects. An overview of the state of the art of the modern concepts of cadastral systems are provided such as cadastre 2014, 3D Cadastre, and Marine Cadastre. The goal behind this approach is to enable decision-makers to have an idea on the existing paradigms of land administration and to master the relationships between humankind and land, infrastructures and business systems, and between the human being and technological issues.

Chapter 14 has the objective is to identify the numerous logics that drive the management of land assets by various actors and to describe the impact of these logics on the land management. The adopted methodology is to analyze the theoretical aspects and functions of the land tenure, study the regulatory framework of the land and the territorial land, and deal with the institutional framework of the land heritage management. The achieved consequences of this study have enabled us to distinguish three logics the conservative logic, the regulatory logic, and the promoting logic. The goal behind is to provide the stakeholders with functional paradigm to take into account for creating the conciliation tools reinforcing the coordination to carry out the urban development of the country.

Chapter 15 is an attempt to develop a benchmark study approach applicable to West African land. It represents a support for land reforms taking into account present and future land issues in the region. West African land legislation is not sufficiently clarified and suffers from a legal dualism that encumbers development and can lead to widespread conflicts. The increasingly visible development of a single regional market and the desire to create conditions for long-term social peace in the region make it imperative to secure the means of subsistence and to harmonize their governance. The adopted approach is based on developing a set of indicators to evaluate and compare the performance of land practices as framework of proposing adequate reforms.

*Moha El-Ayachi*
*Institut Agronomique et Vétérinaire Hassan II, Morocco*

*Loubna El Mansouri*
*Institut Agronomique et Vétérinaire Hassan II, Morocco*

# Acknowledgment

We are grateful to all those with whom we have the pleasure to work during the entire process of this book edition. We would like to acknowledge everyone who played a role in this project accomplishment.

To our book publisher IGI Global for providing us good guidelines of assignment throughout numerous consultations. Thank you for bringing our book to life. We are especially thankful to Mariah for all of her edition assistance throughout this project and particularly in the final steps of the process.

To our chapters' authors who are devoted to enrich the book. We are extremely honored and very motivated by your confidence, and we take it with a great sense of responsibility. This book is also yours.

To the reviewers providing their precious time to read carefully the book' chapters. We are indebted for their valuable contributions regarding the enhancement of scientific quality, coherence, and content presentation of chapters. Most of the authors also served as referees; we highly appreciate their double task.

To many referees and institutions for their valuable help and support in the whole implementation of the book. We would like to thank everyone who provided intellectual assistance, technical help, or special equipment or materials or being a source of motivation. We would also like to expand our deepest gratitude to all those who have directly and indirectly guided us in editing this book.

*Moha El-Ayachi*
*Institut Agronomique et Vétérinaire Hassan II, Morocco*

*Loubna El Mansouri*
*Institut Agronomique et Vétérinaire Hassan II, Morocco*

# Section 1
# Land Governance Concepts and Tools

# Chapter 1
# Concept and Approach to Land Management Interventions for Rural Development in Africa

**Uchendu Eugene Chigbu**
*Technical University of Munich, Germany*

## ABSTRACT

*Unlike urban development, rural development is not getting adequate attention in Africa. This is true in the aspects of development policy and as a research interest. It is still not a priority in the development agenda of many African countries. The consequence is that the governance of rural development in many African countries is still treated as a linear affair. The consequence being a negation of the complexities that impede the improvement of living conditions in rural communities in Africa. This chapter is based upon an explorative study that challenges existing notions of rural development. It uses the structured literature review and focused group discussion conducted with African doctoral researchers in Germany to identify how land management can be used as an intervention for rural development. As a way forward, it presents a land managed-based approach to rural development that can be adaptable or applicable in African countries.*

## INTRODUCTION

In Africa, the search for better living conditions within and across countries is increasing. Within countries, the majority of the population that live below poverty lines (and who move to cities in search of economic opportunities) come from rural areas. Also, the majority of the over 240 million international migrants that move across countries in search of physical security come from rural areas. The most practical way to improve these situations is to improve food security, reduce poverty, and make the rural environment livable for those that dwell in rural areas. That is why sustainable rural development remains the most direct and effective way to improve the lives of the rural people and living conditions in the rural place.

DOI: 10.4018/978-1-5225-5939-9.ch001

The debate on how to deliver rural development in Africa has been going on for more than five decades. It is a debate that has taken theoretical, ideological, conceptual and methodological angles. Despite the varied discourses on rural development in Africa, embracing responsible land management as an intervention to rural development remains unexplored. Scholars who have written about rural development have mostly viewed it from different aspects, depending on their school of thought and development philosophies (or ideologies). Some have written about it from the perspectives of modernization (Klinghoffer, 1973), basic needs (Palmer, 1977), local government administration (Rondinelli, 1979), participation (Cohen and Uphoff, 1980) and agricultural development (Bryceson, 1999). Others who have viewed it from multifaceted approaches (Phuhlisani, 2009), development policy (Alemu, 2012), and cultural development (Chigbu, 2013a). Some have viewed it from the lens of place attachment (Chigbu, 2013c), women's empowerment (Chigbu, 2015a), ruralization (Chigbu, 2015b) and community development (Chigbu et al., 2018).

Considering that many rural areas in Africa are dependent on land-based livelihood options, land management is crucial to rural development. There are scholars who have conceptualized rural development approaches from land management perspectives (see Chigbu, 2013b; Magel, 2014). However, it was only of recent that de Vries and Chigbu (2017) introduced the concept and practice of responsible land management and applied it in the context of rural development in Ghana and Rwanda. In furtherance of the application of responsible land management to rural development, this study argues for a rural development approach that is based on responsible land management as an intervention. The study is based upon a conceptual research that challenges existing notions of rural development and embraces the role of land management in achieving rural development objectives (in African countries) in a responsible way. The essential elements of the study are that it emphasized the theoretical and practical issues that make responsible land management a practicable intervention for rural development in Africa.

## QUALITATIVE EXPLORATIVE METHODOLOGY

This study is best described as an exploratory methodology because it investigates questions that that have not been clearly studied in the context of Africa. It serves as an exploratory effort to show the application of land management as an intervention in rural development. The use of quantitative methods based on exploratory techniques is known to allow for freedom of defining research boundaries and terms of operations (Chia and Lee, 2017). However, it has the limitations of not being "sufficient in practical aspects" because "there is no comprehensive analytical system dedicated to this issue, as well as research regarding this subject" (Czechowski et al., 2018: 1010). The qualitative explorative methodology has been adopted for this study because its objective is on presenting a pathway for action (land management intervention) rather than put emphasis on the specifics of carrying that action. The study emanated from a rural conceptualization forum on Africa with African doctoral researchers in land management at the Technical University of Munich, Germany. The methodological approach to the study involved the use of Focus Group Discussion (FGD) to collect data on the scope, nature, and role of Land Management at the local level in Africa. The FGD specifically dealt with two major questions. (1) What are the developmental definitions of Africa, land management and rural development in African context applications? (2) How can land management serve as an intervention for rural development in Africa? The first question was answered on the verge of exploring a methodology for answering the second question. In answering these questions, the discussants based the FGD on two critical definitions (and models) pertaining to land management and rural development in Africa—the socio-scientific model of land management, the generalized definition of rural development, and the situation of Africa as a heterogeneous spatiality.

## Adapting Socioscientific Model of Land Management

This study views land management as the aspect of land profession that deals with "science and practice related to the conceptualisation, design, implementation and evaluation of socio-spatial 'interventions', with the purpose to improve the quality of life and the resilience of livelihoods in a responsible, effective, efficient, consensual and smart manner" (de Vries and Chigbu, 2017: 66). This definition considers land management as a knowledge-based intervention for enforcing rural development. "Intervention" as has been used here refers to "both the activity of changing a set of physical objects, as well to the process of consultation and deliberation resulting in a legitimate or legal decision" (de Vries and Chigbu, 2017: 66). As a consequence of adopting the above definition of land management, this study adopted a land management model that is based on the socioscientific concept and praxis. The socio-scientific perspective of land management "encompass social dilemmas with conceptual or technological links to science" (Sadler, 2004: 513). This conception of land management entails viewing land management as a set of interventions that can influence all social and technical aspects of the African society. This, according to de Vries and Chigbu (2017: 66) can be represented as a change caused by land management, ΔLM, which can lead to or relate to a combination of 6 types of changes or adaptations:

- Governance → ΔG
- Land, property, real estate, land use Law → ΔL
- Social-spatial relations → ΔS
- Economic opportunities/dependencies → ΔE
- Perceptions/beliefs) → ΔP
- Behaviour → ΔB

These changes or adaptations by land management interventions (listed above) "form a combination of changes that must occur for rural development to happen. Mathematically put, if an intervention in land management is qualified as ΔLM, then ΔLM is a function of (or depends on, or yields) the respective changes in all or any of ΔG, ΔL, ΔS, ΔE, ΔP, ΔB. In short form:

$$\Delta LM = f(\Delta G, \Delta L, \Delta S, \Delta E, \Delta P, \Delta B)$$

This short form can be understood as a conceptual model for creating changes that can lead to rural development. These rural changes can manifest in the form of changes in governance, land (including property, real estate, and land use law), social-spatial relations, economic opportunities and dependencies, perceptions and beliefs, and behavior. Hence, the ΔLM model reflects the need for land management in Africa because its six components—ΔG, ΔL, ΔS, ΔE, ΔP, ΔB—are what drive the development of land issues in Africa. Factoring these six components in rural development is therefore mandatory "for the contextualization, analysis, and understanding of key land issues, the achievements and shortcomings of existing land administration policies, systems and programmes" in rural Africa (African Land Policy Initiative, 2016: 51). This is a fundamental deviation from the perspectives of land management defined by the Bathurst Declaration (1999), Carter (2002), Whitlock (2004), Shepherd et al. (2015), Keesstra et al. (2018), among many others.

## Viewing Africa as a Heterogeneous Spatiality

In order to put the FGD outcomes into context, it was necessary to ascribe a specific description or definition to Africa. This is important because land management and rural development, as human activities, can only be implemented within defined geographical, territorial, sociocultural, political or/and administrative domains. Hence, agreeing on a specific identity (notion or concept) of Africa was necessary in order to avoid the mistakes of the past—situations where "many development initiatives and discourses have treated it (Africa) as a homogeneous unit" (Chigbu et al., 2018: 99). As a geographical unit with a size of more than 30 million km$^2$, Africa could contain the geographic size of the United States, China, and India, "and still leave room enough to accommodate Europe" (Chigbu et al., 2018: 99). This makes the continent massive in terms of spatiality, and diverse in terms of people, cultures and development aspirations. A major point of note is that:

"Africa is not a language, or a religion, or a race, or a government. Consequently, it is not a culture and there is nothing homogenous about it. But we speak of African languages, African religion, and African culture exactly in the same way we speak of European languages, European religion, and European culture. We all know that German culture is not identical with the English… In spite of these discrepancies in European religion, government, and languages as well as in social customs, we still have what we like to call European culture. So it is with Africa". ~ (Ojike (1946: 5)

This study recognizes that Africa is racially, ethnically, linguistically, geographically, economically, culturally and politically diverse. The nomenclature, Africa, "is a term used in identifying all indigenous and migrant ethnicities living in the second biggest continent in the world" (Chigbu et al., 2018: 99). Hence, the study acknowledges that land management (as an intervention) and rural development (as a development process) cannot be uniform or be applied homogeneously across Africa. The recommendations presented in this study are highly generalized for specific adaptability in any part of Africa. The implication of this is that in applying the ΔLM model, all the six components (ΔG, ΔL, ΔS, ΔE, ΔP, ΔB) may differ from country to country, region to region, and community to community. They may, therefore constitute qualitative (and quantitative) confounding variable that can have direct (and hidden) effects on development outcomes.

## Embracing a Highly Generalized Definition of Rural Development

The definition of rural development is already highly contentious. This makes specificity in scholarly definitions more acceptable. However, in a study that takes a cursory look at Africa, it is more realistic to be general because Africa is highly diverse geography. In determining what should constitute a definition of rural development in Africa, a starting point is to grasp what development and rural areas mean in African contexts. This study considers Africa's rural areas as "land-spaces with culturally defined identity; situated within a place statutorily recognized as non-urban, and occupied by settlers predominantly depending on primary sources of labour for their livelihood" (Chigbu, 2013b: 10-11). This definition fits the understanding of rural areas in Africa because the meaning is fluid and can be viewed differently—in legal, social, spatial, administrative and demographic terms (and a combination of these—in different African countries. Development in this study means any improvement in the living conditions of people and/or the environment of a people in qualitative or quantitative terms (or both). This highly generalized description of development is important in Africa because different people, places, and communities may define what constitutes development differently (or indifferently). However, a common element of any definition of development in Africa is that it has to lead to an improvement in the living conditions of people and/or their environment. The description of rural development, as derived from the meaning of development and rural areas, entails an improvement in the living conditions of people in rural areas and the rural areas. Therefore, this study defines rural development as:

*All activities are done for improving the socio-economic conditions of all rural people, and the environmental conditions of their locality (Chigbu, 2013b: 12).*

The above definition provides a perspective of what rural development means in Africa. By socio-economic conditions, this definition covers improvements in the cultural, economic, political, social and technological needs of all rural people. By mentioning environmental conditions of (the people's) locality, the definition caters for all other conditions that improve the rural place. The definition puts into consideration the role land management plays in the rural development process. Rural areas are not developable without encountering various land-related (local) issues. From the perspective of land management, it is deducible that environmental, social and economic sustainability are some of the reasons for developing African rural areas.

## LAND MANAGEMENT BASED APPROACH TO RURAL DEVELOPMENT (RD) IN AFRICA

Land as shown from the concept and praxis of land management as an intervention, land management provides multifaceted opportunities as a primary instrument of rural development (RD). For this study, the FGD with African experts in land management provided an understanding of land management functions in an African context. It confirmed that in Africa, land management has functions that cut across multidisciplinary and multi-sector issues that are linkable to natural resource governance, policy, ownership, and use. This linkage also relates to issues of traditions and cultural value of rural structures and ways of living and the arrangement of physical infrastructure and management of the rural environment. Land management helps in establishing an effective and efficient real estate market, providing natural resource information, and making appropriate development planning. The sustainable use and sharing of natural resources is an important component of land management. When well instituted (planned and implemented), land management has the potential to make rural spaces (such as villages) functional living and working places by balancing the presence of farm and non-farm activities. For details on the best practices in implementing land management (or its aspects) as a tool for rural development, see Wilson (1998), Pašakarnis and Maliene (2010), Chigbu (2012), Asiama et al. (2017 and 2018), Paul and wa Gĩthĩnji (2018), among others.

All the possible functions of land management within a rural place make it a veritable tool for rural development in Africa. For this to be realizable in Africa, a land management based approach to rural development is need to be put into practice. The question this study must answer is, how can land management serve as an intervention for rural development in Africa? In answering this question, it is important to note that land management at provides conceptual, strategisation and implementation approaches for rural development in Africa. In order for it to serve as a tool for rural development in any African country, the success of its application (and form) and adaptation will depend on the institutional structure, social structure and the local development needs of the people in the country in which it is to be applied in Africa. Ideally, land management as a tool for rural development can be best explained by looking at three aspects—the conceptualization, strategisation, and implementation phases—that comprise of seven major activities.

This study refers to this interaction of these phases and activities as a land management based approach to rural development. A land management approach to rural development is a rural development process that incorporates land management functions as a primary base for human settlement improvements. Such an approach to rural development is relevant in Africa because most African communities are dependent on land (and natural resource) based livelihood activities such as agriculture, and water and forest-related dependencies. What this study does is to provide a path through which land management interventions can serve as a conduit to rural improvements beyond agricultural development, and towards the socio-spatial development of rural areas. This is possible in Africa if the following phases (and activities) of rural development are designed to elicit the aforementioned functions of land management within the rural geographies.

A land management based approach to rural development can be described in many formats. However, in this study, it is described as a cyclical process comprising of 3 phases, and beginning with conceptualization to implementation, through a stratification phase. Each of the three phases consists of a collection of activities. This is explained in the following subsections (with reference to Figure 1 and following the numbering of the phases and activities).

*Figure 1. Phases and activities for a land management based approach to rural development in Africa*

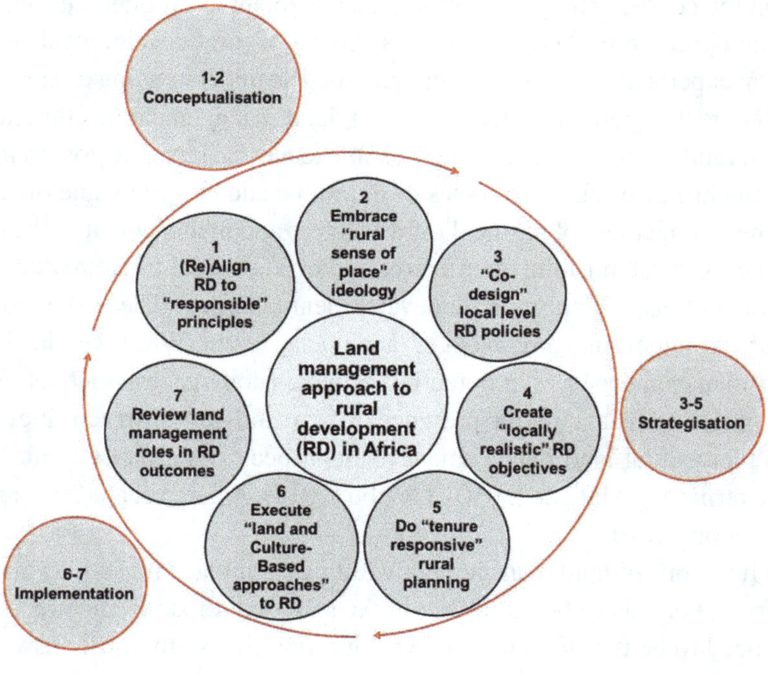

## Conceptualization Phase of Land Management Intervention for Rural Development

The conceptualisation phase of rural development intervention involves a stage of creation that involves aligning rural development ideas with responsible principles and embrace of a rural sense of place ideology. The land is a crucial factor of development, rural livelihood and national development in Africa. That is why the conceptualization of rural development has to be based on principles that are responsible. A responsible principle is a belief (and behavior) channeled to words improving the living conditions of people (de Vries and Chigbu, 2017). Responsible principles should support the promotion of an ideology of rural sense of place is "the relationship people share with their homeland (place of habitation), usually based on their feelings to land-based practices, cultural patterns, values, attitudes, traditions, and customs they share in connection with their environment" (Chigbu, 2013c: 266). As an ideology, adhering to a sense of place will help engender strong people-to-place relationship, as well as people's attachment to place. When rural development is conceived this way, then there is need to go into strategizing towards

## Strategisation Phase of Land Management Intervention for Rural Development

The strategisation phase of rural development intervention involves a stage choice-making in terms of what should be done and what should not be done in the course of achieving rural development outcomes. A land management based approach should involve three major actions. They are co-designing local level rural development policies, creating locally realistic rural development objectives and doing tenure responsive rural planning respectively.

Strategisation of rural development in the land management context should begin with a co-design of local level rural development policies. Co-designing or co-producing rural development objectives is important because people in rural areas are the key stakeholders in the rural areas. People are also the primary victims of underdevelopment—and the primary beneficiaries of development—in rural areas. Land is the primary asset in rural areas. Decisions on the development of rural areas (or rural land) should be made together with the people that matter most (that is, the rural people). It is particularly important to co-design policies (rules, regulations, and bye-laws) that embrace land and natural resource tenure security; and which can lead to set realistic objectives for rural development.

Creating locally realistic rural development objectives should be one of the off-shoots of co-production. Rural development objectives are those outcomes that rural development actions sought to achieve. An ideal objective of rural development in Africa would be the improvement of rural peoples' livelihoods (and living conditions). Considering that Africa is largely land and natural resource dependent, a realistic path to achieving this objective should be through sustainable land and natural resource management (backed by environmental protection). However, specific objectives of any rural area in Africa should focus on fulfilling local needs. When a local need is pursued in accordance with priorities agreed by all, and if it is basic for making the greatest positive impact on peoples' lives, then it constitutes a realistic local objective.

After a realistic objective is made, there is a need to conducting tenure responsive rural planning. A tenure responsive rural planning involves designing strategic forethoughts and strategies that are based on improving land tenure security of people in rural areas. "Land tenure defines the way people hold, own and enjoys rights to land, so it reflects local realities (regarding people-to-land relationship) in any society in legal and social terms" (Chigbu et al., 2017: 1624). In the course of strategising for rural development, the challenges encountered as result of relationships (especially, tenure insecurity) should be targeted for improvements. Examples of land tenure issues in rural areas that must be factored into rural planning include loss of land and natural resource uses, rights, privileges and ownership and various rights people exercise over land. This is why land management and land tenure related issues should be factored in the planning of rural areas to limit these challenges. If these crucial land issues are factored into planning, it is important that they become implemented for them to have desired impacts on rural people.

## The Implementation Phase of Land Management Intervention for Rural Development

The execution phase of rural development intervention involves a stage of converting strategies into concrete activities that can create direct impact on people and the rural environment. This can be done in different ways and approaches. However, in the context of land management interventions, a direct approach would be to follow land and culture-based approaches to rural development; and then continue with reviewing (and assessing) land management impacts after execution activities.

A major aspect of rural development implementation in Africa is the provision of infrastructure, social services, utilities and maintenance of amenities. However, a missing link in rural development in the continent tend to be the lack of efforts to preserve rurality; disappearing rural façade; women insensitive land tenure; lack of protection of forests, open spaces, agricultural land, grazing land, markets, rural landscapes and water resources. The consequence of these has been the loss of rural culture, degradation of rural environmental resources and unattractiveness of the rural areas as places of unstable economic situations. Land and land-based activities are embedded in rural cultures in Africa. Hence tackling rural development with a focus on the land challenges can lead to realigning cultural practices that are counter development to become developmental, while promoting existing development oriented cultures that affect how people live in rural area. The implication of this is that infrastructure provision (and other rural development activities) have to be made to satisfy rural needs and be made to fit into the living and physical landscape of the rural environment. After this, the next stage should be to keep reviewing and updating the process of rural development.

This means that if these rural outcomes are achieved through land management interventions, it is important to keep reviewing the impact of those outcomes by collecting information about its impact on people. This is the best way to monitor the functions of land and ensure that any anomalies, low impacts or weak results being experiences are rectified through realignment of rural development principles or establishing more realistic objectives. This is what makes a land management based approach to rural development a cyclical process. However, it is necessary to mention that it must be an iterative process. This means that, even as a cyclical process, the decision-making embedded in the process is not linear. They are complex, and any change in the decision should be introduced at any point, rather than wait until the end of the process (because the process has no end). Each phase (and activity within phases) are as important as others. Rural development principles, needs, objectives, and strategies are ever-changing factors in a land management based rural development approach.

## MAJOR IMPLICATIONS ON CONCEPT, THEORY, POLICY, AND PRACTICE

The concept and practice of land management based approach to rural development, if embraced by academics, practitioners and policymakers will have very positive implications in the academic, development practice and policy arenas in Africa. It will fill the existing gaps already discovered concerning rural development in the continent in the aspects of concept, theory, policy, and practice. In terms of theory, de Vries and Chigbu's (2017) land management intervention or change concept, which is mathematically symbolised in the format of $\Delta LM = f(\Delta G, \Delta L, \Delta S, \Delta E, \Delta P, \Delta B)$, constitutes a new way of theorising land management in academic and development domain in order to explore the best ways possible to use land management as a development intervention—whether in rural-urban or peri-urban settlements. As a notion or theory, it captures the ideal mindset that practitioners and academics (especially geodesists) should bear in mind in the course of defining the boundaries for the use of land management as an instrument for development.

Concerning the academic perspective of rural development in the continent, Chigbu et al. (2016) investigated curricula in land education from 10 notable African countries (in ten African countries) to establish the extent to which they address the current needs in Africa. They (Chigbu et al. 2016) evaluated these curricula based on the report of Curricula Needs Assessment in Africa (African Land Policy Initiative, 2016). Their findings showed that rural development was only a marginal aspect of capacity development on land issues in the continent. Even in the less than two universities where rural development was taught, it was not taught from a land management perspective. This led to Chigbu et al. (2016: 3) to make recommendations for improving curricula in land education in Africa by introducing "land interventions" in the "capacity development" effort to ensure appropriate "governance of land in Africa in the future". The Framework and Guidelines on land policy in Africa (AUC-ECA-AfDB Consortium, 2010), which was endorsed by the African Union's joint conference of Ministers in charge of Agriculture, Land, and Livestock in 2009, recognizes the importance of rural development to the African continent. For details of other African Union documents that either proposes land management based approaches to development practice in Africa, see AUC-ECA-AfDB Consortium (2011, 2012 and 2014). This study presents land management based rural development approach as a concrete example of a subject that should be included in upcoming (and new) curricula in land management (land governance and land policy) programmes in the continent. Doing this will further enrich the concept and theories of rural development from (for and in) an African perspective.

In terms of policy practice, existing rural development practices have focused on either agricultural or infrastructural development (or both) in Africa. This is in itself not bad because the key missing links in rural development in Africa are food security and infrastructure. Most importantly, rural Africa is mostly agrarian, so it is not surprising that agricultural development dominates the elements of the continent's approaches to rural development. The concept of land management based approach presented in this study is an effort to broaden (rather than to stop) the existing approaches beyond disciplinary and sectoral lines. This is exactly what land management (as a development intervention) can do. It provides an interaction mechanism for critical land matters—related to agriculture, governance, planning, culture, management, spatial, land-use technical matters, policy, geodetic, and anthropological and sociological matters—within the rural space to produce the best development possible results. Adopting a land management based approach to rural development in many African countries will have an effect on policy practice because it will push policies away from strictly sectoral to multi-sectoral lines.

## CONCLUSION

This study, recognizing that land management activities can be implemented as standalone measures in within any spatial unit in Africa, sought to show how it can be factored directly into a rural development process. This is in accordance with the notion that since many African local communities are plagued with challenges related to land, any rural development measure is bound to be more people-centred when it addresses land related problems (Chigbu and Klaus, 2013; Teshome et al., 2016: Dube, 2016; de Vries and Chigbu, 2017; Liu, 2018). This is because rural development measures can benefit from land management based measures, principles, objectives, and strategies. For instance, it is necessary to have processes for defining land rights (including restrictions and responsibilities), spatial planning, land-use planning, gender mainstreaming in land tenure, land consolidation and land adjudication or conflict management (to mention a few) as integral aspects of rural development. However, this study argues that when these land-based processes drive the rural development process (that is, land management based rural development), the rural development process has higher potential to produce outcomes that directly have positive impacts on rural people and their environment.

These arguments have been painstakingly presented in this study because the future of rural Africa will largely depend on how the continent's land question is answered. In this regard, structures have been put in place to for continental land governance guidelines, and it is important for all (especially academics and practitioners) to ensure that rural development is included in the fundamental implementation of Africa's land policy discussion and practice. The best way to do this is to frame rural development within the context of land issues. That is why this study has evoked the land management based approach to rural development as a starting point.

## ACKNOWLEDGMENT

This study is based on a Focus Group Discussion with doctoral (PhD) researchers from Africa at the Technical University of Munich, Germany. The author will like to acknowledge the knowledge that emanated from constant brainstorm with these doctoral researchers. It helped in shaping the concept of rural developed prescribed in this study.

## REFERENCES

African Land Policy Initiative. (2016). *Guidelines for the development of the curricula on land governance in Africa*. Addis Ababa: Land Policy Initiative.

Alemu, Z. G. (2012). *Livelihood strategies in rural South Africa: Implications for poverty reduction*. Paper prepared for presentation at the International Association of Agricultural Economists (IAAE), *Triennial Conference*, Foz do Iguacu, Brazil.

Asiama, K. O., Bennett, R., & Zevenbergen, J. (2017). Land consolidation for sub-Saharan Africa's customary lands—The Need for Responsible Approaches. *Am. J. Rural Dev*, *5*, 39–45.

Asiama, K. O., Bennett, R., Zevenbergen, J., & Asiama, S. O. (2018). Land valuation in support of responsible land consolidation on Ghana's rural customary lands. *Survey Review*, 1–13. doi:10.1080/00396265.2018.1467672

AUC-ECA-AfDB Consortium. (2010). Framework and guidelines on land policy in Africa: A framework to strengthen land rights, enhance productivity and secure livelihoods. AUC-ECA-AfDB.

AUC-ECA-AfDB Consortium. (2011). Land policy in Africa: West Africa regional assessment. AUC-ECA-AfDB.

AUC-ECA-AfDB Consortium. (2012). Tracking progress in land policy formulation and implementation in Africa. AUC-ECA-AfDB.

AUC-ECA-AfDB Consortium. (2014). Guiding principles on large-scale land-based investments in Africa. AUC-ECA-AfDB.

Bathust Declaration. (1999). *The BucharestDeclaration on land administration for sustainable development. FIG Publication no. 21*. Copenhagen, Denmark: FIG.

Bryceson, D. F. (1999). African rural labour, income diversification & livelihood approaches a long-term development perspective. *Review of African Political Economy*, *26*(80), 171–189. doi:10.1080/03056249908704377

Carter, M. R. (2002). Soil quality for sustainable land management. *Agronomy Journal*, *94*(1), 38–47. doi:10.2134/agronj2002.0038

Chia, J., & Lee, B. (2017). An exploratory methodology to incorporate transfer location in transit spatial coverage quantification. In *Proceedings of the Australasian Transport Research Forum (ATRF) 2017*. Australasian Transport Research Forum (ATRF).

Chigbu, U. E. (2012). Village renewal as an instrument of rural development: Evidence from Weyarn, Germany. *Community Development (Columbus, Ohio)*, *43*(2), 209–224. doi:10.1080/15575330.2011.575231

Chigbu, U. E. (2013a). Rurality as a choice: Towards ruralizing rural areas in sub-Saharan African countries. *Development Southern Africa*, *30*(6), 812–825. doi:10.1080/0376835X.2013.859067

Chigbu, U. E. (2013b). *Territorial development: suggestions for a new approach to rural development in Nigeria* (PhD thesis). Technical University of Munich, Germany.

Chigbu, U. E. (2013c). Fostering rural sense of place: The missing piece in Uturu, Nigeria. *Development in Practice*, *23*(2), 264–277. doi:10.1080/09614524.2013.772120

Chigbu, U. E. (2015). Ruralisation: A tool for rural transformation. *Development in Practice*, *25*(7), 1067–1073. doi:10.1080/09614524.2015.1071783

Chigbu, U. E. (2015). Repositioning culture for development: Women and development in a Nigerian rural community. *Community Work & Family*, *18*(3), 336–352. doi:10.1080/13668803.2014.981506

Chigbu, U. E., Izugbara, C. O., & de Vries, W. T. (2018). Land, culture, culture loss and community: rural insights from sub-Saharan Africa. In S. Kenny, B. McGrath, & R. Phillips (Eds.), *The Routledge Handbook of Community Development: Perspectives from Around the Globe* (pp. 98–114). London: Routledge.

Chigbu, U. E., & Klaus, M. (2013, May 6). Insecurity-Generating System of Land Tenure and its Impact on Rural Development: Evidence from Uturu, Nigeria. FIG Working Week.

Chigbu, U. E., Schopf, A., de Vries, W. T., Masum, F., Mabikke, S., Antonio, D., & Espinoza, J. (2017). Combining land-use planning and tenure security: A tenure responsive land-use planning approach for developing countries. *Journal of Environmental Planning and Management, 60*(9), 1622–1639. doi:10.1080/09640568.2016.1245655

Chigbu, U. E., Tenadu, K., & Mwasumbi, A. (2017). Curriculum Reform in Land Governance Education: The Need for Transforming Existing Curricula in Africa. *World Bank Conference on Land and Poverty*, Washington, DC.

Cohen, J. M., & Uphoff, N. T. (1980). Participation's place in rural development: Seeking clarity through specificity. *World Development, 8*(3), 213–235. doi:10.1016/0305-750X(80)90011-X

Czechowski, P. O., Owczarek, T., Badyda, A., Majewski, G., Rogulski, M., & Ogrodnik, P. (2018). Preliminary comparative assessment of PM10 hourly measurement results from new monitoring stations type using stochastic and exploratory methodology and models. *E3S Web of Conferences, 28.*

de Vries W. & Chigbu U. E. (2017). Responsible land management – concept and application in a territorial rural context. *Fub –Flächenmanagement und Bodenordnung, 79*(2), 65-73.

Dube, K. (2016). Implications of rural irrigation schemes on household economy. A case of Lower Gweru Irrigation Scheme, Zimbabwe. *South African Journal of Agricultural Extension, 44*(1), 75–90. doi:10.17159/2413-3221/2016/v44n1a371

Keesstra, S., Nunes, J., Novara, A., Finger, D., Avelar, D., Kalantari, Z., & Cerdà, A. (2018). The superior effect of nature based solutions in land management for enhancing ecosystem services. *The Science of the Total Environment, 610*, 997–1009. doi:10.1016/j.scitotenv.2017.08.077 PMID:28838037

Klinghoffer, A. J. (1973). Modernisation and political development in Africa. *The Journal of Modern African Studies, 11*(1), 1–19. doi:10.1017/S0022278X00008053

Liu, Y. (2018). Introduction to land use and rural sustainability in China. *Land Use Policy, 74*, 1–4. doi:10.1016/j.landusepol.2018.01.032

Magel, H. (2014). *Rural-urban partnership: key for a better land and water management?* International Symposium on occasion of IFAT 2014 on Intelligent Peri-Urbanisation through Decentralisation, Munich, Germany.

Ojike, M. (1946). *My Africa.* New York: John Day.

Palmer, I. (1977). Rural women and the basic-needs approach to development. *Int'l Lab. Rev., 115*, 97.

Pašakarnis, G., & Maliene, V. (2010). Towards sustainable rural development in Central and Eastern Europe: Applying land consolidation. *Land Use Policy, 27*(2), 545–549. doi:10.1016/j.landusepol.2009.07.008

Paul, M., & wa Gĩthĩnji, M. (2018). Small farms, smaller plots: Land size, fragmentation, and productivity in Ethiopia. *The Journal of Peasant Studies*, *45*(4), 757–775. doi:10.1080/03066150.2016.1278365

Phuhlisani, S. (2009). *International and Local Approaches to Rural Development: Key Issues and Questions: A Review of the Literature for the Drakenstein Municipality*. Phuhlisani Solutions.

Rondinelli, D. A. (1979). Administration of integrated rural development policy: The politics of agrarian reform in developing countries. *World Politics*, *31*(3), 389–416. doi:10.2307/2009995

Sadler, T. D. (2004). Informal reasoning regarding socioscientific issues: A critical review of research. *Journal of Research in Science Teaching*, *41*(5), 513–536. doi:10.1002/tea.20009

Shepherd, K. D., Shepherd, G., & Walsh, M. G. (2015). Land health surveillance and response: A framework for evidence-informed land management. *Agricultural Systems*, *132*, 93–106. doi:10.1016/j.agsy.2014.09.002

Teshome, A., Graaff, J., Ritsema, C., & Kassie, M. (2016). Farmers' perceptions about the influence of land quality, land fragmentation and tenure systems on sustainable land management in the north western Ethiopian Highlands. *Land Degradation & Development*, *27*(4), 884–898. doi:10.1002/ldr.2298

Whitlock, C. (2004). Land management: Forests, fires and climate. *Nature*, *432*(7013), 28–29. doi:10.1038/432028a PMID:15525968

Wilson, O. J. (1998). Village renewal and rural development in the former German Democratic Republic. *GeoJournal*, *46*(3), 247–255. doi:10.1023/A:1007028409123

## ADDITIONAL READING

Chigbu, U. E. (2013). Rurality as a choice: Towards ruralising rural areas in sub-Saharan African countries. *Development Southern Africa*, *30*(6), 812–825. doi:10.1080/0376835X.2013.859067

Chigbu, U. E. (2015). Ruralisation: A tool for rural transformation. *Development in Practice*, *25*(7), 1067–1073. doi:10.1080/09614524.2015.1071783

Chigbu, U. E., Izugbara, C. O., & de Vries, W. T. (2018). Land, culture, culture loss and community: rural insights from sub-Saharan Africa. In S. Kenny, B. McGrath, & R. Phillips (Eds.), *The Routledge Handbook of Community Development: Perspectives from Around the Globe* (pp. 98–114). London: Routledge.

Chigbu, U. E., Schopf, A., de Vries, W. T., Masum, F., Mabikke, S., Antonio, D., & Espinoza, J. (2017). Combining land-use planning and tenure security: A tenure responsive land-use planning approach for developing countries. *Journal of Environmental Planning and Management*, *60*(9), 1622–1639. doi:10.1080/09640568.2016.1245655

Zevenbergen, J., de Vries, W., & Bennett, R. M. (Eds.). (2015). *Advances in Responsible Land Administration*. London: CRC Press. doi:10.1201/b18988

## KEY TERMS AND DEFINITIONS

**Africa:** A term used in identifying all indigenous and migrant ethnicities living in the second biggest continent in the world.

**Land Management:** The science and practice related to the conceptualization, design, implementation, and evaluation of socio-spatial interventions on land; with the purpose to improve the quality of life of people and their environment, and the resilience of their livelihoods in a responsible, effective, efficient, consensual, and smart manner.

**Land Management Intervention:** Any change caused by land management in order to improve the living conditions of people and the environment. It can occur in the form of land issues changes related to governance, law, policy, property rights, politics, administration, economics, culture, beliefs, perceptions, and behavior (to mention a few).

**Rural Area:** Land-spaces with culturally defined identity; situated within a place statutorily recognized as non-urban; and occupied by settlers predominantly depending on primary sources of labor for their livelihood.

**Rural Development:** All activities are done for improving the socioeconomic conditions of rural people, and the environmental conditions of their locality.

# Chapter 2
# Land Governance and the Information Society

**Mohamed Timoulali**
*GTOPIC sarl, Morocco*

## ABSTRACT

*In a globalized world evolving into knowledge societies and economies, geographic information plays a very important role, as illustrated by the spatially enabled society (SES) concept. In this context, the role of a modern cadastre in the good governance of land for sustainable development is worth noting. The international and regional agendas and various initiatives related to these themes are examined to deduce the emerging global trends and the possible transformation of the national statistical. One of the prerequisites for an integrated information system is the availability of harmonized data and the interoperability between the various dedicated systems. This depends on the effectiveness of the national statistical system and the establishment of a national spatial infrastructure. The institutional and organizational arrangements for better land governance are discussed. The resulting technological infrastructure to be implemented is then addressed.*

## INTRODUCTION

Among the international agendas, the countries are engaged in the World Summit for the Information Society (WSIS), the Agenda 2030 for Sustainable Development, and also following the Open data movement, the UNGGIM activities.

In Africa, others regional agendas were implemented: the Agenda 2063, the Land Policy initiative and the Africa data Revolution.

The paper is discussing the link between national land policies, National Geomatics Strategies, the information/Society, knowledge / Economy. and implication on the land Management and provide recommendation for the mobilization of the Geospatial sector for the national development (Economic Commission For Africa, 2007), (African Union, 2010).

DOI: 10.4018/978-1-5225-5939-9.ch002

Morocco is referred to, as the country is implementing several sectoral strategies in agriculture, fisheries, industry, tourism, energy, environment, town planning, which all face a problem of land availability and ambiguous legal environment.

The country has a long history of legal pluralism. Parallel systems coexist in customary and modern law, formal and informal land administration systems, and face the absence of a land policy that recognizes the range of various rights and interests

For example, the inadequacy of the legal system governing the property of ethnic communities remains one of the constraints that hamper the optimal valorization of this important national heritage.

Obstacles also relate to problems of the nature of agricultural land structures, and to the impact of urban development on agricultural land.

The development of this sector is also hampered by rising land prices in areas characterized by high attractiveness, high demand for investment, and the high cost of the infrastructure.

## A MODERN CADASTRE FOR GOOD LAND GOVERNANCE

The role of a modern cadastre in good land governance is highlighted by Williamson, Enemark, Wallace, and Rajabifard, 2009, While the role of geographic information in the knowledge society, and Spatially Enabled Society (SES) Is discussed in (FIG, 2012).

The publications addressed the following pillars to achieve the SES vision:

- The legal framework supporting the institutional structure of sharing discovery and access to data, the concept of multi-source data integration and interoperability, the positioning infrastructure
- The concept of multi-source data integration and interoperability
- A positioning infrastructure to benefit from the possibilities of available precision
- A spatial data infrastructure to facilitate data sharing, reduce duplication and link producers, suppliers to users on the basis of common goals
- Information on ownership as a dominant topic between government, business, and citizens about land and water resources
- Data and information in order to respect the basic principles to increase availability and interoperability of freely reusable data from different actors and sectors.

In the 2016 edition of United Nations e-Government Survey conducted every two years, the Survey featured the global and regional trends, and provided new evidence that e-government has the potential to help support the implementation of the 2030 Agenda and its 17 sustainable development goals (SDGs).

In line with the call of United Nations General Assembly in 2015 for a close alignment between WSIS and SDG processes, the WSIS Action lines facilitators have been working together to further implement the WSIS-SDGs Matrix (ITU, 2017).

Hence, the availability of sectoral updated statistics data on applications related indicators is a major challenge.

*Figure 1. WSIS-SDG Matrix*
*ITU, 2017.*

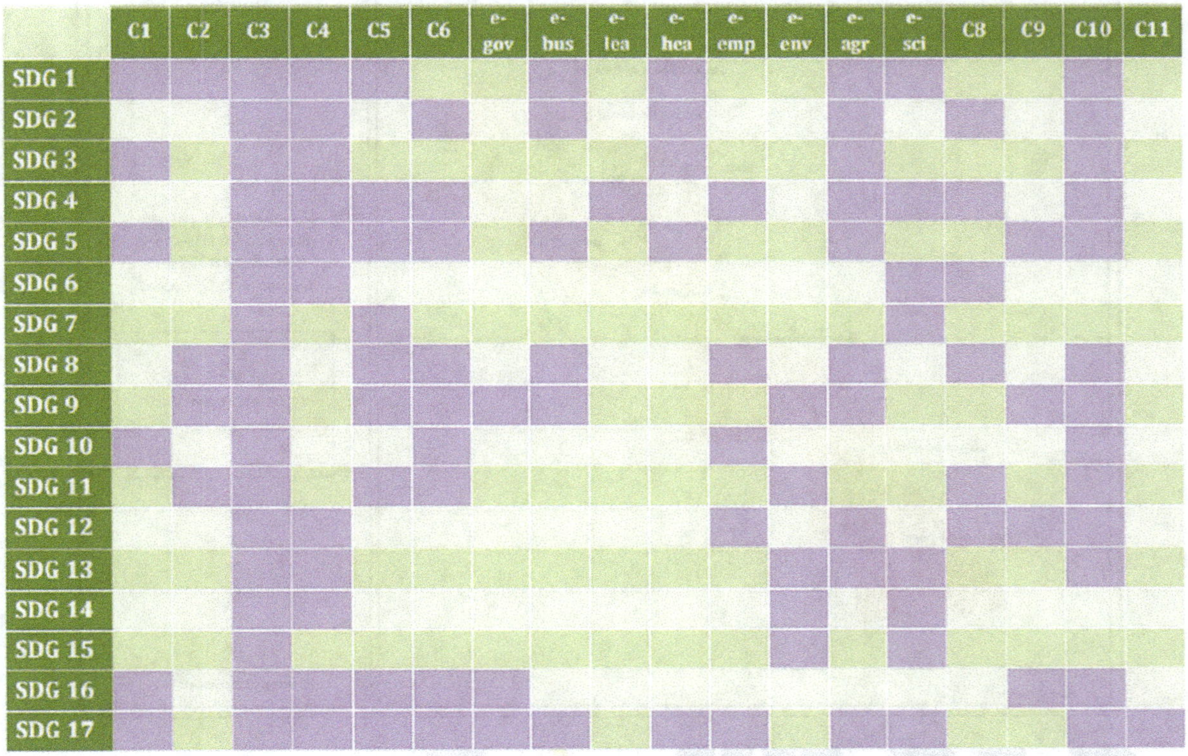


## THE ABSENCE OF A LAND POLICY IMPACTS NEGATIVELY DEVELOPMENT

The lack of a state land policy is one of the obstacles to achieving socio-economic development objectives, due to the lack of good governance of land resources, which are essential to all sectorial strategies.

The absence of a strategic vision concerning the private domain of the state was highlighted by the report of the Public Finance Court (Cour des comptes, 2015).

The report indicated the difficulty for the state to control its land holdings, as only 53% of these areas are registered with land conservation, while 41% are in the process of being registered. The remaining 6% are not registered at all.

The consultations launched by the Head of Government on the state land policy in Skhirat on 8-9 December 2015 provided the following roadmap:

- For a comprehensive reform of the land sector, with the adoption of a vision encompassing the legal, institutional, organizational and procedural dimensions.
- Focusing on the adoption of adequate mechanisms for streamlining the interventions of all stakeholders in order to ensure the effectiveness of the intervention at national and territorial levels.

Another result of the consultation, is the creation of the Permanent Ministerial Committee on Land Policy chaired by the Head of Government, by Decree No. 2-16-263 of 24 May 2016.

*Figure 2. ICT statistics: Households with mobile phone*

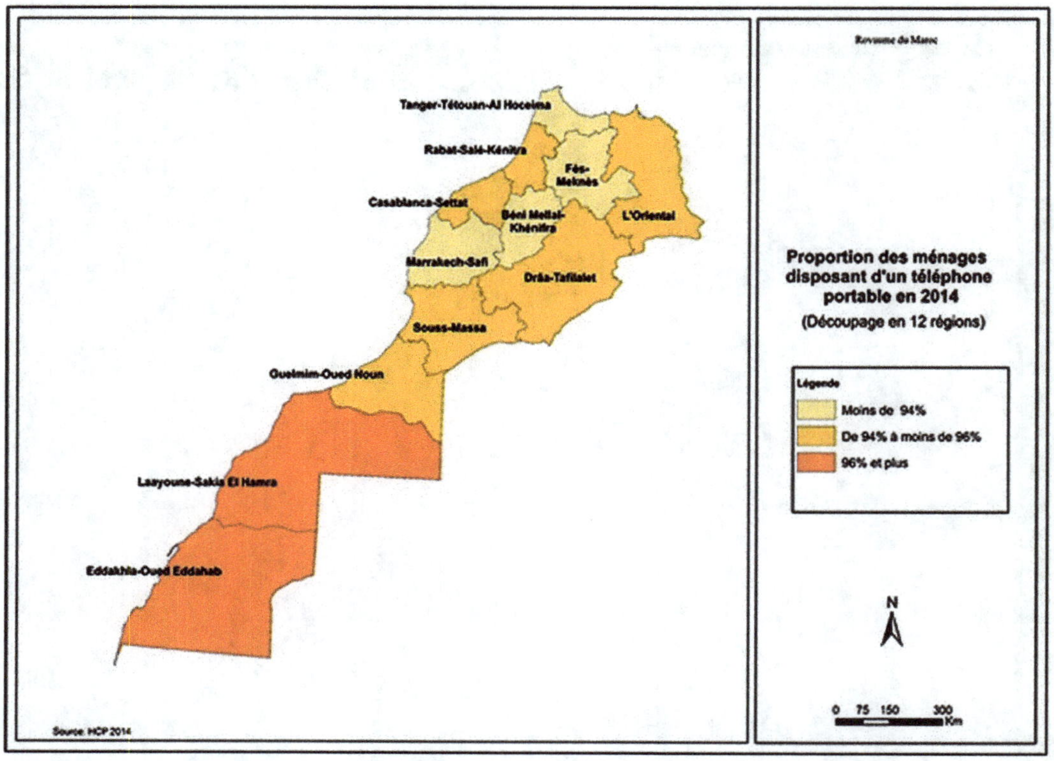


## THE LAND AS A SHARED STRATEGIC RESOURCE

The report of Public Finance Court cited above, deals with the lack of a land policy, and addresses also the need for a dedicated observation and knowledge on land markets.

The purpose of land observation is to produce statistical or cartographic data contributing to a more precise knowledge of the territories. This observation aims not only to share information, but also to evaluate past decisions in order to anticipate and to prepare future ones.

As an illustration of the shortcomings in this area, we refer to (Cour des comptes, 2015). The report devoted a chapter to the Rural Development and Mountain Areas Fund (FDRZM), and to the organizational flaws in project monitoring. In this regard the court noted that the Geographic Information System (GIS) whatever available, does not cover the main areas of activity, and is not integrated. It does not, therefore, promote the flow of information and its sharing between central and decentralized services for good decision-making.

In this connection, reference is made to the opinion of the Social and Environmental Economic Council (EESC) on cross-cutting policies for the management of shared strategic resources, such as land tenure and its importance in public policies (CESE, 2014) and (UN- Habitat., 2014).

*Figure 3. Map of Search mining license per regions in 2014*

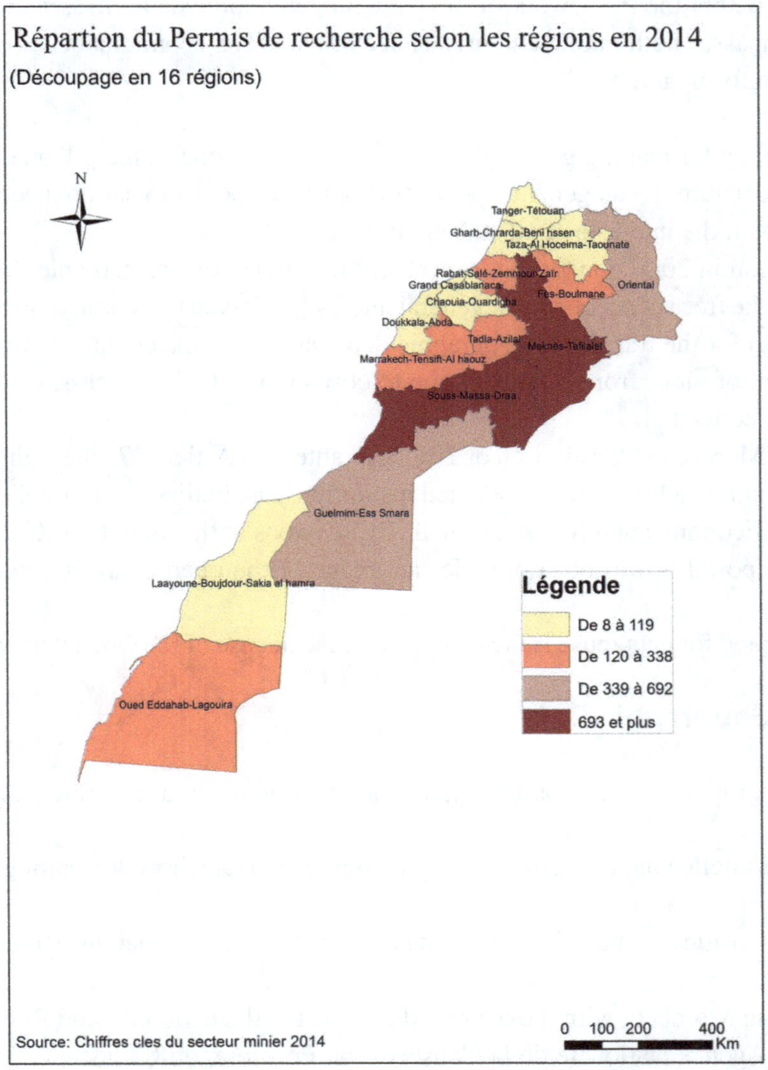


## LEGAL STATUS OF GEOGRAPHICAL DATA AND TRANSFER POLICIES

### Open Data Agenda

Open data in the field of geographic information vary from one country to another. In the fields of cadastre and geoinformation, the following examples could be cited:

- **Spain:** Users have free access to data free of charge, except for those protected by law.
- **Australia:** The distribution of geographic data is promoted under the Creative Common License (Fitzgerald, Hooper and Cook, 2013).

- **South Africa:** The COFUR (Cost of Fulfilling User Request) model is used to make the data available to users within the framework of the national geographic data infrastructure. South African policy encompasses both public and private sector data. The economic benefits of this policy are provided for information only.

An adequate policy for making geographical data available is one of the pillars of the national geographic data infrastructure. It can generate positive benefits, responds to national economic conditions, and takes into account the interests of all stakeholders.

Since its inception in 2009, the Open Government has undergone remarkable development, giving new dimensions to the free access of data of general interest, collected by central or local administrations.

National policies for the transfer of geographic data vary from one country to another, and thus go from a pricing point of view, from a contribution to cost support to free access. Open Data is also becoming part of this ecosystem.

In this area, the Moroccan Constitution of 2011 guarantees in Article 27 “the right of access to information held by the public administration, elected institutions and bodies with a public service mission”. The opinion of the Economic and Social Council (EESC) goes in this direction (CESE, 2013).

The Open Data portal in Morocco provides an image of the open data situation of the Moroccan administration.

The license adopted for data reuse is the Open Data base license of the Open Knowledge Foundation.

## Ownership of Geographic Data

The creation of geographical data is part of intellectual production and therefore subject to intellectual property law.

Thus, the law of intellectual property, copyright applies to data whose formatting must have a character of originality.

Economic law also allows data to be protected because of the substantial investment that has allowed them to be produced.

This is the case in Morocco, with the data produced by the the national Land Registry, cadastre and Mapping Agency (Agence Nationale de la Conservation Foncière, du Cadastre et de la Cartographie: ANCFCC).

## Access to Cartographic Data

The access to cartographic data in Morocco is regulated by the following legislation:

- Decree of 30 December 1975 relating to consultation, reproduction and use
- Decree of 4 May 2000 amending that of 30 December 1975 to extend the protection of cartographic material against the duplication and illegal reproduction of digital data and maps produced by ANCFCC.
- Order of 8 July 1996 fixing the sales and research fees for the on-the-spot consultation of documents and technical works published or reproduced by the ANCFCC
- Order of 31 October 2001 supplementing that of 8 July 1996 by including the tariffs for digital cartographic products.

Access to digital land and cadastral data is possible through the e-land services of the ANCFCC. Access is free for some data, and paying for others.

## GEOGRAPHIC DATA AND LEGAL RESPONSIBILITIES

At this level, responsibility and liability is examined in the use of geographical data applies for both the producer and the user side.

### Protection of Personal Data

It is necessary to comply with Law 09-08 on the protection of individuals with regard to the processing of personal data of 18-02-2009, which created the National Commission for the Protection of Personal Data (CNDP). This commission was officially installed on August 31, 2010. One of the objectives of the law is to strengthen and consolidate the protection of fundamental rights and individual and collective freedoms.

### Importance of Quality and Accuracy

Geographic data are parts of many decisions that can have important economic and social implications. The quality of these data and their accuracy must therefore be taken seriously and face the possible degradation of this quality by the fusion of data from various sources and diverse qualities. Quality information becomes a fundamental element, as described by (Jakobsson & Tsoulos., 2007), (Servigne, Lesage, & Libourel, 2006). (Lush, Bastin & Lumsden, 2012).

Available standards from the International Organization for Standardization (ISO) describe the elements of the data quality to be considered, including: completeness, logical consistency, Positional accuracy, temporal accuracy, and thematic accuracy. (Talhofer, Hošková-Mayerová & Hofmann, 2012).

Although the continuous improvement of the data quality and accuracy thanks to the developments in the geospatial technology sector, a 'Fit For Purpose ' approach is recommended for a choice responding the user real needs (FIG, 2015).

### Challenge of Cybersecurity

The safety aspects are very acute considering the uses that could be made of the geographical data. The necessary cybersecurity measures must therefore be scrupulously respected in order to ensure the security of critical infrastructures and information systems, and to ensure the security, integrity and confidentiality of the digital transaction, In conformity with the Law 53-05 on the electronic exchange of legal data.

The producers and users of the digital data bear the penal consequences of non-compliance with the measures enacted, in accordance with Law No. 07-03 supplementing the Criminal Code with regard to offenses relating to automated data processing systems. This law sanction all unauthorized intrusions into an automated data processing system, including those that could result from the disclosure of the personal login to a third party.

## RECOMMENDATIONS FOR A NATIONAL GEOMATICS PLAN

As international benchmark, a review was undertaken of several geomatics plans and:

- The Pan-Canadian Geomatics Strategy is a collaborative and open-ended effort of the Canadian Geomatics Community Round Table to determine the desired future of the Canadian geospatial sector from a collective and inclusive perspective.
- Evaluation of 10 years of regional geomatics policy in France
- Basic plan for the use of geospatial information in Japan
- Topography and mapping policy formulated on the basis of the national land policy project in Nepal
- The national plan of Geomatics of Senegal developed with the support of the Government of Canada
- The NSDI strategic plan of the USA
- Strategic geomatics plan for Wallonia
- Rwanda's action with the availability of several integrated strategies

Consequently, proposals are made here to mobilize all stakeholders to formulate a national transversal geospatial information strategy and plan, to support the sectoral policies and strategies such as the land policy.

The national Geomatics Strategy and Plan will address the institutional, legal, technical, human and financial constraints; and also integrate international developments and best practices:

- Focus on geo information requirements for sustainable development developments (Espinoza, 2011).
- Adopt an integrated vision for the realization of a Decision Making Information System on monitoring and evaluation of public policy (Richard. 2014).
- A realistic implementation plan of the pillars of the strategy, resulting from inclusive open consultations with interested parties.

*Figure 4. Conceptual Integrated global view*

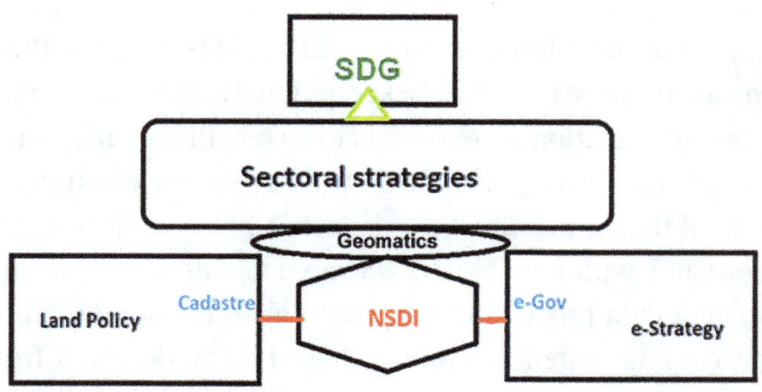

# DEDICATED SPATAIL DATA INFRASTRUCTURE TO SUPPORT TERRITORIAL DECISION-MAKING

From the presented analysis, the following recommendations can serve to align the national Geospatial sector with the national development priorities.

## Adapt the National Institutional Arrangements

The national institutional arrangements provide the adequate platform for dialog, coordination and co-operation between the producers and users of geospatial data, (UNGGIM, 2017).

The most common form is a national independent transversal geoinformation structure, with the authority to guide the national consultations for the formulation, monitoring and implementation of the national Geomatics Strategy and Plan.

In the absence of such structure, arrangement can be made with the newly created Agency for digital development, and the inter ministerial commission for the national and Policy.

The consultations will guide the process and actions to be taken in institutional and legal frameworks, Technology and Standards, Human Resources and Capacity Development, and finance resources mobilization.

## Moving From GIS to SDI as Part of a National Geomaticss Strategy

The MORGEO 2017 congress organized at the EHTP on 16-17 May 2017 offered the opportunity to have of the following figures on geomaticss research in Morocco:

- 10 plenary lectures including two related to Spatial Data Infrastructure (SDI) (20%)
- 74 oral communications with three related to SDI (4%)

These figures reflect the mood of the operators in Morocco, in relation to the concerns that led to the evolution towards SDI, such as the need for data openness / sharing, system interoperability, and standardization.

In the absence of a favorable framework for the establishment of a National Spatial Data Infrastructure (NSDI), we see emerging projects oriented thematic or territorial SDI.

Territorial decision-making is becoming increasingly complex, given the multitude of issues to be addressed, the disparity of data sources, the quality and the volume of data to be handled.

This is pushing for the creation of national observatories and the use of technologies to support policy and strategic decisions, based on geospatial statistics data from interoperable systems and a standardized model (Peters, 2014), (Roth & al. 2013).

The design follow the Spatial Data Infrastructure (SDI) approach, where harmonization and interoperability of components make it possible to ensure that geographic data and services can integrate and combine in a consistent manner. (Hintz, 2012).

The international benchmark shows the need for interoperability of systems, and modeling within the framework of an open architecture to allow proper sharing of information, decision-making and / or communication (Idrees, Vahideh, Yusuf; & Shariff, 2012).

## Adapt the International Standards

The international Organization for Standardization (ISO) and the Open Geospatial Consortium. (OGC) have already published more than sixty standards and continue to work on complementary standards. ISO focused on real world modeling based on the concept of geographic entities including their definitions, attributes, associations and operations. The OGC was interested in services and developed specifications to access and download the data. The work of the two bodies is complementary. ISO has adopted OGC standards and the OGC has taken over the methodology and modeling aspects of ISO in its core documents, George Percivall.(2006), Linbda Peters.(2014).

ISO and OGC standards are important for SDI, as they govern interoperability, and sharing of the geographic information.

A commission In the Institut Marocain de normalization (IMANOR), is in charge of the Geographical information standards (CN 094) as a part of the Electric and Electronic Industry sector. IMANOR participates in the ISO committee on the geographic information: ISO/TC 211.

*Figure 5. Distribution of land surveyors Engineers by regions in Morocco 2015*

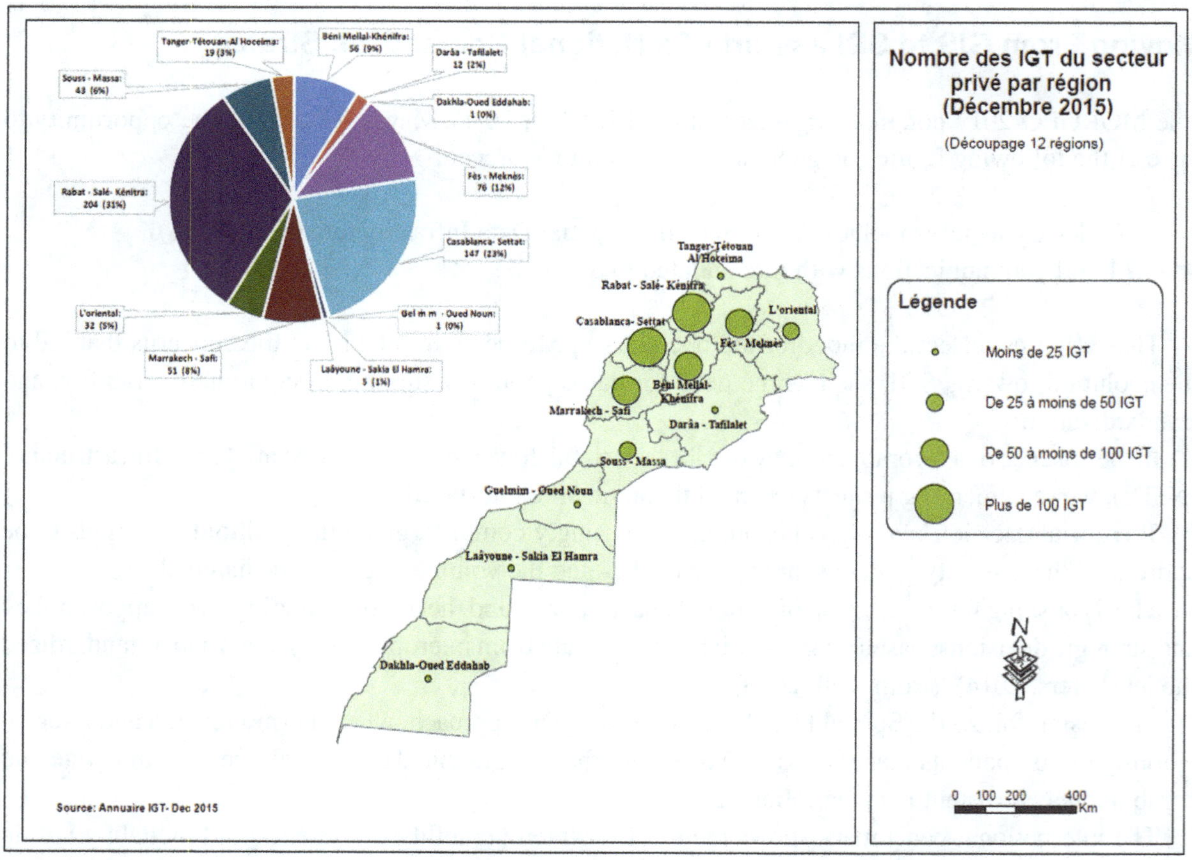

## Reinforce the Capacity Development in Land Governance

- Promote an interdisciplinary training approach that combines technical, legal and social, and links the various fields of land management science
- Respond to the need for continuous professional development, aimed at maintaining and developing professional skills
- Develop the interaction between education, research and professional practice and benefit from regional and international cooperation

The equitable territorial distribution of the professionals has also to be considered.

## CONCLUSION

Since Morocco launched the national land policy process, it was important to recall the need to make it an inclusive task, with an imperative cohesion with sectoral strategies and sustainable development objectives.

Building on the best practices, and technological developments, the success of this land policy will be based on improved land governance through the availability of reliable and shareable information.

The foundation for this would be a national geospatial data infrastructure to be realized as part of a national geomatics strategy. The outline of the development process and content of the strategy is provided based on an international experiences benchmark, and the singularities of the national context.

The pillars of the Strategy include 1- Revision of the legal Framework for the access and security of Data. 2- Adapt the institutional arrangements for a better cooperation and exchange of geospatial data. 2- Adoption of the Geospatial information international standards. 3- Regular revisions of the Curricula on Geo Science and Technology education. 4- Research program in cooperation with regional and International institutions.

## REFERENCES

African Union, African Development Bank, Economic Commission for Africa. (2010). *Framework and Guidelines on Land policy in Africa*. Addis Ababa: ECA.

Banque Mondiale. (2008). *Marchés fonciers pour la croissance économique au Maroc*. Washington, DC: Word Bank.

Bouderbala, N. (1990). *Les systèmes de propriété foncière au Maghreb: le cas du Maroc*. Montpelier: CIHEAM.

Conseil Economique Social et Environnemental. (2013). *Open data. La libération des données publiques au service de la croissance et de la connaissance*. Rabat: CESE.

Conseil Economique Social et Environnemental. (2014). *Cohérence des politiques sectorielles et accords de libre échange*. Rabat: CESE.

Cour des comptes. (2014). *Evaluation de la stratégie Maroc Numeric 2013*. Rabat: Cour des Comptes.

Cour des comptes. (2015). *Annual report of the Public Finance Court for the year 2015*. Rabat: Cour des Comptes.

Economic Commission for Africa. (2007). *e-Strategies - Best Practices and Lessons Learnt*. Addis Ababa: ECA.

El Ghrib, A. (2016). *Privatization of collective agricultural Land in Morocco. World Bank Conference on Land and poverty Conference*, Washington, DC.

Enemark, S., Clifford Bell, K., Lemmen, C., & McLaren, R. (2014). *Fit-For-Purpose Land Administration. International Federation of Surveyors*. Copenhagen: FIG.

Espinoza, J. (2011). What is required to promote Good Land Governance in the 21st Century. *Word Bank Conference on Land and Poverty*, Washington, DC.

Féderation Internationale des Géomètres. (2012). *Spatially Enabled Society*. Copenhagen: FIG.

Féderation Internationale des Géomètres. (2015). Fit for purpose land administration. Copenhagen: International Federation of Surveyors (FIG).

Fitzgerald, A., Hooper, N., & Cook, J. (2013). Implementing open licensing in government open data initiatives: a review of Australian government. practice. *Proceedings of the 9th International Symposium on Open Collaboration*. 10.1145/2491055.2491094

Guigoz, Y., Giuliani, G., & Ray, N. (2013). *State of the art in Open Source GIS and SDI*, ARPEGEO project Symposium on Environmental Data Sharing for the Benefit of the South Caucasus Region.

Hintz, D. (2012). *Data Harmonization Principles and Development Approaches as Applied to INSPIRE SDIs*. INSPIERE - GMES Information Brochre. Technische Universität München.

Idrees, M., Vahideh, Y. A., & Shariff, R. (2012). Comparing Approaches and Strategies for NSDI Implementation Between the Developed and Developing World. *The International Journal of Spatial Data Infrastructures*.

International Telecommunication Union. (2017). *WSIS-SDG Matrix-linking WSIS Action Lines with Sustainable development Goals*. Geneva: ITU.

Jakobsson, A., & Lysandros Tsoulos, L. (2007). The Role of Quality in Spatial Data Infrastructures. *Proceedings of the 23 rd International Cartographic Conference*.

Land Policy Initiative. (2016). Needs assessment and mapping of training institutions on land governance in North Africa. Consultancy report UNECA.

Lush, V., Bastin, L., & Lumsden, J. (2012). *Geospatial data quality indicators*. International Spatial Accuracy Research Association.

Percivall, G. (2006). OpenGIS International Standards for GEOSS Interoperability Arrangements. *IEEE International Symposium on Geoscience and Remote Sensing*.

Peters, L. (2014). *Statistical analysis of geospatial information relevance of SDI's and international standards*. ESRI.

Richard, R. (2014). *Geospatial Land Information Management: Addressing the Data and information needs*. Addis Ababa: UNGGIM HLF.

Roth, E., Donohue, G., Sack, M., Wallace, R., & Buckingham, M. (2015). *A Process for Keeping Pace with Evolving Web Mapping Technologies*. University of Wisconsin–Madison.

Salvemini, M. (2009). *The Infrastructure for Spatial Information in The European Community Vs Regional SDI: The shortest Way for reaching Economic and Social Development. Ninth United Nations Regional Cartographic Conference for the Americas*, New York, NY.

Servigne, S., Lesage, N., & Libourel, T. (2006). *Quality Components, Standards, and Metadata. Rodolphe Devillers*. London: Robert Jeansoulin. Fundamentals of Spatial Data Quality.

Talhofer, V., Hošková-Mayerová, Š., & Hofmann, A. (2012). Improvement of digital geographic data quality. *International Journal of Production Research, 50*(17), 4846 - 4859.

UN-Habitat. (2014). *Good Land governance for the 2030 Agenda*. UNGGIM HLF.

United Nations Committee of Experts on Global Geospatial Information Management. (2017). Compendium of good practices for national institutional arrangements. New York: UNGGIM.

United Nations Committee of Experts on Global Geospatial Information Management. (2017). *National Institutional Arrangements: Instruments, Principles and Guidelines*. New York: UNGGIM.

Williamson, I. E., Wallace, J., & Rajabifard, A. (2009). Land administration for sustainable development. ESRI Academic Press.

# Chapter 3
# Comprehensive Ontology Model of Moroccan Land Administration Domain Applied to the National Land Governance

**Moulay Abdeslam Adad**
*ANCFCC, Morocco*

**El Hassane Semlali**
*Institut Agronomique et Vétérinaire Hassan II, Morocco*

**Fatiha Ibannain**
*ANCFCC, Morocco*

## ABSTRACT

*Moroccan land and real estate are characterized by a large number of properties and an advanced and varied typology. Among the related shortcomings is how to identify various land statuses and the stakeholders in the sector. These characteristics of land are multidimensional, including urban, suburban, and rural; land use plan; lands under formal and informal registration systems and multiple regulatory frameworks; the importance for the state, the public, and private individuals; regarding its heritage, urban planning, and taxation; the development projects in each sector; etc. The complex character and distinct fields of land administration in Morocco necessitates the creation of a unified and standardized land data model for use by all public and private actors who have a relationship to land. The present study contributes to the perception of an applied taxonomy and aims to define the basic concepts of the land domain ontology. Such expertise is a prerequisite for land governance and land policy that integrate land and real estate into national development processes.*

DOI: 10.4018/978-1-5225-5939-9.ch003

## DEFINITIONS AND CONCEPTS

The International Organization for Standardization (ISO) describes land administration as the process of determining, recording and disseminating information about the relationship between people and land (ISO, 2012). In this context, the term "land" may refer to "real estate", "immovable property", "parcel" and other terminologies used in private and professional fields related to land. Land and real estate can also refer respectively to soil and property resulting from land use like building and housing.

"Individuals" are natural and legal persons, under public or private law, who can relate to land through any kind of property right. They can also be entities belonging to the public sector and the private sector, individually or in a group, which are involved in land administration. Additionally, it could be a property linked to another property in terms of servitude or easement for example.

In a more specialized context, land administration can be defined as the set of instruments, rules, and institutions that make it possible to manage land in a given legal, social, political, economic and environmental context. It refers more specifically to processes aimed at regulating the development, use and conservation of land, the collection of income from its sale, concession and taxation, as well as the resolution of conflicts over its appropriation. It can be provided by state services, but also by various public, territorial, customary, and associative institutions (C.T., 2009).

## INTRODUCTION

The Moroccan land administration system, as a process of determining, recording and disseminating information about the relationship between people and land, provides a foundation for land governance and land policy, and is a prerequisite to these functions. This information is at the heart of any process of land administration and therefore it is vital to identify and understand the concepts of "lands", "individuals" and "land-individual relationship".

It must be noted that without a unique land data model that takes these concepts into consideration, drawbacks arise in manipulating this information, such as data redundancy and difficulties in sharing data between all the major users and stakeholders involved in these domains. To agree on a unified model is a necessary objective, which in turn requires the identification and description of the entities and concepts of these domains.

The Moroccan land administration domain has widened as its typology has diversified and the area administered has increased. This diversity is perceived in the context of this study as richness and as the achievement of historical efforts made in describing land and recognizing its peculiarities, not in the sense that a multiplicity of land statuses makes the holding and manipulation of land ownership problematic. This perception speaks to the informative function of land administration and requires an exhaustive approach to address our topic. No aspect of land governance should be overlooked or underestimated.

First, we will give an overview of land policy and land governance. The volume and typology of Moroccan land will then be highlighted, and a description of the individual-land relationship will be proposed. The domain of Moroccan land administration will be detailed after. We will end with a discussion, some Cadastre-2014 aspects', international perceptions, proposal and conclusion.

## LAND POLICY AND LAND GOVERNANCE

Land policy is defined as the set of rules and administrative practices governing, in the framework of public intervention on private rights, the rights of ownership, the forms of access and use of land, and the transfer of rights to land. A land policy includes conceptions of citizenship, the public interest, and land governance. The latter is the set of rules, procedures, and structures that define and organize access to land, its use and transfer, and the resolution of land conflicts (C.T., 2009).

The Department of the Head of Government organized a conference at the national level on state land policy in late 2015 in Skhirat under the theme "National Conference on State Land Policy and its role in economic and social development". The debates were articulated on the one hand on land as a resource: collective lands, agricultural lands of the State, Habous (ADAD et al., 2015), Guich (land transferred in the past to tribes for use in return for a military-type service rendered), the public domain of the State, the rental of real estate, etc. On the other hand, it focused on administrative aspects: dual structure and diversity of legal regimes, land tenure security, access to decent housing, revision of the property tax base, protection and valuation of land, flexibility in drafting and implementing planning documents, urbanization, establishment of a global and integrated policy, the key role of land in town and urban planning, in promoting agriculture and rural development, in economic and social development, and constraints of the Moroccan land sector (Al Baouaba, 2015).

Currently, the head of the government requests the Economic, Social and Environmental Council of Morocco to prepare a study on the national strategy for state's land policy and to plan for actions for its implementation. This strategy aims to increase the efficiency and effectiveness of the state's land policy, by doing a comprehensive diagnosis of the situation of the land in Morocco, identifying the main existing constraints preventing its contribution to the economic and social development and making proposals of an integrated and efficient land and real estate policy (P.M., 2018). The Millennium Challenge Corporation (MCC) and the Government of Morocco are partnering to implement a $450 million compact to address two major constraints to economic growth; land productivity is one of them (MCC, 2018).

## VOLUMETRIC ASPECTS AND TYPOLOGY OF THE MOROCCAN LANDS

The Kingdom of Morocco comprises more than 71 million hectares (710 850 $km^2$). It is composed of local authorities who are legal persons of public law. The territory of Morocco has 12 regions, 75 prefectures and provinces, as well as urban and rural communes (PNCT, 2017). The population of Morocco is distributed over 1538 communes: 256 urban and 1282 rural. The following paragraphs present indicators on the amount of data and property typology of the Moroccan land administration domain.

The flows of land transactions recorded for properties subject to land registration at the time of their registrations in land registers testify to the order of magnitude of the volume of Moroccan land in general. The National Agency for the Land Registry, and Cadastre and Cartography presents its key figures (ANCFCC, 2017), which testify to the level of geographic and land infrastructure and the quality of its information assets. See Figure 1.

With regard to the housing stock, the results of the general population and housing census in 2014 show that the legal population of the Kingdom reached 33,848,242, including 33,762,036 Moroccans and 86,206 foreigners, as of 1 September 2014. The number of households is 7,313,806. The Figure 2 shows more indicators on population, household, and dwelling in urban and rural areas.

*Figure 1. ANCFCC key figures in 2017*

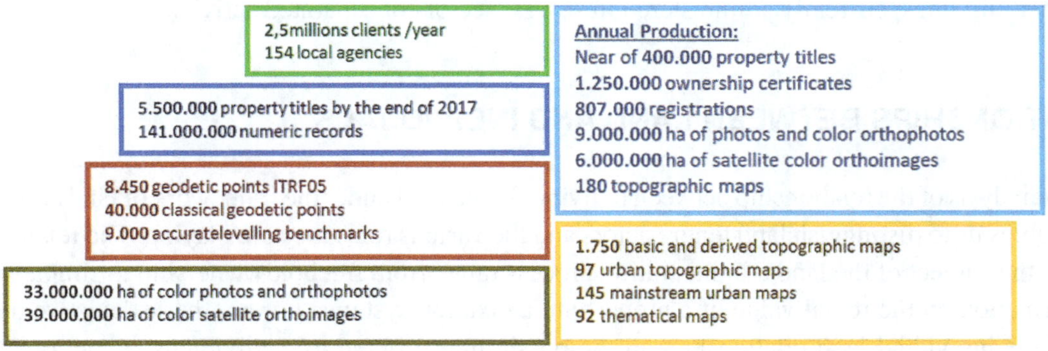


*Figure 2. RGPH indicators of 2014, population, households and dwelling*

| | Population and Households | | | Housing Type | | | | | | Occupancy Status | | |
|---|---|---|---|---|---|---|---|---|---|---|---|---|
| | Municipal Population | Household | Average size | Villa | Apartment | Traditional House | Light Dwelling | Rural Dwelling | Others | Owner | Tenant | Other |
| Urban | 20 284 436 | 4 807 743 | 4,2 | 4,5 | 17,5 | 70,4 | 5,2 | 1,3 | 1,0 | 62,7 | 27,6 | 9,6 |
| Rural | 13 325 648 | 2 506 063 | 5,3 | 0,8 | 0,3 | 30,7 | 3,1 | 64,1 | 1,0 | 89,6 | 1,9 | 8,5 |
| Total | 33 610 084 | 7 313 806 | 4,6 | 3,2 | 11,6 | 56,8 | 4,5 | 22,8 | 1,0 | 71,9 | 18,8 | 9,2 |

For agricultural units, the 1996-1997 general census of agriculture identified 1,431,633 farm units, with a useful agricultural area of 8,732,200 Ha. The number of parcels was 9,528,685 with 6.7 parcels per farm on average. The dominant legal statuses are Melk (which is privately owned land) and assimilated Melk, collective land, and the property of the state (sometimes referred to as the domain), with their shares in area being 75.79%, 17.69% and 3.09%, respectively. The useful agricultural area includes the Melk status, which covers almost one million hectare, about 270,000 Ha of state land, 240,000 Ha of Guich land, and 59,000 Ha of Habous land (Akesbi, 2001).

Commercial real estate is made up of office space, retail properties, industrial premises, warehouses and storage facilities. Its size is expressed in the total square meters covered by each of these types. By way of example, in 2010 Casablanca had 2,000,000 m$^2$ of office space, 7,000,000 m$^2$ of industry, 2,300,000 m$^2$ of warehouses and storage rooms, and 2,300,000 m$^2$ of retail space (El Bouanani, 2011). The number of units, on the other hand, is more revealing in terms of land administration.

One example of a land category characterized by large areas is the collective lands. These lands cover an area of 15 million hectares and are inhabited by 4563 ethnic collectivities. The ethnic collectivity enjoys legal personality and is known as a tribe, tribal party, douar, or any ethnic group (CETC, 2017). Another example is forest lands, which cover 9,037,720 hectares, and are composed in order of decreasing importance of hardwoods, Alfa, conifers, reforestation and matorral. (HCEFLCD, 2017).

The volumetric and typology indicators of the field of Moroccan land administration presented above demonstrate the large scale of this domain. On the one hand, the urban and rural population characterizing the user potential of land, and on the other hand, the quantity and diversity of land parcels and dwellings and other buildings, currently administered in several sectors of separated activity.

## RELATIONSHIPS BETWEEN LAND AND INDIVIDUALS

A deep analysis of the relationship between individuals and land under the framework of land administration enables us to distinguish land tenure actors and the various roles they are playing. The relationships that are the subject of the land administration process range from freehold ownership to simple interest in information on the rental value of a property to a taxation systems. For a rather exhaustive description of the individual-land relationship, all the possibilities offered to "individuals" to act on "lands" should be considered. These relationships are wide and of various consequence and depend on several parameters. We distinguish between impact and non-impact relationships through a given relationship link. See Figure 3.

The so called impact occurs when some property features are changing, namely the delimitation, the content, the ownership right, and the zoning and planning designation. Here are some applications in several sectors of activity.

- In the domain of property right, managed by the National Agency of Land Conservation, Cadastre, and Mapping (ANCFCC), courts, statutory notaries, customary notaries (adouls), surveying engineers, lawyers, etc., alterations have impacts on the identity of the property owner, on the type of right, on the extent and object of ownership and on the delimitation and contents of land ownership.
- The development of land, urban planning, setting land zoning regulations, change in the use of real estate, ensure land and dwelling equipment, etc. (with impact).
- For professionals such as engineers and construction contractors (without impact), identifying the state of the property at a given moment for a specific purpose: drafting of property deed, registration of the property in registers, delimitation or demarcation of the property, estimate of market value, etc.
- Taxation (without impact), operation of collection of property tax.

*Figure 3. Relations INDIVIDUALS - LANDS*

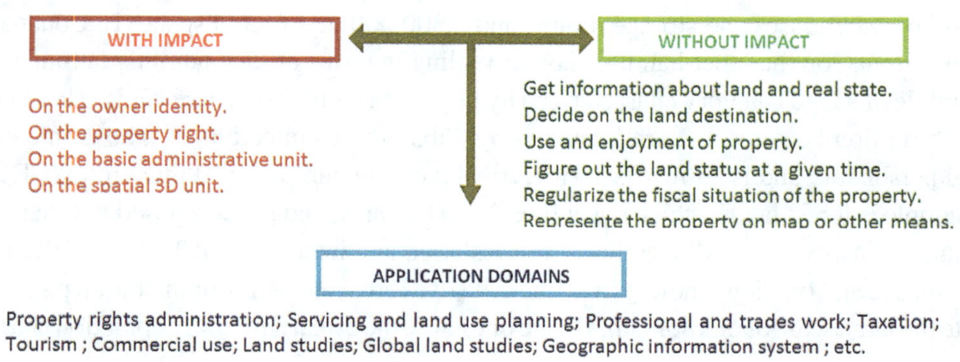

- In commercial use (without impact), having information on the property and the owner for commercial purpose in real estate agencies (sale, lease), banks or financial institutions (mortgage, foreclosure).
- At the tourist service (without impact), offering orientation or tracking information.
- In global studies (without impact), identifying for a given territory its parcel status and its legal status with a view to land consolidation, overall registration, urban development, agricultural development, national agricultural register, etc.
- In land studies, identifying the extent of any land status at the national or regional level to decide in relation to any problem in connection with a sectoral project of upgrading or development of land.
- Presenting thematic maps by manipulating and analyzing land data using geographic information systems, as a system for decision making.

## MOROCCAN LAND ADMINISTRATION DOMAIN

"Individual-land" relations highlight areas of activity which we will briefly recall in order to detect the most important aspects of land administration. Each of these areas is governed by general or special legislation and is framed by either empirical procedures or informal modes of management.

The reasoning on the nature and the priorities of the land administration process leads to this encapsulated sequence: Property - Owner - Shaping - Regulation - Protection. In the first place, this process is based on the raw material that is land, private or public. Secondly, this land is differentiated by the nature of its owner, the state or the non-state, natural or legal person. Thirdly, this property is subject to shaping (creation, change, use, enhancement, etc.). Fourth is taxation and urban and agricultural land regulations, which have a regulatory effect on land. Finally and fifth, the protection of the property is ensured by the regulatory provisions of property rights, either through the land registration system or other systems, governed by the code of real rights, special provisions, the principles of Malekite-rite Muslim law, the local customs, the administrative delimitation law, the expropriation law, the inalienability of public land, the Habous law or other provisions. See Figure 4.

Hence, we will approach the field of Moroccan land administration in five themes: land (state and non-state), the shaping of land, land and real estate as a tax base and urban and agricultural asset, the management of property rights outside of the land registration system and the registration system.

### State and Non-State Lands

The Moroccan Constitution guarantees the right of property; it also provides for the limitation of its extent and its exercise by law if the requirements of the economic and social development of the country require it (C.M., 2011). Private property not belonging to the state is relatively straightforward, unlike public or private state land (forest, highway, dam, lake, etc.) or lands intended for public interest (the collective domain, the Habous, the communal domain, etc.). The latter is protected by specific laws and special procedures, involving the local authority and public departments and prescribing precise and relatively well-defined deadlines to carry out the delimitation and publicity of landed property. These include the procedures for administrative delimitation, expropriation in the public interest and special procedures for land registration, which are particularly helpful in securing state property or that under its supervision.

*Figure 4. Land administration model*

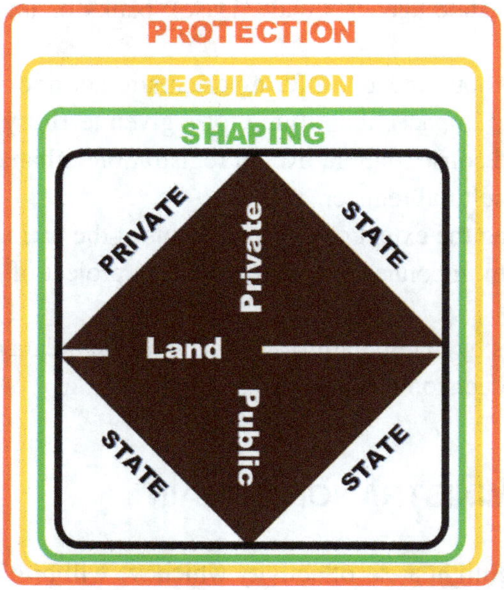


Land assets of the state and its dismemberments represent a national land store that can be used for housing, industry, tourism, socio-cultural facilities, etc. This led to the creation of the national land agency, launched in December 1979 in response to a royal directive at the Colloquium on Urbanism, Habitat and National Authenticity in Marrakech. The aim was to create an information center within the ANCFCC, responsible for collecting and disseminating technical and legal information related to this land heritage (El Guerraoui, 1986).

The Moroccan legislature has regulated procedures of administrative delimitations of three types of property of public interest and great potential in terms of area: the public domain of the state by the DAHIR of July 1, 1914 (BO n ° 62 of the 10 / 07/1914), the private domain of the state (including forests) by the DAHIR of 2 February 1916 (BO n ° 141 of 10/01/1916) and the collective domain by the DAHIR of 18 February 1924 (BO n 596 of March 25, 1924).

The municipal domain (B.O No. 470 of October 25th, 1921) and the domain of the rural communes (B.O No. 2177 of July 16th, 1954) are property of public interest. They are formed by the public domain (streets, gardens, monuments, slaughterhouses, souk, etc.) and by private domain (all the property owned by the municipality that has not been formally assigned to their public domain: buildings, plots of land, etc.).

The expropriation of land for the public interest (BO 3685 of 15 June 1983) is carried out by the state and local authorities as well as other public and private legal persons or natural persons to whom the public authority delegates its rights in order to undertake works or operations declared of public utility. Religious buildings of various denominations, cemeteries, buildings in the public domain, and military works cannot be expropriated.

## The Land and Real Estate Shaping

We refer to "land shaping" as any human action of voluntary transformation of the primary resource land and real estate or its zoning and planning designation for a given usage. The territory of the Kingdom of Morocco is divided by zones on the basis of planning documents, namely urban areas delimited into communes, suburbs, and urban planning groupings, in addition to rural. The main urban planning documents in force are the general plans for planning, the zoning plan, the development plan, the alignment orders and the alignment orders giving transferability to the municipal authority, and the development plans of agglomerations. Urban and rural planning in Morocco is essentially based on the guiding scheme for planning (strategic development orientations) and on the development plan (land use regulation).

As a land-shaping actor, the competent municipal or territorial authority authorizes the division, the subdivision, or the construction of single buildings or housing complexes. It issues a permit to inhabit or a certificate of conformity when the building is intended for a use designation other than the dwelling (industrial, tourist, commercial or craft lot) as it issues a provisional acceptance of the equipment and servicing work for the allotments. The favorable opinion of the Urban Agency ensures the conformity of the applicant's request with the planning documents. From the private sector, the participation of architects, surveyors and specialized engineers is usually mandatory. Together, they work in the process of shaping lands, either in defining their limits or in stating the kind of buildings or activities that are appropriate to it.

The general building regulations applicable since February 1, 2014 set the form and conditions of filing, examination of applications, and authorization issuance of allotment, creation and fragmentation of dwelling groupings, issuance of building permits, permit to inhabit, and certificate of conformity (DECREE 2-13-424). This regulation uses the designations "Project", "Petitioner" and "Authorization" in accordance with the provisions of DAHIR No. 1-60-063 relating to the development of rural agglomerations, to the law n ° 12-90 relating to the urbanism and the law n ° 25-90 relating to fragmentations, allotment and groups of dwellings. The processing of the applications for authorization is carried out according to a procedure of "big project" (public use building, commercial equipment of large surfaces, etc.) or a procedure of "small project" (individual buildings, etc.) by commissions of instruction composed of the representatives of the local and communal authorities and multiple administrations and actors.

The land redistribution (Dahir of may 31, 1938) and the land consolidation (Dahir of july 25, 1969) characterize particular cases of shaping and redefining land parcels. These are total revisions of the properties, starting from a given situation and changing to another that is totally different from the point of view of delimitations, areas and consistencies, although owners and property rights are retained. The former limits and locations of parcels are deleted from plans, databases and field and replaced by new ones. With the exception of these cases, operations are normally limited to dividing parcels or merging one to another. Therefore, it is relevant to note these special procedures and take them in consideration when building land data models.

## The Real Estate Taxation

Real estate taxation and fees play a revenue generating and regulatory role on land, as the annual tax encourages the full use of land and discourages land hoarding and claim rights on lands that are not valued. The taxes and fees due to land transfer serves to formalize these and cover the operating costs of their registrations (CT, 2009). Figure 5 summarizes the main direct and indirect property taxes and levies in Morocco (CESE, 2012).

A summary of real estate tax case studies will be discussed below (SEDDIK, 2010) as well as an overview on the agricultural tax, all are subject to any update of the finance law.

## Due Taxation on the Agricultural Incomes

The accounting and tax legislation has been constantly evolving since the exemption in 1984 from any present or future direct tax on agricultural income, which was previously subject to the agricultural tax introduced in 1961. In 2014, the government redefined agricultural income and began a staged system of gradual taxation, and in 2015 a new accounting plan for agriculture was developed to facilitate taxation (LAAROUSSI, 2017). The progressive taxation is expected in 2014, 2016, 2018 and 2020 respectively for gross income exceeding 35, 20, 10 and 5 million dirham.

## Registration Duty

Authenticated or private deeds (notarial, adoulary, hebrew, judicial or extrajudicial) which establish agreements on immovable property are subject to the registration duty process. This aims to acquire a certain date in the private agreements by means of their recording in a register called "register of entries" and to keep the records (assets transfer inter vivos, lease, and transfer of a right to the lease or benefit of a promise of lease, constitution or release of mortgage, partition, antichresis or pledge). The

*Figure 5. Immovable property tax scope*

**GENERAL TAX CODE**

- Income tax
- Registration fees
- Corporation tax
- Value added tax

**SPECIAL LEGISLATIONS (PARAFISCALE TAX)**

- Mining exploration permit
- Operating license
- Transfer tax

**FISCALITE AGRICOLE**

**LOCAL COMMUNITIES TAXES**

- Professional tax
- Housing tax
- Municipal services tax
- Tax on urban undeveloped land
- Tax on construction operations
- Tax on subdivision operations
- Tax on the sale of forest products
- Tax on mining operations
- Temporary occupancy charge of the communal public domain

**OTHER CASES**

- Property acquisition case involving registration fees (current rate of 6%) and Land registration fees (current rate of 1.5%).
- Acquisition of shares and shares of real estate and property-related companies (registration rate of 6% and 3% depending on the case).
- Assignment cases involving local taxes on housing, bare rent, professional assets and undeveloped land.
- Cases of sale of buildings.

taxable base for the settlement of duty is the value of the property, the bare ownership, the usufruct and the enjoyment of the immovable property. Proportional tariffs are at a rate of 6%, 4%, 3%, 1.5% and 1% depending on the purpose of the act. Reduced rates of 4% and 1.5%, fixed fees, and a minimum of collection are provided under specific conditions.

### Taxation of Land Profits

Land that generates profit is subject to the professional tax on land profits. This tax is payable by any natural or legal person of Moroccan or foreign nationality who practices a professional activity in Morocco. The tax base is the annual gross, normal, and current rental value of stores, shops, factories, workshops, sheds, yards, depots and all premises, locations, and facilities used for the exercise of professional activities. They are taxable according to their class, at 10%, 20% or 30% of the rental value. This value is determined for hotel establishments according to their price.

The habitation tax applies annually to all buildings occupied as dwellings, in urban or suburban areas, as well as resorts. Exemption cases are provided, in particular for public utility buildings. This tax applies to the annual rental value (average rent) with a revision every five years (+ 2%). A tax abatement of 75% is provided for primary dwellings. The rate applied to the rental value varies between 10%, 20% and 30% for values ranging from 5,000 to more than 40,000 DH.

The municipal services tax applies to all buildings considering the total amount of rent or rental value, at the rate of 10.50% for properties located in the perimeter of urban communes, delimited centers and resorts, and 6.50% for properties located in the peripheral areas of urban communes.

For self-delivery, consisting in taking charge of building construction operations on one's own, the tax base for VAT is constituted by the cost of the construction; the exemption of this VAT requires that the covered area of the building is less than 300 m2 and a minimum duration of primary personal residence greater than 4 years.

## MANAGEMENT OF REAL RIGHTS NOT UNDER LAND REGISTRATION

### Real Rights

The law No. 39-08 relating to the code of real rights (rights in rem) is presented as the general legislative framework governing land ownership and the real right, as long as it does not conflict with special land legislation. The aspects and cases not covered by this law would be governed by the Dahir forming the Code of Obligations and Contracts (BO, Sept. 12, 1913), failing that, by the customs and precepts of the Malikite school of Islamic jurisprudence.

The real right allows the use, enjoyment and disposal of the property, above and below, unless it is subject to limitations imposed by law or agreement. This property can only be expropriated for reasons of public utility. We distinguish between collective ownership in joint ownership (property, walls, and private roads), easements (natural, legal, or born of agreement between owners), land charges (right of usufruct, right of Omra, right of enjoyment, right of surface area, emphyteusis right, Habous right, Zina right and Houa and Elevation rights). There are also ancillary rights (liens, conventional mortgages, and official mortgages).

## Customary Rights

These constitute dismemberments of the right of property, originating in the customary Muslim right of Menfâa (enjoyment). The study of the various customary rights brings out the following rights (DECROUX, 2007, P.450): right of Guelza or Kholou or Halaoua, right of Gza or Istidjar, right of Zina, right of houa, and right of Hazaka. Added to this list the customary status of the Berber tributes (LAFUENTE, 1999), object of the Dahir of May 16, 1930, according to which the civil or commercial, movable or real actions are judged by special customary courts (BO No 918 of May 30, 1930, page 652). The deeds and inscriptions of registers produced in the spirit of this regime continue to have the effect of proof of ownership even now, possibly against other evidence.

## Holding of Melk Property Not Submitted for Land Registration

Under the real rights code and even under its promulgation, the courts keep records of real estate properties that are outside of the land registration system, including the notarial register and the register of conservatory foreclosures. The Adular acts before the promulgation of the Dahir of July 17, 1914 regulating justice and the transfer of real estate property are given to the parties concerned without registration in a particular register. This Dahir, as subsequently amended, notably in 1944, 1983 and 1993, set up the notarized register, to be kept by the judge to record the acts in chronological order of arrival. In addition, the 1974 Civil Rights Code established the register of conservatory seizures of immovable not subject to the land registration. This register also serves for the publicity of the seizure, of which a notice is also posted for 15 days in court at the expense of the beneficiary of the seizure (Driouch, 2003).

From the context of the general cadastre, we note the provisions of the Dahir No. 1.73.163 of 31 May 1973, establishing the National Register of Land Property (NRLP), which provides for registration of all rural properties including those outside of the land registration system. This register is to be kept for each rural commune, and must provide information on all of its properties and their apparent owners or holders of real rights (Adad & Semlali, 2015). These works has not been carried out so far, in our opinion, due on the one hand to the legal background and the optional nature of the Moroccan cadastre (no urgency to be generalized), and on the other hand to the difficulty of recovering back the cost to spend for establishing a widespread cadastre, having a non-tax purpose though and not auto financed consequently.

In the same context, we also note that the Ministry of Agriculture and Maritime Fisheries (MAPM, 2016) launched the National Agricultural Register (NAR) through a general census of agriculture in 2016 in support of the Green Morocco Plan, which aims at developing a high-performance and competitive agricultural sphere while preserving the activity and the income of small vulnerable farmers working in difficult conditions. It is a five-year program of periodic surveys designed to update data on economic and social structures. The NAR will be the first legally-based farmers register and will provide the list of farmers, with a unique identifier (1.5 million), and their farms (useful agricultural area of nearly 9 million hectares).

## Management of Real Rights Submitted to the Land Registration

The scope of the Dahir of August 12, 1913 as modified and completed by subsequent amendments, relating to the land registration was extended to the entire national territory in 1966 (Royal Decree, no 114-66,

BO no 2817, October 26,1966), thus unifying various regimes applied by regions of Morocco before that date (DGUA, 2007). This system consists, on the one hand, of registering an immovable property, creating a land title which purges all prior rights, and on the other hand to subsequent formalities of land registration that record on the land title thus established any deed or act constituting, transmitting, modifying, recognizing or extinguishing real rights or land charges relating to the immovable property concerned. Each property getting a title is represented individually on cadastral plan at appropriate scale, with the property's area in square meter and tied to the national geodetic network based on the Lambert Conformal Conic projection. The preliminary formalities of registration can be ordinary (publication of the extract of requisition to the BO, boundaries delimitation, drawing up of the plan, closure of the operations of demarcation, publication of the notice in the Official Bulletin and registration) or special (governed by special provisions), with publicity of the boundary closure notice to the BO or without. The principal acts and facts subject to registration on land titles are transfers, inheritances, mortgages, subdivisions, concordances, condominiums, amalgamations, civil status, change in property name, issuance of new duplicates, release of mortgages, foreclosures, seizure releases, and prenotations. Any holder of a right in a property in the process of being registered may request its publication in the BO by means of an excerpt (art. 83) or its deposit pending its purification in concomitance with that of the parent requisition (art.84).

Cases of grouped land registrations are provided by special provisions namely: the general land registration (Dahir No. 1-69-174 of July 25, 1969), the land registration of consolidation (Dahir No. 1- 69-32 of 25/07/69) and the group land registration (art. 16 amended by the Dahir of April 5, 1938). Obligatory Registration (OR) areas are provided for by Law 14.07 (Dahir No. 1-11-177 of November 22, 2011), although so far this possibility has not been exploited.

With regards to the field of development and encouraging investment in the agricultural sphere, Morocco is proceeding in the "melkization", or conversion to private individual property, of agricultural lots in two contexts. The first concerns subdivisions of the agrarian reform regulated in 1972 (notably by the Dahir n ° 1-72-277 of December 29, 1972 relating to the attribution to farmers of agricultural lands forming part of the private domain of the state) and in 2005 (Dahir 1-04-251 of January 7, 2005 relating to the transfer of agricultural lots to regular farmers). The second concerns the subdivisions of collective lands located in the irrigated areas (Dahir n ° 1-69-30 of July 25, 1969 relating to the melkisation in favor of entitled beneficiaries belonging to ethnic communities). These properties, initially collective domains under the authority of the Ministry of Interior, extends over hundreds of thousands of hectares and benefits tens of thousands of private beneficiaries. These procedures have undergone notable simplifications, namely: the lifting of prohibitions and binding obligations for the subdivisions of the agrarian reform in 2005 and the exemption from the fees of the land registration of the collective lands located in the irrigated perimeters in 2016.

Mining rights are classified by the mining code in the category of immovable property and are subject to the registered property regime, even if they are located offshore in the ocean. Following the granting of a mining right by the minister in charge of mines, the property registrar simply establishes the mining title, without any particular procedure and without effect of purging all other rights (DECROUX, 2007, P.119).

Let us note the cases of buildings subject to the regulations of condominium. The law 18.00 as modified and completed by subsequent amendments on the status of the co-ownership of buildings and law 01.07 on the status of real estate tourism promotion (hotel, guest house, convention center, etc.) highlight as a unit of ownership a lot composed of a private party and a share of the common portions.

This unitary property acquires a very specific land title, derived from the ground title and located in the 3D space of the condominium building. Another case of co-ownership is provided by Law 25.90, that of land subdivisions with common equipments and common spaces such as streets, gardens or parks.

## DISCUSSION

The above presentation of the Moroccan land administration domain highlights a multitude of sub-areas of interest, each with its own group of stakeholders, its relative significance in terms of area, and its informational supports, both in paper-based systems and computerized systems. They proceed separately to the administration of the land according to their jurisdictions and perspectives. Difficulties arise whenever we want to set up a single and unified Moroccan land administration system in order to integrate each of these sub-domains. These difficulties are due upstream to the multitude of adopted analog and digital data models and a lack of agreed and standardized terminologies of the concepts and terms in use, and downstream to the difficulty of interoperability between the data coming from the different domains and sub-domains. The cost of acquiring, handling and maintaining land data is thus multiplied: multiple production of similar data and redundancy, lack of compatibility, impediments to synergy during dematerialization efforts carried out by each domain all over the country, etc.

Take the example of land information related to an apartment and involving various stakeholders (owner, ANCFCC, court, tax and registration services, bank, municipality, urban agency, etc.). It was partially or totally duplicated in the files, registers and databases of each of these stakeholders, named and held in different ways, and required the same efforts of acquisition, registration and updating. There is a lack of symmetric and standardized data disseminated among the various stakeholders.

Generally speaking, any immovable property raises a field of interaction for any intervener; see Figure 6; characterizing the process of access to land by a stakeholder, by means of part or all of the dismemberment of the right of ownership, through a role, relying on a document, and highlighting data. See Figure 7.

In this context of interaction, to access land and build structures on it, their uses and their transfers require the control of the land's information, and this requires cross-checking the data coming from several sources and holders. The consensus on the concepts, the relations between entities, and the technical and legal details of the land domain is essential to operate and rationalize land governance, which is by definition based on information about it. Some additional prerequisites are the standardization approach, the use of new information and communication technologies, and modeling as a tool for reengineering systems and a basis for communication between different systems as well as developing integrated information systems. That is helpful to master identifying immovable properties and its components, transactions, relationships between properties and persons, classification of land statuses, land use, land value, and finally to map representations of land and real estate.

## SOME CADASTRE-2014 ASPECTS

In the spirit of the conceptual framework of 'Cadastre 2014' of the International Federation of Surveyors (FIG), especially its first, second and sixth statements, which are respectively "Cadastre 2014 will show

*Figure 6. Types of interactions involved in access to immovable property*

| STAKEHODER | ROLE | DOCUMENT | DATA |
|---|---|---|---|
| **Owner, Objector, Tenant** | Usus, abusus, fructus | Deeds, land register copy, plans, certificate, rental agreement | Common information between stakeholders and parties |
| **Registration** | Registration fees | Register of entry, copy of deed and convention | Value, Price et Charges of formalities and transactions |
| **Imposition and tax** | Professional tax, Housing tax, municipal tax, profit tax, VAT | Real estate price reference data | Rent value, professional works, location |
| **Cadastre** | Demarcation, Survey and handling cadastral information | Minute of demarcation, calculation, plan, cadastral map | X and Y Coordinates, positioning, size and measurements, areas |
| **Land registry** | Land registration, land ownership, land charges and easements | Land register, certificate, zoning of venal value | Owner identity, real rights, charges and easements |
| **Cartography** | Representation at small scale | Map, urban plan | Position, dimension, thematically |
| **Bank** | Housing loan | Mortgage | Identifying the owner, loaner, property, and bank loan value |
| **Rural/Urban Commune** | Subdivision/construction approval, rural lands national register | Register, Permit, license, project files, maps, plans | Situation affectation et utilisation du bien, équipements |
| **Urban Agency** | Town planning, Instruction and Issuance of authorization permit | Information sheet, urban planning regulations | Land use designation, urban planning regulations |
| **Urban Planning** | Planning and managing the land | Planning policy documentation | Land designation |
| **Court** | Judgment on conflict about land or eviction of a tenant | Judgment, notarial register, register of provisional seizures | Identifying parties, lands, deeds et contracts, acts outside land registry |
| **Real estate agency** | Marketing of the immovable property | Plan, certificate, sheet | Venale value, price, seller, buyer |
| **Water, electricity services** | Servicing, utility, water, electricity and sewerage services | Files, bills, statistics | Identification and positioning of real estate, Consumption, statistics |
| **Agriculture** | Agricultural development | Census, maps, statistics | Soil class, agricultural activity |
| **Professionals** | Professional works : engineers, Architects, Adouls, Lawyers | Document, maps, plans, censuses, expertise, measurements | Common Information between stakeholders |

*Figure 7. Model of the access process to the immovable property*

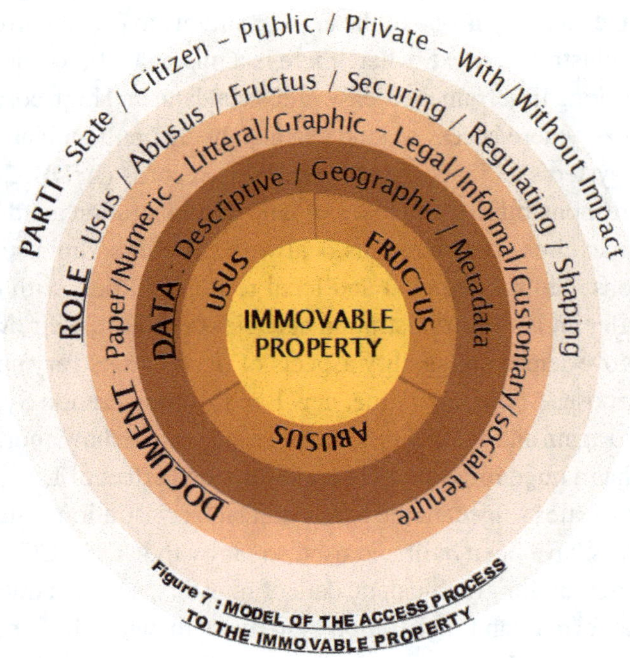

the complete legal situation of land, including public rights and restrictions" and "The separation between 'maps' and 'registers' will be abolished" and "Cadastre 2014 will be cost recovering"(KAUFMANN, 1998); the Moroccan's provisions and good practices previously cited have helped in achieving decent results so far, given the bases of comparison of the area of Morocco and the useful agricultural area.

- The boundaries of properties within large areas of collective land and forest land have been fixed by means of the procedure of administrative delimitation. These lands are covered by the formal registration system (titling system) in majority and will be totally in few years following the ongoing works.
- Hundreds of rural communes are covered by the overall registration. The formal registration of these lands is in rapid progress. An ongoing ambitious program aims to cover annually more and more rural communes by the overall registration procedure.
- The 5.5 million property titles cited previously produced so far guarantied the land registration of the quasi majority of urban areas, most of valuable rural areas and lands cited above.
- The OR, the NRLP and the NAR cited previously could widespread the cadastre works; some comments are given before for the first and the second one, the last one is being implemented.

## INTERNATIONAL PERCEPTIONS

According to the study entitled "Morocco's Growth Diagnostic", resulting from the joint commitment of the Government of the Kingdom of Morocco, the African Development Bank and the Millennium Challenge Corporation, the Government of Morocco expressed the wish to have an objective diagnostic analysis of economic growth. The analysis noted that in Morocco, land remains a key determinant of domestic and foreign investment and that land tenure plays a significant role in boosting economic activity. It also points out the inefficiency of Morocco's land administration, which aggravates the constraints in many land domains. This inefficiency is one of the key constraints to private investment development in Morocco, whether in the industrial context, where there is a mismatch between the quality of industrial land and the needs of firms, long timeframes to lease industrial land and high costs for obtaining permits, or in the agricultural context, where access to land can also be difficult and land rights can be insecure due to overlapping tenure systems and incomplete ownership records (AfDB, 2014).

In many countries the responsibilities and tasks of land administration are distributed among different organizations. Sometimes those organizations deal with different administrative territories, which may have other subdivisions: central, regional and local responsibilities, with either public or private roles. Quite often these different land administration agencies each have their own information systems and data. As a result, the governance and quality aspects of the data sets vary and can be incomplete or outdated and not fit for purpose. At the same time, new Land Administration Systems (LASs) are being developed all over the world again and again. Sometimes countries even have more than one IT-system for land administration. This has a huge impact on the continuity and effect of LASs (Lemmen et al, 2013).

Actually, just for the properties subject to the Moroccan land registration system (which dates back to the year 1913), several LASs have been built and used so far by the ANCFCC to manage cadastral and land registry information, mainly for graphic data, data of registers, scanned documents and transaction data. Properties outside the formal land registration system are managed by other systems within many other departments and agencies, or by customary notaries (ADAD et al., 2015).

There is a need for domain specific standardization to capture the semantics of the land administration domain based on the agreed foundation of basic standards for geometry, temporal aspects, metadata, and observations and measurements from the field. A standard is required for communication between professionals, for system design, system development, and system implementation for data exchange and data quality management. Such a standard will enable GIS and DBMS providers and/or open source communities to develop products and applications for land administration purposes. In turn, this will enable land registry and cadastral organizations to use the components of the standards to develop, implement and maintain systems in an even more efficient way (Lemmen, 2012).

This International Standard Land Administration Domain Model (LADM: ISO 19152) as a conceptual model brings a formal language for describing existing systems and provides a reference model which serves two goals:

- To provide an extensible basis for the development and refinement of efficient and effective land administration systems, based on a Model Driven Architecture (MDA), and
- To enable involved parties, both within one country and between different countries, to communicate, based on the shared vocabulary (that is, an ontology), implied by the model (ISO, 2012).

The LADM supports the full continuum of land rights as well as a continuum of accuracy, of land recordation's, of types of spatial units, of types of parties involved, and of data acquisition approaches. As a result, LADM allows for a flexible, step by step approach in the development of a Land Administration based on the needs, priorities and requirements of users and society. This can be combined in a natural way with organizational development with a proper alignment to ICT development. This makes the concept of LADM a basis for strategic development in land administration (Lemmen et al, 2013).

Following are a few examples in which LADM model based studies have contributed to the integration of multiple agencies concerned by land administration, the upgrading of the cadastral and registry systems from 2D to 3D and its improvements, the resolution of redundancy of data systems, and to avoid the LASs drawbacks cited above.

In Malaysia, the integration of the systems of two agencies, the Land Office and the Department of Survey and Mapping systems is proposed using LADM model (Amalina 2014). A prototype was developed for the conceptual model consisting of several novel aspects such as 3D representations, full version management, linking of database information and source documents, and basic property units (Amalina et al. 2014). The model of 3D topology was introduced to enhance the volumetric description approach adopted in cadastral registration system in Malaysia (Amalina, 2015).

A 3D land administration model is proposed using a cadastral resurvey form, including the representation of 3D physical properties and 3D rights. This study followed the proposition in 2012 of the country profile of the Land Administration Domain Model (LADM) to the ISO/TC211 by the Korea and the announcement of the 'the Korean special Cadastral Resurvey Act' announced in 2014. This law includes the representation of 3D partitioned areas for buildings and underground facilities into the land information system. This new version of Korean country profile was designed to represent all types of cadastre information (Bo-Mi Lee et al, 2015).

To resolve the redundancy within the Croatian land administration-related official registers and the weak or non-existing linkage between them (registers of persons, registers of properties and registers of rights), a linking based on the Land Administration Domain Model (LADM) was introduced. This research aims to create a strategy for improving the effectiveness of the overall system of registers (Mario Mader et al, 2015).

The derivation of the Portuguese model and the creation of an integrated legal cadastre was supported by a system development methodology based on the generic and worldwide nature of the Land Administration Domain Model and taking in consideration the interrelations between the technical, legal and organizational aspects (Hespanha, 2012).

As a new data model, the LADM was introduced to enhance the existing Cyprus Land Information System (CLIS), to facilitate the management of data, to reduce data duplication, and to improve data integrity. This investigation concluded that the adoption of LADM was a great opportunity for the Cyprus Department of Lands and Surveys to introduce an ISO standard model, improving and expanding the services provided by CLIS and the broader land administration system to the Cyprus community (Elia et al., 2013).

The LADM is a generic domain model. It is expandable and adaptable.. Additional attributes, operators, associations, and additional user-defined elements and classes are required and can be added for specific regions or countries. Country profiles are used for customizing the LADM to meet specific needs. Annex D of the LADM standard document mentions eight country profiles using the LADM, namely for Portugal, Queensland and Australia, Indonesia, Japan, Hungary, The Netherlands, Russian Federation and Republic of Korea. Each of those profiles demonstrates specific usages of the LADM model (ISO, 2012).

## PROPOSAL

The solution proposed as a result of this study is based on two points:

- The development of a unified conceptual model of data of Moroccan land administration model, supported by an ontology model and covering a comprehensive taxonomy of this domain. This model to be developed for the Moroccan context can be formalized later in the Land Administration Domain Model (LADM - ISO 19152: 2012); in order to benefit from the efforts made in this field by the International Organization for Standardization.
- The establishment of a government structure in charge of holding and sharing this model with the various holders of geographic information in general and land in particular. Broad national consultation on this issue is necessary to ensure that all areas and sub-domains of Moroccan land administration adhere to this goal and enable them to take the necessary measures to transition to the new unified model.

## CONCLUSION

The Moroccan land domain is rich in its many facets. This testifies to its importance for its owners, beneficiaries and users. The efforts made by the many public and private actors in the land sector for its control and management are felt in the regulatory and procedural systems put in place. Any rationalization in the field of Moroccan land administration, especially that introducing a data conceptual model should take into account the concepts and entities presented in this article explaining the cases of formation of unitary land ownership according to its status, its stakeholders, and its regulatory and procedural contexts.

The unified and standardized handling of information about the relationship between "individuals" and "land" is of great relevance, given the context of a highly diversified and asymmetric system of land administration, as detailed above. A standardized perception of the Moroccan land domain will facilitate informational and economic flows of exchanges and interoperability between the various actors of the land and increase the synergy in any project of land valuation.

Our contribution above in the Moroccan land governance distinguishes the ontology of the Moroccan land administration domain, through the identification of the sources of formation of the Moroccan land ownership and in relation with the multiple stakeholder and actors of its administration. A land administration model and a land tenure model are proposed in the general context for a conceptual land data model.

## ACKNOWLEDGMENT

Our sincere thanks are dedicated to professionals in the field of land, authors of opinions, advice, participation, and writing, who have supported the drafting of this paper.

## REFERENCES

Adad, M. A., & Semlali, E. H. (2015). *Etude d'un modèle pour les données foncières et cadastrales marocaines à travers le modèle du domaine d'administration des terres*. Istanbul, Turkey: WCS-CE - The World Cadastre Summit, Congress & Exhibition.

African Development Bank (AfDB). (2014). *Morocco's Growth Diagnostic: Identifying Morocco's Binding Constraints to Broad-Based Growth*. Author.

Akesbi, N. (2001). Les exploitations agricoles au Maroc: Diagnostic à la lumière du recensement Général Agricole. *MEDIT, 4*.

Al Baouaba & Al Watania. (2015). *Portail national des collectivités territoriales*. Ministère de l'Intérieur.

Amalina, N. A. (2014). *Adoption of land administration domain model for land administration in Malaysia*. (Thesis).

Amalina, N. Z., Abdul Rahman, A., Hassan, M. I., & Tan, L. C. (2015). Conceptual Modelling of 3D Cadastre and LADM. WCS-CE - The World Cadastre Summit, Congress & Exhibition, Istanbul, Turkey.

Amalina, N. Z., Abdul Rahman, A., Jamil, H., Teng, C. H., Tan, L. C., & Looi, K. S. … van Oosterom, P. (2014). *Development of a Prototype for the Assessment of the Malaysian LADM Country Profile*. FIG Congress 2014 Engaging the Challenges – Enhancing the Relevance, Kuala Lumpur, Malaysia.

ANCFCC. (2017). Retrieved from www.ancfcc.gov.ma

CESE. (2012). *Le système fiscal marocain: développement économique et cohésion sociale*. Conseil Economique, Social et Environnemental, rapport du conseil économique et social, Auto-Saisine n° 9/2012.

C.M. (2011). *Constitution Marocaine. Royaume du Maroc*. Secrétariat Général du Gouvernement, Direction de l'Imprimerie Officielle.

C.T. (2009). *Gouvernance foncière et sécurisation des droits dans les pays du Sud*. Comité Technique du Foncier et Développement, Agence Française de Développement internationale.

Decroux, . (2007). *Ouvrage Droit Foncier Marocain*. Editions la Porte.

DGUA. (2007). *Rapport final de l'étude portant sur l'analyse de la problématique foncière sous le régime du DAHIR khalifien de 1914, et son impact sur le développement urbain dans les provinces du nord, Direction générale de l'urbanisme et de l'architecture*. Retrieved from www.muat.gov.ma

Driouch. (2003). *Ouvrage « les origines du système d'immatriculation foncière »*. Les publications de la série de la connaissance juridique. DECRET n° 2-13-424..

El Bouanani. (2011). *Immobilier d'Entreprise au Maroc*. ICH Rabat INAU.

Elia, E. A., Zevenbergen, J. A., Lemmen, C. H. J., & van Oosterom, P. J. M. (2013). The land administration domain model (LADM) as the reference model for the Cyprus land information system (CLIS). Survey Review, 45(329).

Guerraoui. (1986). *Les Problèmes Fonciers et l'Agence Foncière Nationale*. Actes du Symposium sur la Gestion Foncière au Maroc et en Tunisie, Barcelone.

HCEFLCD. (2017). *Inventaire Forestier National. Haut-Commissariat aux Eaux et Forêts et à la Lutte Contre la Désertification*. Retrieved from http://www.eauxetforets.gov.ma

Hespanha. (2012). *Development Methodology for an Integrated Legal Cadastre Deriving Portugal Country Model from the Land Administration Domain Model* (PhD thesis). Geboren te Lisboa, Portugal.

ISO. (2012). Modèle du Domaine d'administration des Terres (LADM). ISO 19152: 2012(F), Information géographique - première édition 2012-12-01, Projet de Norme Marocaine 2016 (Indice de classement IC 17.8.417).

Kaufmann & Steudler. (1998). Cadastre 2014. A vision for a future cadastral system. FIG Commission, Brighton, UK.

Laaroussi. (2017). *Expert-comptable, commissaire aux comptes*. Cabinet EurO-CAF Consulting. Retrieved from oec-casablanca.ma/img/uploads/IR_agricole_10_01_17.pptx

Lafuente. (1999). *La politique berbère de la France et le nationalisme marocain*. Academic Press.

Lemmen, C. (2012). *A Domain Model for Land Administration* (PhD thesis). Faculty of ITC, The Nederland.

Lemmen, C., van Oosterom, P., Uitermark, H., & de Zeeuw, K. (2013). *Land administration domain model is an ISO standard now*. The World Bank.

Mader, M., Matijevi'cb, H., & Roi'ca, M. (2015). Analysis of possibilities for linking land registers and other official registers in the Republic of Croatia based on LADM. *Land Use Policy*, *49*, 606–616. doi:10.1016/j.landusepol.2014.10.025

MAPM. (2016). *Le recensement agricole, une étape essentielle pour la création du Registre National Agricole*. Retrieved from http://www.agriculture.gov.ma

MCC. (2018). *Millennium Challenge Corporation of the United State of America*. Retrieved from www.mcc.gov

P.M. (2018). Letter to the head of the Economic, Social and Environmental Council of Morocco. No. 159. January 22, 2018.

PNCT. (2017). *Portail National Des Collectivités Territoriales*. Ministère de l'Intérieur.

Seddik. (2010). *Expert-Comptable DPLE*. Fiscalité Immobilière. ICH Rabat INAU.

# Chapter 4
# A Conceptual Framework for Implementing a Multipurpose Cadastral System in Morocco

**Moha El-Ayachi**
*Institut Agronomique et Vétérinaire Hassan II, Morocco*

## ABSTRACT

*For a decade, many countries have re-engineered their cadastral systems focusing closely on organizational and technological aspects in order to meet requirements of the new digital era. Appropriate configurations of the cadastral systems have been designed. The cadastral system, as a public institution, has a large relationship with all kinds of customers and includes a number of structural, operational, and functional processes that must be improved. This chapter has the aim to illustrate a new conceptual framework by describing the requirements and the technical perspectives of the Moroccan multipurpose cadastre. Basic concepts are discussed in accordance with various assumptions to provide a deeper understanding of the vision and the new structure. The development of an integrated system for administering land and the establishment of the cadastral information systems are carried out to highlight the fundamental aspects of the multipurpose cadastre.*

## INTRODUCTION

The driving forces behind re-engineering the cadastral systems are the new customer needs, the new global changes around the world, the technology evolution, and the state requirements. The cadastral system customers are of two kinds: individuals and group of individuals. The needs of both kinds of customers are significant and should be dealt with great and specific attention. The major customer demand is the system efficiency, which means the ability of the cadastre to determine exactly their properties and provide in a short time the related ownerships. The global changes affect, on the other hand, the nature of the cadastral system to develop. Many efforts may be conducted to improve the certainty of land rights and the security of tenure. Under the globalization, the cadastral system should serve the on-going reforms around the world and serve the urgent requirements of land and territory planning to support

DOI: 10.4018/978-1-5225-5939-9.ch004

international companies that will use land inside Morocco. A new system must be developed to increase security on land and to allow its use with respect to local land formalization regarding obligations and restrictions. Through the development of a multipurpose cadastre, various benefits will arise and will have direct impacts on economic and social life. The current technology has seen major increases in different domains. The cadastral system is concerned by these new technological improvements. Varieties of advanced tools were recently introduced that will accelerate, in the future, the processes of surveying, land titling, registration, and mapping. The State, as a primary landowner, needs to establish an efficient system of cadastre to increase the security on land and facilitate land administration. This allows for monitoring of land market, the improvement of planning in urban and rural areas, the regulation of legal framework and statutes of land, and the introduction of new technology to maintain land reform such as redistribution, consolidation, valuation, and assessment. In addition, the multipurpose cadastre will ensure within the society an optimum level of quality and establish a uniform land information system based on properties and parcels.

The implementation of a multipurpose cadastre requires various technical steps to collect data from and for different institutions. This new system will provide an extensive inventory of land in both rural and urban areas. The establishment of regular cadastral sections will serve as a fundamental framework to generalize the land inventory process and the creation of large-scale maps. These will be used as references for the implementation of national land information system. All thematic resources can be added in an easy, progressive, and flexibly manner.

This chapter describes the technical perspectives needed to implement the components of the multipurpose cadastre at different levels. The technical procedures will not be dealt with this study. They must be developed in a national commitment gathering the specialists of various domains. A number of competencies and experts will carry out great efforts to improve technically the development of this new system.

## THE TECHNICAL REQUIREMENTS FOR THE MULTIPURPOSE CADASTRE

### Introduction

The fundamental network is a geometric foundation that allows a proper spatial linkage between features across an area. All the locational data will be shared and integrated to serve the multiple purposes of the cadastre. The cadastral data and related information will integrate numerically the modern system. To serve these needs, all surveyed properties and parcels must be tied to the cadastral sections to permit adding safely all natural details to the existing cadastral maps.

While this foundation is established for a long-term use, its control points must be compiled accurately and progressively. The ongoing GNSS network is tied to the first and second order of the existing geodetic network and to the International Reference Frame (ITRF2005). When the transferring parameters are compiled, we can densify the GNSS network and provide locational information of the new control points in both the Lambert and Cartesian systems. Once the required density is achieved, the cadastral sections are tied to the new control points. In several projects, various technical issues merit special considerations. One distinguishes the spacing density required between the framework control points and the accuracy levels (Figure 1).

*Figure 1. The process of creating the fundamental geodetic frame*

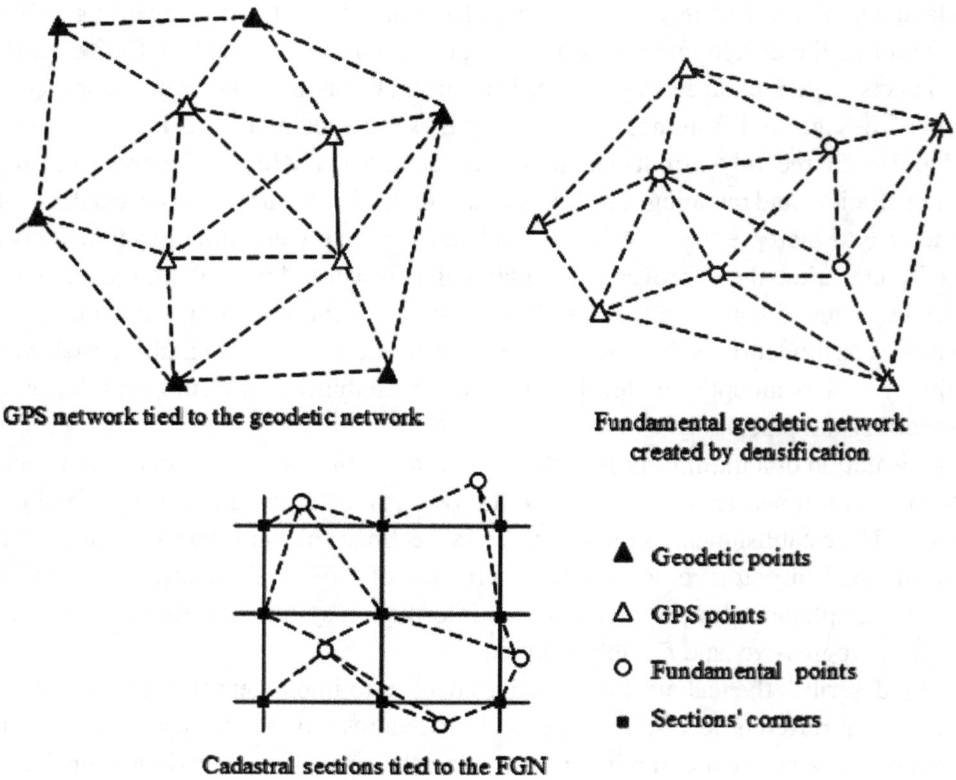


The Moroccan geodetic network has already been created. The new GNSS network is ongoing to be finalized to avoid the shortcoming of the former system. The most important decision to make now is concerned with the utilization of the two systems. They are considered as a reference framework to define the fundamental geodetic network for the multipurpose cadastre. They should be updated and maintained to provide current, accurate, and appropriate locational information. The GNSS network provides either relative or absolute positions of all natural and physical features in the WGS84 as a geometric framework for this system. The national coordinate system provides positional information in a rectangular system based on the Clark spheroid frame. The GNSS network will uniformly cover the whole country to facilitate the definition of mathematical models permitting the transformation of position coordinates to and from the national system based on the Lambert Conformal Conic projection.

## The Density

Several studies recommend that the spacing between the monumented points ranges from 0.3 to 0.8 km in urban areas and from 1.6 to 3.2 km in rural areas (McLaughlin, 1977). These densities are costly and time consuming if they are determined by means of the geodetic measurements. Then, a minimum density can be required within the fundamental network and a uniform density as a practical framework will be suggested to serve the cadastral matters.

A sufficient densification of the fundamental geodetic network will be realized to achieve at least 16 km of distance separating the monumented points (NRC, 1983). These minimum spacing points will serve as well as a basis for implementing the cadastral sections. This will be a solid foundation from which a multipurpose cadastre could be built upon.

## The Accuracy

The GPS as a recent satellite technology has improved the limits of the relative positioning accuracy. Three highest orders and three lowest orders of the geometric positioning accuracy are specified as in the Table 1 (FGCC, 1989). The three highest orders are called AA, A, and B. the lowest orders ranged from 1, 2, and 3 that constitute the category C.

The accuracy standards presented in the table 8.1 are not based on the technical ability of a technician surveyor, but on the capabilities of the GPS measurements. According to these standards, the maximum allowable linear error in centimeters can be reckoned by:

$$s = \pm\sqrt{e^2 + (0.1 \times p \times d)^2}\,cm$$

where,

s = maximum allowable error in centimeters at 95% of confidence level,
d = distance in kilometers between any two stations,
p = the minimum geometric relative position accuracy standards in ppm,
e = base error in centimeters.

For example, if we have a distance d=10 km and the baseline measurement is specified as first order, the maximum allowable relative error is 10 cm.

*Table 1. Geometric relative positioning accuracy standards*

| Survey Categories | Order | 95% Confidence Level | | |
|---|---|---|---|---|
| | | Minimum Geometric Accuracy | | |
| | | Base Error | Line-Length Dependent Error | |
| | | e(cm) | p(ppm) | a(1/a) |
| Global regional geodynamics | AA | 0.3 | 0.01 | 1/100.000.000 |
| National geodetic reference system "primary" for regional local geodynamics | A | 0.5 | 0.1 | 1/10.000.000 |
| National geodetic reference system "secondary" for local geodynamics (high precision engineering surveys) | B | 0.8 | 1 | 1/1.000.000 |
| National geodetic reference system for: mapping, land information, property, and engineering, ..etc. | 1 | 1.0 | 10 | 1/100.000 |
| | 2I-2II | 2.0-3.0 | 20-50 | 1/20.000 -1/50.000 |
| | 3 | 5.0 | 100 | 1/10.000 |

FGCC, 1989.

When surveying operations are conducted using GNSS in the relative positioning static mode, the results are better for long periods of observations. More accuracies that are reliable are achieved with a good geometric distribution of satellite, an accurate ephemeris, and a number consisting of four or more satellites (FGCC, 1989). The GNSS relative positioning techniques lead us to perform geodetic surveys at decimeter level or better.

The relative accuracy of the primary order geodetic network is 1/250.000. The relative position accuracy is estimated to 10 centimeters. Then, when the coordinates for a series of points are determined within both the GNSS and geodetic network, the coordinates for new points can be compiled. One must adjust the systematic differences between the two reference systems.

## THE REGULAR CADASTRAL SECTIONS

### The Concept of the Cadastral Sections

The design of a geometric framework of a series of regular cadastral sections is a necessary task to implement the multipurpose cadastre. These sections permit to collect cadastral data and both natural and physical features. At the beginning of implementing the new system, we can use the same range from 3.o to 4.5 km between monuments in both urban and rural areas. Progressively, we may improve the density in the urban perimeters.

The corners of each cadastral section should be well monumented on the ground and tied to the fundamental geodetic network. The densification within urban areas can be conducted by the traversing techniques. The new technology for surveying will be utilized to curry out positional data. The basic cadastral map at scale 1/20 000 has for dimensions 18 km by 12 km. The basic cadastral section will be at scale 1/5000 and have for dimensions 4.5 km by 3 km. It is 13.5 square km and can be divided into 25 sections of 0.9x0.6 sqkm each, which has the scale 1/1000 for the use in urban and suburban areas (Table 2).

The cadastral sections' indexes are derived from the cadastral base map's indexes at the scale of 1/20000. The subsection index will be derived from the base section a below:

- The base section is divided into 25 subsections,
- The following field of the former index ranges from 1 to 25.

For example, if a cadastral section has its index as 32-21-a, the number of the subsections we can create is 25. Their indexes, with respecting the same procedure of creating the map cadastral, will be 32-21-a-1, 32-21-a-2, …….., 32-21-a-25 (Figure 2).

*Table 2. The structure of the proposed cadastral framework*

| Areas | Frame | Dimension | Scale | Coverage | Index |
|---|---|---|---|---|---|
| Rural | Cadastral sections | 4.0 x 3.0 km2 | 1/5000 | 1350 ha | cadsec=##-##-a,...,p |
| Urban | Subcadastral sections | 0.9 x 0.6 km2 | 1/1000 | 54 ha | cadsec-1,2,….,25 |

*Figure 2. The process of implementing the cadastral and sub-cadastral sections*

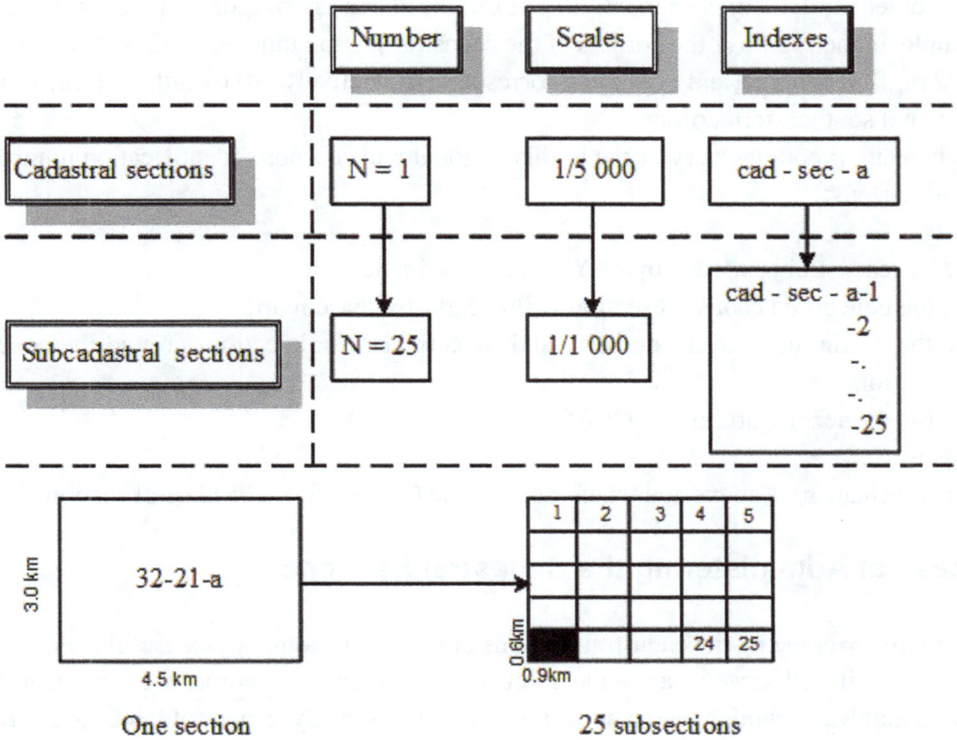


The number (nb) of the corners needed to monument a number (ns) of the cadastral sections can be computed as nb1=3/2*(ns+2) or nb2=2*(ns+1). The numbers nb1 and nb2 depend on the order in which the sections are covering a zone. For a zone covered by ns=94 cadastral sections, the case of Temara, a city close to Rabat, the minimum corners are nb1=144 and the maximum corners are nb2 = 190. The covered area is about 1300 sq.km (130000 ha).

Once the sections are created and monumented, each Office of Land Administration System must create a provincial repertory for archiving and maintenance tasks. The repertory must be at the disposal of the public and private users. The densification of the cadastral sections may be enhanced to meet the users' needs. The entire surveying traverse must be tied to the section corners.

## The Corner's Identification Number System (CIN)

The system for Corner's Identification Number is advantageous for many users and technicians because it permits to avoid problems that happen when the same CIN is assigned to corners used in contiguous areas. It simplifies the amount of the indexed corners differently in a large area. The system enables the developer of cadastral systems based GIS technology to deal with a national wide CIN, instead of customizing indexation system for each situation in a region, province, commune, or city. This system will uniquely identify every corner that will provide long-term benefits to national, regional, and local users.

The proposed CIN is a character field type composed of five characters for the cadastral sections and seven characters for the subsections. The five characters consist of the five characters of the cadastral section's index. This index is affected to the corner situated in the south west of the concerned cadastral section. Then, each cadastral section will have its southwestern corner as its first corner coded by its index. On the other hand, every corner has the index of the northeastern cadastral section as its CIN code.

For example, in the Figure 3, the corners of the cadastral section indexed 32-21-f will have their CIN as 3221f, 3221a, 3221b,3221c, and 3221g that correspond respectively to the southwestern, northwestern, northeastern, and southeastern corner.

The mechanism of coding every borne in this national wide corner's identification number consists of the following steps:

- Locate the cadastral base map by its Y and X coordinates,
- Locate the concerned cadastral section within cadastral base map,
- Locate the northern, eastern, southern, and western cadastral sections around the concerned cadastral section,
- Codify the corners regarding the CIN format.

The same mechanism is suggested when creating the CIN for the subcadastral sections.

## The Process of Administering the Cadastral Sections

The 4.5 x 3 sqkm coverage by the cadastral sections of the whole country is a useful frame for the multipurpose cadastre. It will serve as a unique geometrical reference for property assessment, urban and rural planning, highway planning, and land administration. Currently, a piece of land, parcel, or property

*Figure 3. The national wide corner's identification number*

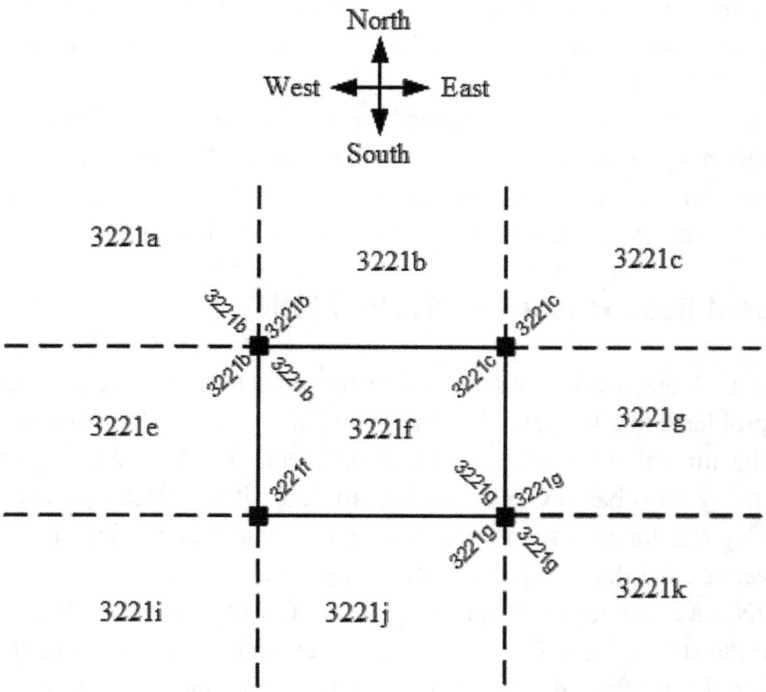

is described within a convenient cadastral section according to its geographic location. As well as every cadastral section is monumented as each natural or physical object on land is easily and accurately located.

The regional agencies will ensure the mission of coordinating activities and establishing guidelines and standards for creating and maintaining the cadastral sections. The activities of linking the cadastral sections to the fundamental geodetic network are scheduled and sent to the Office of Land Administration to be monumented. All the digital or graphical data describing the cadastral sections are archived and maintained by the local level. A copy of each file is sent to the Regional Agency to be used, controlled, and for creating large-scale maps.

### The Beneficial Uses of the Cadastral Sections

The new concept of the cadastral section is helpful for referencing, maintaining, and updating parcels or properties in a jurisdiction. Each section has its unique index and its specific scale. This mechanism permits to identify easily properties and related land data.

For several reasons as mentioned below, this mechanism must be retained and reinforced. Major technicians, engineers, and lawyers are familiar with the former forms of the cadastral maps. A unique index is affected to each property and this index is associated to the corresponding section index. The covered area by the basic cadastral section (1/5000) will be the basic unit for all natural and physical data overlays. The basic cadastral sections have several advantages. Firstly, the local level will assume only the responsibility of monitoring the monumentation of the cadastral section corners. The implementation is done by the licensed surveyors from the private sector. The monitoring procedures will be easy within regular and monumented sections. Second, the possibility of surveying lands within a cadastral section as a local system and converting data coordinates to national system is easier.

Third, the cadastral sections will be a standard reference for every institution at different levels to describe their geographic data. The private companies prefer using a unique framework for surveying because it saves time and costs. Fourth, the cadastral sections are useful for creating Digital Elevation Models (DEM) for engineering design, deriving contours, landscaping, route planning, computing slopes and intervisibility, and thematic mapping. However, there are many obstacles in the establishment of these sections. Some additional activities such as creation, update, and maintenance of the cadastral sections need continuous efforts and efficiency. The monumentation of their corners is expensive and time consuming. Indeed, they should be conducted in a gradual and progressive way.

The multipurpose cadastre will thus serve all institutions to locate their data easily when they use the regular cadastral sections. For urban purposes such as planning and building, the responsible agencies should use the subsections derived from basic cadastral sections.

## A SERIES OF LARGE-SCALE MAPS

### The Art of Mapping Within the Cadastral Sections

In the framework of the multipurpose cadastre, the value of land information depends of the costs of producing the basic data. To save time, money, and energy it is a necessary mission for cadastral system to conceive best practices and to develop convenient technical solutions. This is assured by developing large-scale maps presenting natural and physical features related to land data with respects to the various user needs. Generally, similar range of maps was developed across the world to represent data (NRC, 1983). First, we distinguish large-scale maps, which include cadastral maps at scales between 1/100 and 1/5000 and maps for zoning and planning at scales between 1/500 and 1/10000. Second, we find medium-scale maps called topographic base maps at scales between 1/20000 and 1/100000. The small-scale maps constitute the final category, which has a scale smaller than 1/100000.

In case of Morocco, the basic topographic maps at scales 1/25 000, 1/50 000, and 1/100 000 have covered the whole country. The maps at 1/25000 scale are created throughout the costal zones and certain developed cities. However, a wide percent of these maps are not up-to-date. Thus, the creating of large-scale maps at a scale of 1/5000 and more will have its significant justifications regarding the new requirements of both the society and the government. These maps will be the foundation upon which most, or all, other geospatial information overlays are built. Secondly, we observe intensifying pressures on land information for planning and development projects especially at the local levels. Thirdly, for the so-called Regional Centers of Investment in Morocco, they require a large inventory of land to assist policy makers and to evaluate land information needed in various development projects.

The actual needs, the living standards, and the modern techniques available are basic factors to consider for creating an efficient but simple multipurpose system. Some government institutions generate some large-scale maps but they only concern thematic features such as urban planning, tourist activities, or territory management. The responsibility of cooperating and creating these maps belongs to the cartographic units within the regional agencies. These unites will make efforts to set up an integrated framework that allows several governmental and non-governmental agencies for being concerted around a standardized system. The result will be the eliminating of duplication in gathering data, mapping, and developing GIS applications.

The basic unit of the large-scale maps shows the contents of the basic cadastral section that cover an area of 4.5x3.0 sq.km. However, a realistic approach should be made based on the nature of each zone and on the kind of projects to be conducted. For every urban zone, maps at a scale of 1/1000 must be generated. In case of populated cities, maps at a scale of 1/500 are needed. These maps are added to the cadastral overlays to display the properties and parcels they cover, then to be submitted to the local rules in term of restrictions and obligations regulating land uses.

In the areas including suburban zones, towns, and villages, the use of maps at scales from 1/2000 to 1/5000 are practically sufficient. The local conditions concerning the social and economical aspects of lands determine on the other hand the choice of the scales to adopt. In forests and widely cultivated areas, the corresponding maps scales are from 1/5000 to 1/10000.

All natural or physical resources are linked legally to the interests concerning lands and land ownerships. When special surveys are required such as in constructing railroads, dames, pipelines, waterways, bridges, and highways, accurate large maps should be established with high accuracy. Further, the cadastral overlays are indispensable for legal purposes.

## The Technical Considerations

The technical solutions to use for surveying areas to be covered by large-scale maps depend on the natural and economic characteristics of each region. The natural issues consist of the physical characteristics of every region. The types of terrain determine the most appropriate technique to use. The economic issues include the major people activities of each region. They determine where mapping process is appropriate to ensure that the large-scale maps permit to achieve social outcomes. The investment in land within a region is conducive to increase economical opportunities. For these reasons, the establishment of accurate and exact large-scale maps in such region requires specific technical solutions. However, in regions with lowest level of economical activities, the technical solutions should be adapted to regional conditions and modest financial possibilities unless an internal specific funding is expected to support the use of high technology. Despite the non-homogenous economic conditions from region to region, the principle of making large-scale maps for natural and physical features is a priority task.

The existing powerful techniques should widely be used to create large-scale maps. We distinguish two major ways of applying new technology, namely, direct and indirect surveying techniques. The indirect surveying includes photogrammetry techniques, while the direct surveying encompasses field-surveying techniques such as digital theodolites and GNSS. Currently, the ANCFCC uses various photogrammetric techniques such as stereo-photogrammetry, orthophoto techniques, satellite photogrammetry, and remote sensing (Ettarid et al., 2003). The GNSS is also utilized as a fundamental tool for surveying operations in both cartography and cadastre. A single technique cannot serve as a unique method for establishing large-scale maps (Table 3).

A large commitment is recommended, especially with the Regional Centers of Investment, Urban Agencies, Services of Geology, and Representative unites of the Agricultural Department. Further agencies such as private companies, universities, social services, educational and professional institutes, bureau of lawyers, non-governmental organizations, and local communities will play a major role in creating large-scale maps.

All these organizations will support the initiatives of developing such system by providing skilled specialists as aided designers and consultants for the government. The most important role to consider is the role of both surveying technicians and engineers. Their actions as specialists enable them to deal with legal and technical aspects of the new system in a best manner. They have a significant tradition of carrying out surveying, cadastral, juridical, and operational processes. In Morocco, the existing of an organized professional corps of surveyors is for a great interest because it will ensure efficiency and flexibility in dealing with multipurpose cadastral tasks.

The basic professional standards and legal aspects are supervised by the ONIGT organization to achieve cadastral works while the ANCFCC as a public institution will concentrate on supervising and controlling all projects.

*Table 3. The technical tools to create large-scale maps*

| Zones | The Nature of zones | Technical Procedures |
|---|---|---|
| Z1 | Urban | Field measurements<br>Ortho-photogrammetry<br>Stereo-photogrammetry<br>Satellite photogrammetry |
| Z2 | suburban<br>developed villages | Field measurements (GPS)<br>Ortho-photogrammetry<br>Stereo-photogrammetry<br>Satellite photogrammetry |
| Z3 | Rural zones<br>forests<br>large cultivated areas | Ortho-photogrammetry<br>Stereo-photogrammetry<br>Satellite photogrammetry |

## The Data Exchange Between Different Large-Scale Maps

When compiling data from maps to maps with different scales, we eliminate duplication of surveying and mapping. However, problems of precision and consistency come up because of the magnitude difference in scales at which data were compiled firstly. Our concern in this paragraph is how to ensure the needed accuracy and currency in transferring data between the existing maps and the new large-scale maps. The problems linked to the exchange of data between existing cadastral maps and large-scale maps are for two kinds. First, the transfer of data from the map scales between 1/5000 and 1/10000 to a map scale of 1/1000 or 1/2000 means an enlargement that implies unwarranted higher accuracy. This can produces erroneous information regarding properties or land parcels.

Second, transferring data from cadastral maps to large-scale maps means a reduction that produces unnecessary details. The volume of data at the cadastral scale may not be needed for maps showing natural and physical features at scales smaller than 1/5000 (NRC, 1983). The solution to this issue can be summarized in the following propositions. At first, the cadastral section that covers an area of 4.5x3.0 sq.km at a scale of 1/5000 must be the basic unit for developing standards in terms of accuracy. Detailed cadastral maps take more time than large-scale.

The nature of cadastre in Morocco as a sporadic system will not efficiently permit to develop natural resources' maps at large-scales based on cadastral maps. The solution is to reinforce and promote the national cadastre at the local level to establish a general land inventory regarding the standards and specifications of each zone within a region. The national land agency must be supported at the regional level to improve the general inventory of the public lands within urban and rural zones in a region. Once the large-scale maps of natural and physical resources are developed, the local offices of land administration system will furnish systematically cadastral details concerning their jurisdictions for regional uses (Figure 4). Cadastral details, physical and natural resources are compiled within the same frame, which is the cadastral section. Cadastral sections are tied to the global geodetic network, which facilitates the transfer of data within a unique reference framework. Cadastral maps can be rescaled easily as references for natural and physical resources.

The overall features must be represented in separates overlay that enable the mapmaker to establish new maps through recombination, reduction, or enlargement. The separating of different kinds of information will also help in developing digital maps. For a multipurpose cadastre, the large-scale maps may display in separates at different scales the following layers:

- Geodetic control points and cadastral sections,
- Property, parcel, block, and zone boundaries,
- Buildings and utility services (schools, hospitals, pharmacies, etc.),
- Streets, electric/communication lines, sanitary/storm sewers, hydrographic flows, drainage, and transportation,
- Surfaces such as parks, forest areas, lakes, and reserved zones, etc.,
- Hypsographic information in form of contour lines and vegetative information.

These layers will be used in a combined manner or separately. This ensures the consistence of large-scale maps with cadastral maps. The development of a cadastral information system at each office of land administration system will be a fundamental tool to recompile and adjust spatially other land information related to properties and parcels.

*Figure 4. The compilation process of large-scale maps*

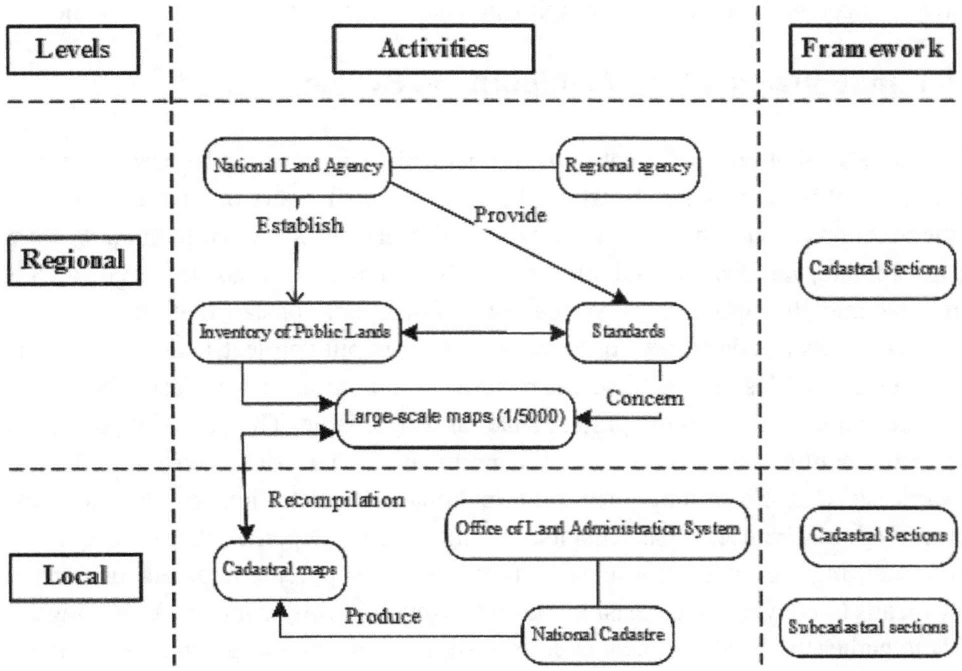


## AN INTEGRATED LAND ADMINISTRATION SYSTEM

### The Wide Partnership at the Local Level

Without local leadership in managing lands and land ownerships, there will be little chance of success to implement a multipurpose cadastral system. The former Services of Cadastre shall be upgraded as Offices of Land Administration System (OLAS) at the executive level. They will play a major role in coordinating systematically the local functions of the government. This will help to develop standards according to local and regional requirements and to national guidelines. The so-called Land Facility will include both the OLAS and the Registry Service within each province. Every local activity gathering municipalities, towns, and cities will be a fundamental key to advance in implementing the multipurpose cadastre. The spirit of gaining such initiative is through a large partnership approach between many providers and users of the land information. As a pivotal point around which the performance of the project is assured, this approach will provide a ready access to land information and avoid to waste time and money. This approach will naturally lead to a form of interconnecting data sources that support operational steps towards a multipurpose cadastral information system.

Various partners can be distinguished at the local level. They can be categorized into land administrators, land use planners, land managers, and land policy makers. Those groups must operate in a comprehensive and integrated approach to deal locally with the economical, environmental, and social considerations. The powerful analytical functionalities associated with the modern information technology based GIS/LIS are needed to better respond to the public inquiries. The goal of different partners of the multipurpose cadastre is to handle successfully all forms of the land information. They need operational and a friendly-user tool based on GIS/LIS technology to conduct easily their strategic studies and planning. This involves generally a long and complex process, which often costs much money.

## Managing Functionalities for a Multipurpose System

The use of a unique kind of land information is mechanically easy and can be based on manual process or just on a unique software package. However, because of the diversity of land data, data sources, and data users, the functions can not be easily designed. With enhanced system functionality we will be able to produce, manipulate, and display land information through the use of advanced GIS/LIS technology. This system is technically supported by various sets of analytical phases that are getting started with potential partners, assessing needs and requirements, running a pilot project, and developing an operating system. The first step enables us to achieve the benefits of a multipurpose cadastre by cooperating with a minimum of participants to keep the project small and affordable. The potential partners at the local level are the municipalities and certain service departments such as forest, agricultural, transportation, and public work services. For a long-term, other national administrations or agencies can operate at the local level; they may be useful participants. During the first step, the OLAS must provide detailed information concerning the multipurpose cadastre. It also must prepare booklets, demonstrations, and all necessary means to convince the decision makers to invest in this program. A working committee is indispensable to gather the local technical experts, design necessary equipment, and assess the preliminary costs of the project. The next step consists of analyzing the current and future related land activities at the local level. It encompasses both the assessment of people needs and the requirements to develop such system. The needs assessment covers the identification of the potential uses of land information, the forms of managing and maintaining land data, the needed personal resources, and the possible prospective advantages through an introduction of a GIS/LIS technology. The requirements consist of defining the software functionalities, data models, and the manner in which the modern technology is integrated for the benefits of the partners. These assessments are generally conducted by surveys and interviews with users and partners.

The third step "pilot project" provides, in a limited geographic area, a manner of testing the designed procedures and methods for developing the multipurpose system. It permits to identify problems and test various solutions before initiating a large project. Other guidelines are determined and several lessons are learnt from a pilot project. The last step consists of developing an operational system around enough partners and users. During this step, responsibilities and functionalities and are defined and work plan is designed to perform each task. These analytical phases require significant financial supports and institutional commitment to be achieved. By operating together, all partners take advantage of the system, profit from the economies of a joint effort, and share the assumed costs. Then, a new legislation is required to motivate individuals from various partners and generate enough resources supporting the project.

## The Conceptual Scheme of the Startup Project

A conceptual scheme of conducting a pilot project is needed in order to develop a final diagram for implementing the Moroccan multipurpose cadastral system. The conceptual scheme is largely linked to the contribution of the national, regional, and local levels. In this diagram, we distinguish four essential phases (Figure 5).

*Figure 5. Conceptual framework of the startup project*

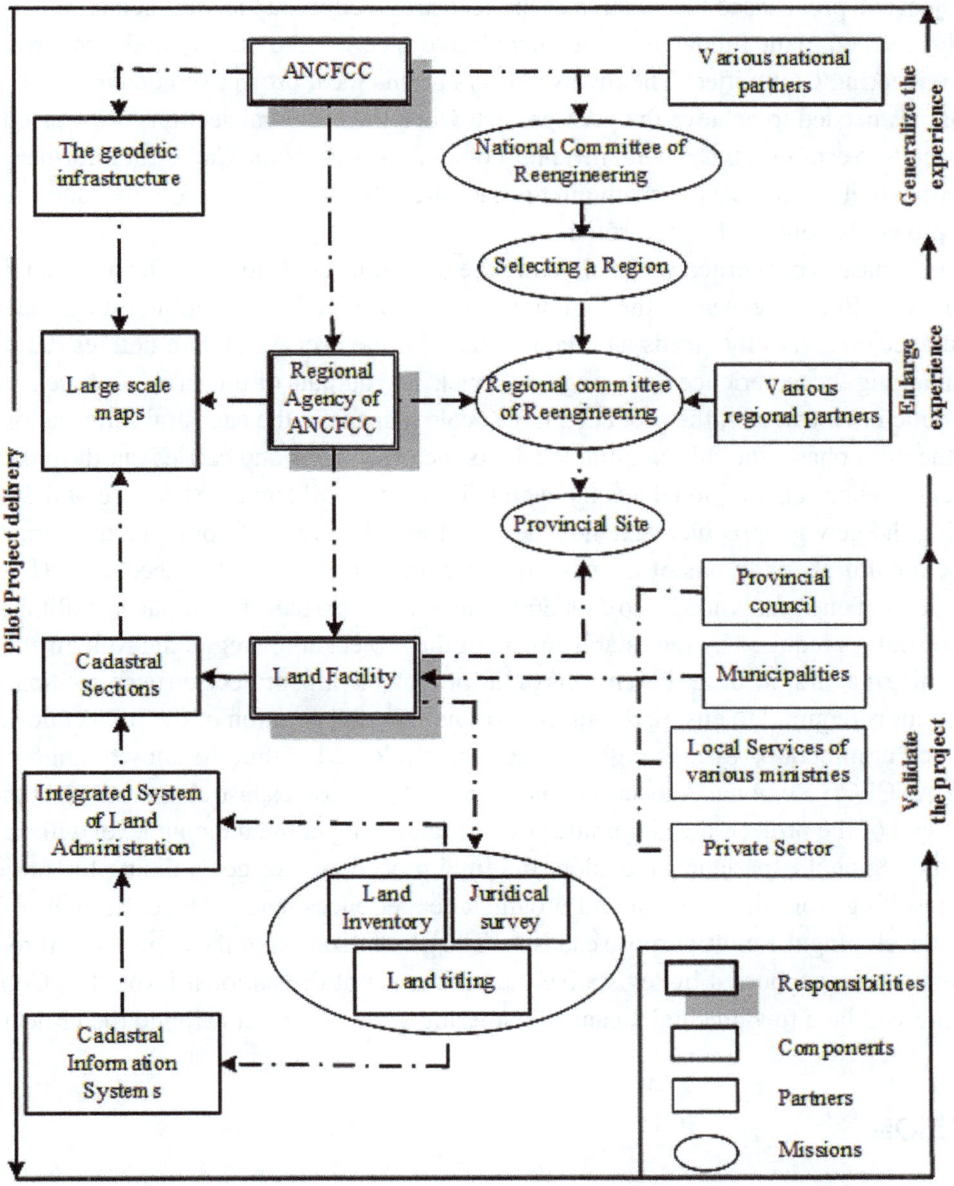


The first phase consists of making the organizational structures that enable monitoring and controlling the advancement of the pilot project. Throughout these strategies, the ANCFCC is the only responsible for conducting tasks relating to the five components of the multipurpose cadastre. The Province as a local level is considered the best site for conducting the pilot project. The Province is selected among the provinces belonging to a Region characterized by huge economic and development activities in the fields of urbanization, agriculture, and industrialization. The Land Facility, located within the selected province, proceeds to the creation of the Users Group of the multipurpose cadastral products. This Group involves the representatives of provincial council, municipalities and towns, local services of the various ministries and the representatives of the professional sector.

The Land Facility collaborates with the Users Group to establish the operational steps of conducting the pilot project and proceeds to the creation of the cadastral sections at the provincial level. The sections are normally attached to the fundamental geodetic frame according to the national standards developed by the Reengineering Committee. The involved users in this local project should support all costs and material charges needed to achieve the pilot project. Once the cadastral sections are created, the Land Facility proceeds to establish the general inventory of lands as well as the Cadastral Information System. On the other hand, it is necessary to maintain the cadastral operations relating to the land titling as well as the pilot project is going to be performed.

The second phase is concerned with validating the procedures developed at the provincial level. The sub-Committee of Reengineering at the regional level supervises the validation, establishes a study of quality assurance, analyzes the needs and requirements of the users, and then defines the new adjustments for enlarging the experience to the regional level. The outputs of this project phase consist of the created geodetic infrastructure, the produced large-scale maps, and the cadastral information system.

During the third phase, the sub-committee adopts the procedures and carries out the enlargement of the project experience at the regional level. Several difficulties will come up because of the new factors characterizing the new geographical extent. These factors relate to the geographical characteristics of the area, socio-cultural conditions of the new involved provinces, and to the needed costs to cover the Region. Duration of one year is needed to cooperate with various partners for preparing utilities, technical, and human resources required by the establishment of the project at the regional level. Furthermore, we should recognize that in spite of the best results of the provincial pilot project, an even political consensus and engagement is required to ensure the success of the second operation of the project development.

Once the commitment of various regional partners is released within the sub-committee of reengineering, the ANCFCC Regional Agencies launch the multipurpose cadastral operations at the regional level. At the end of the project, the Committee of Reengineering at the national level will carry out the analysis of the regional experience and adopt the final procedures for generalizing the project. These procedures resulting from the regional and provincial experiences lead to the definition of Moroccan standards in developing the multipurpose cadastre. The generalization of the experience throughout the fourth phase must be supported by legislative measurements at the national level. The Committee of Reengineering will be a fundamental means of gathering various partners around the project.

## CONCLUSION

The multipurpose cadastre for Morocco will be a comprehensive system that offers many advantages in managing all kinds of land information. Its development does not mean the creation of a new system, but our perspective is to re-engineer the existing frame and upgrade it to improve managerial, functional, and operational procedures. Naturally, a progressive approach is necessary to make the first steps of building the new system less expensive. The ANCFCC should focus on the enhancement of its technical procedures by the mean of developing guidelines and standards and introducing new technologies. The primary tasks necessary for future system implementation is to establish systematically and gradually regular cadastral sections and large-scale maps.

The same sections are tied to the global geodetic network and will serve technically as the basic reference for cadastral and surveying operations. The large-scale maps will permit as a basis for generalizing the titling process. Once a region or a province is surveyed and has all its natural and physical features presented on maps, the process of surveying and titling parcels or properties for cadastral concerns will be conducted easily and rapidly.

Normally, if the guidelines and standards are respected in establishing large-scale maps, the cadastral maps can be compiled from. A strict control should be conducted on the ground especially in the period of adjudication process to demarcate definitively parcel limits and corners. The progress of implementing the new system will depend on the natural and economical conditions of each region. Regions with high level of economical activities are primarily selected. The implementation will on the other hand be proceeded area by area according to national priorities. A flexible approach is required to determine in which region and in which area, the multipurpose system will have a huge impact in improving land management and planning.

The new technology using digital theodolites, GPS, photogrammetry, and satellite imagery play a major role in performing the multipurpose cadastral tasks. As the most promising techniques, they will improve both the acquisition and transfer of data to digital platforms based on geographical information systems to facilitate processing, management, update, and storage.

## REFERENCES

Ettarid, M., El-ayachi, M., Ettaqy, E., & Elyamani, M. (2003). L'Orthophotographie numérique au service du cadastre national. *Proceeding of the 2nd FIG Regional Conference. Session TS14 Cadastre in the Mediterranean*. Available from: https://www.researchgate.net/publication/238073176_L'Orthophotographie_numerique_au_service_du_cadastre_national

FGCC. (1989). Geometric geodetic accuracy standards and specifications for using GPS relative positioning techniques. Version 5.0. preliminary document. National Ocean Service. NOAA.

McLaughlin, J. D. (1977). *Maritime accuracy study, Land Registration and Information Service technical report*. Fredericton, Canada: University of New Brunswick.

NRC. (1983). *Panel on a Multi-Purpose Cadastre. Procedures and standards for a multi-purpose cadastre*. National Academy Press.

## KEY TERMS AND DEFINITIONS

**Cadastral Map:** A piece of plan established in a paper or digital format to describe a piece of land, which is parcel. It is the fundamental tool of a cadastre that ensures monitoring data on land properties in a territory.

**Multipurpose Cadastre:** A developed cadastral system to meet the requirements of both juridical and fiscal cadastre. It deals with various land data to produce information for land administration, investment, and sustainable development.

# Chapter 5
# Spatial Data Infrastructures in Morocco (From the Diagnosis to the Success of the Implementation)

**Elmostaphi Elomari**
*Institut Agronomique et Vétérinaire Hassan II, Morocco*

**Hassan Rhinane**
*Hassan II University of Casablanca, Morocco*

## ABSTRACT

*A spatial data infrastructure (SDI) is a platform for coordinating the exchange and sharing of spatial data between several producers or users of spatially referenced data. In Morocco, there is a massive production of spatial data and several generally public administrations are starting to feel the need for geographic information governance through a mechanism of exchange and management of data to optimize their efforts and avoid a redundant production. The purpose of this chapter is to draw up an inventory of the state of the art of geo-spatial data, systems, and tools existing in the central administrations in Morocco in relation with the collection, management, storage, and dissemination of geographical information. Through this study, it was found that the problem is more a question of global governance, and that the current context has assets for the establishment of a spatial data infrastructure in Morocco.*

## INTRODUCTION

This paper is part of a general assessment to prepare the necessary elements of an academic research that seeks to build a spatial data infrastructure in Morocco's Rabat-Salé-Kenitra region (RSK region). The SDI aims to promote data sharing through a common platform that takes into account the producers and users of spatial data. The study area is located in the north-west of the country and involves several data producers who have created their own geographic information system to meet their specific needs.

DOI: 10.4018/978-1-5225-5939-9.ch005

However, the existing systems are neither compatible nor interoperable. In order to attain the objectives of the study, it is necessary to touch upon the current state of the art and the available knowledge on spatial data infrastructures and establish a national benchmark on the implementation of spatial data platforms.

For the time being, spatial information in Morocco is used in many fields. After being used primarily to produce paper maps, techniques and representation tools of the geographical space have evolved over time. In this respect, cartography is now facing new challenges mainly related to technological advances in the field of geomatics (Ibannain and Donnay, 2012).

The analysis of the current situation made it possible to identify new needs in terms of the administration of geographic information. In other words, attention should be directed to establishing the necessary arrangements for the possible pooling of geographic information, which could lead to the establishment of a national data infrastructure. But the question is: do we have in Morocco the main components of SDI as defined in the literature (Ibannain and Donnay, 2012).

The SDI concept (Spatial Data Infrastructures concept) continues to develop as it becomes a basic infrastructure that supports economic emergence, environmental management and social stability in developed and developing countries. Because of its dynamic and complex nature, it is still a fuzzy concept for many, with practitioners, researchers, and governments adopting different perspectives according to their needs and circumstances (Williamson, et al, 2003).

Since the 1990s, the development of SDI concept has been underlining the need expressed at the international level to have access to, exchange and share spatial information held by many producers in order to best use and manage it. The SDI-recommended information sharing between organizations is essential to understand the complexity of environmental and human phenomena. “Taken in isolation, no organization has the information and the know-how to reach this goal” (Craglia, 2010).

## Presentation of the Study Area

The indispensable and ambitious project of “advanced regionalization”, in which Morocco has recently embarked, is characterized by the advent of two major concomitant factors as far as country planning and socioeconomic development are concerned.

The adoption of a new division of the national territory, which has increased the number of Regions from 16 to 12, gave rise to various spatial reconstructions of varying importance in the configuration of new regional entities.

In this redistricting operation, the case of the new RSK region is rather peculiar, since it results from the pure merger of the two former regions of Rabat-Salé-Zemmour-Zaer(RSK), on the southern side, and Gharb-Chrarda- BniHsene, on the northern side.

The new RSK region, which extends over almost 17,800 km$^2$, extending over some 200 km from North to South and more than 100 km from East to West, enjoys a privileged geographical location that locates it in the heart of North-West Morocco. In other words, the region, by its economic dynamism and its population density, comes second, after the Casablanca-Settat region (HCP, 2016).

Made up of three prefectures (Rabat, Salé and Skhirate-Témara) and four provinces (Kenitra, Khemisset, Sidi Slimane and Sidi Kacem), the new RSK region is currently home to 4 580 866 inhabitants (HCP, 2016). It represents the second mass regional population of Morocco, after that of Casablanca-Settat (about 7 million), and also 13.5% of the population of the country as a whole, while it covers only 2.5% of the total area of the country.

*Figure 1. Map Location of RABAT-SALE-KENITRA region according to 16,12 regions of national territory*
*Source: Internal realization.*

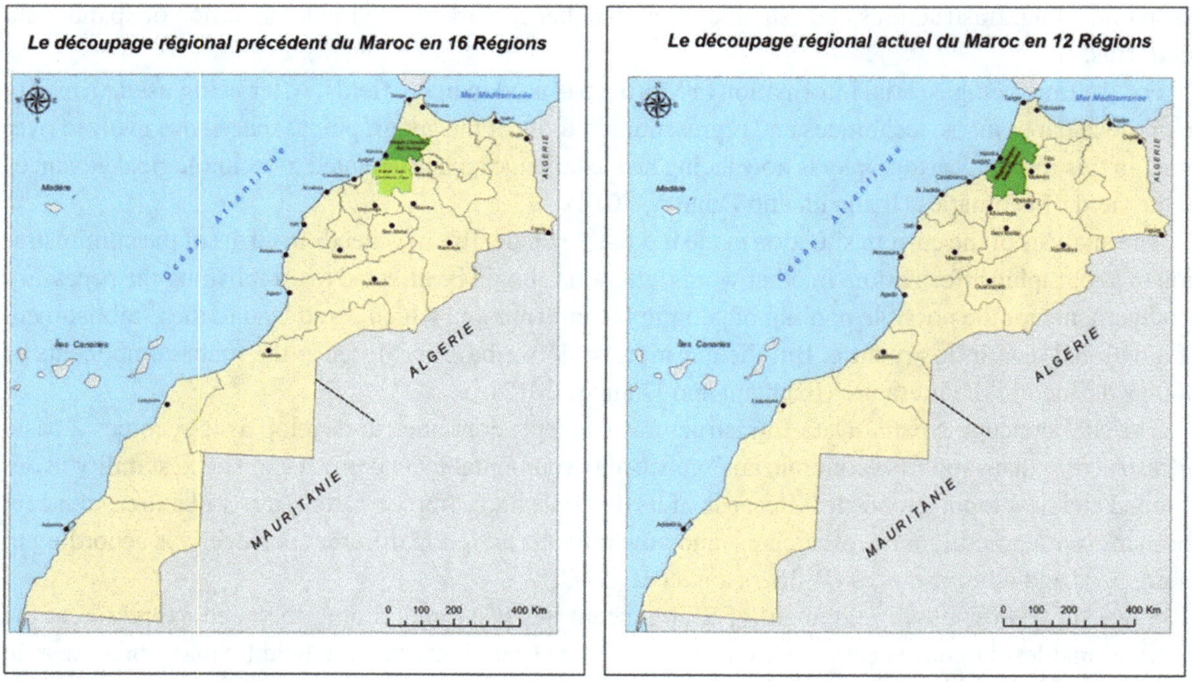


Clearly, RSK region is favored by its exceptional geographical situation and privileged by the presence of the political and administrative capital of the country. Furthermore, it is strong of its multiple and diverse natural resources as well as its human potential quantitatively and qualitatively considerable. Thanks to this, the RSK region possesses great capacities for growth and development at all levels.

The availability of spatial data in this study region is due to the concentration of administrations in this zone. In addition, the politic and economic importance of geo-referenced information stems from their daily requirement.

The use geographic information goes beyond governmental planning and decision-making, to include electoral campaigns, economic operations and scientific research. In fact, it is a prerequisite for the proper functioning of democratic regimes. And because of its enormous politic and economic potential, geographic information constitutes a first-class asset for this area. Within the public institutions of this region are a set of spatial data as well as geographical information systems.

Other than the above-mentioned advantages, the study area is hampered by many spatial data-related constraints which hold back its development. These constraints are reflected particularly while preparing the regional planning tools (Regional Development Program RDP, The Regional Spatial Planning Scheme RSPS) and implementing the sectoral development policies.

From its geographical location and its specificities, the study area has both strength and enormous constraint in terms of major spatial data having an impact on its development. This aspect is mainly reflected in the preparation of regional planning tools (RDP, RSPS) as well as the implementation of sectoral development policies.

It is therefore imperative to devise, now, a strategy for the establishment of a regional spatial data infrastructure that can be used by the region's actors and partners.

## Scientific Contributions and Methodology

The wide variety of spatial information and the absence of a common policy of standardization of technologies among the users of spatial data impede the efficient exploitation of spatial data in an interoperable environment.

Compared with the rest of the country, RSK region is relatively favored to be an example to tackle this problem.

The SDIR project will ultimately lead to a qualitative leap in both regional and national territorial administration. Its significance lies in the fact that spatial data promotes knowledge and analysis of territories, and hence their development. It will be equipped with the necessary skills and tools to, on the one hand, produce and create databases on telecom networks, and on the other hand, animate a network of actors in order to strengthen the interoperability of information systems.

The SDIR project will be an incentive to well-organized and efficient investment that seeks to improve the planning, productivity and attractiveness of the RSK region.

The scope of this paper is to provide answers for the following questions:

- What is the current state of the spatial data infrastructure in Morocco?
- What are the spatial data and SDI concepts required for the development of a spatial data infrastructure?

The purpose of this study is to:

1. Study and analyze national experiences of the construction of spatial data infrastructures in order to shed light on the best practices.
2. Identify all the producers and users of spatial data in this region with their different exchanges, types of information and GIS applications used.
3. Analyze and model the different components of SDI to define an infrastructure implementation strategy. The aim of this step is to define the norms to be applied to spatial data, and to specify the methods, tools and services of spatial data management.
4. Define the work environment and the technical setup and management of an exemplary spatial data infrastructure.

# MOROCCAN EXPERIENCES OF SHARING SPATIAL DATA

The challenge that the Moroccan Information Technology (IT) sector will face in the foreseeable future is not only strengthening the progress already achieved, but also enabling the country to find its place in the field of the global knowledge economy. This is attainable only by an amplified and widely distributed integration of IT at the level of all the actors of society: State, administrations, businesses and citizens (Plan Maroc Numeric, 2013).

Indeed, development in the vast majority of the world's States is conditioned by access to information and its appropriate use. Modern and prosperous nations exhibit a high production capacity and use of information. In this regard, Morocco has adopted several strategies and programs that seek to give impetus to the development of the information technology sector, and particularly spatial information sharing.

## Law 31-13 on the Right of Access to Information

In application of the provisions of article 27 of the Moroccan Constitution, which provides that "Citizens have the right to have access to information held by the public administration, elected institutions and organizations. The right to information can only be limited by law, in order to ensure the protection of all matters relating to national defense, the internal and external security of the State, as well as private life of individuals, to prevent the infringement of the rights and freedoms set forth in this Constitution and to protect sources and areas expressly determined by law". (MC,2011)

The Moroccan Government Council approved on 31st July 2014 Bill 31-13 on the right of access to information.

Thus, Morocco has embarked on a path of openness and progress with the inclusion of the right of access to public information in Article 27 of its new Constitution, to place itself among the most progressive countries in the region.

However, it still faces many challenges to make this right a reality. And while the upgrading of the national legislation and administrative functioning remain a priority, another important aspect is the change of mindset. Openness and transparency, trust between the constituent and the administration, as well as the dissemination of information for the better management must replace the culture of secrecy and the retention of information at the service of individual powers (MichaelT, 2014).

It is a lengthy process that requires the joint commitment of all stakeholders, government, administrations, civil society organizations, journalists and individual citizens. This commitment may, over the medium to long term, enable Morocco to better develop its infrastructures (MichaelT, 2014).

The NGO Transparency Morocco believes that access to information is a fundamental right and an effective means of improving fairness, preventing corruption and consolidating the National Integrity System. In addition, the dissemination of information can only raise public awareness and identify the stakeholders that can constitute a social movement in favor of the effective enforcement of the right of access to information.

In application of this law, Morocco is ready to implement information sharing initiatives, including spatial information, in a global development of the Information Technologies sector.

## Moroccan E-Government Program

Public services and their effectiveness have a significant effect on economic and social life across the country. Thus, it is increasingly important to provide, transparently and safely, these service so that the Moroccan society can be gradually transformed into an information society, through and effective use of IT and in accordance with the needs and expectations of citizens and companies.

Launched in October 2009, Morocco's 2013 Digital Plan seeks to bring the administration closer to citizens. To achieve this goal, a vast E-Government program is set up, which consists of a portfolio of 89 projects and services, under the responsibility of different organizations and administrations, for a total investment of more than 2 billion MDH (E-GOV Program).

The program aims to exploit information and communication technologies to deeply reconfigure processes, make them efficient, effective and totally oriented to the service of citizens and companies. This requires an inter-sectoral collaboration between public administrations and a shared vision between the entities that collect and use information at the different territorial scales. (E-GOV program).

A multitude of projects and services are being carried out within the framework of this E-Gov program. This includes: the General Interoperability Framework which constitutes a reference for the Moroccan administrations. This Framework contains the rules of conformity for the use of standards, norms and references in the development of information systems in the public sphere. Moreover, the portal data.gov.ma, which has been online since May 2014, is developed by the Ministry of Industry, Trade and New Technologies in order to bring together in a single instrument all public data, in a directly exploitable format.

## National Observatory of the Territories Dynamics (ONDT)

The Department of Territorial Planning (DAT) is the government structure responsible for anticipating territorial issues, designing intervention strategies, and stimulating and supporting regional actions in terms of development of the territory. It fulfills this objective by ensuring coherence with national priorities, helping in the decision-making process and establishing homogeneity among sectoral actions related to spatial planning.

The spatial planning policy is, in essence, cross-cutting and multi-sectoral. As a result, this monitoring approach must be shaped by mechanisms and tools that constitute a unifying framework of strategies for promoting a common vision on the state of our territories. Thanks to the reinforcement of cohesion and the integration of public policies, the Department of Regional Planning has implemented an observatory of territorial dynamics (ONDT). It constitutes since 2015 a division within the DAT.

It is a tool to make decisions and define public policy actions to develop a common vision of territorial planning. It aims accompanies public policies and measures their impact on the territory, on the basis of reliable and relevant data.

The construction of the ONDT was done through a roadmap indicating that the institution shall be a mechanism to observe, evaluate and accompany the dynamics Moroccan territories. Another task is anticipate the future evolution of these territories and to ensure a functional and operational link with all partners. Among the guidelines of the construction roadmap of the ONDT is to "monitor a platform of reflection on the adoption of a common infrastructure of Moroccan spatial information".

The first concrete project set up by ONDT is the Geoportal of the national territories (www.geoportail.gov.ma) which aims at sharing with local and national decision-makers the necessary tools (cartographic supports, statistical dashboards, etc.) to let them observe the changes and territorial dynamics and their impact on development policies. This state of affairs requires the democratization of access to information and the sharing of territorialized, up-to-date and centralized data, which is essential for the formulation of development policies, strategies and decision-making.

A data collection protocol: Considered as cornerstone of the ONDT, this project is an information system that ensures a regular and permanent supply of databases. In other words, it allows the establishment of a data collection and exploitation protocol with all producers able to ensure the life of the observation device (collecting, processing and publishing territorialized statistical information) while establishing the necessary ramifications with the regional levels in order to be able to trace the information and to allow the regular update of the data. This study is now in its final phase.

## The Experience of the National Observatory for Human Development (ONDH)

The main objective of the National Observatory for Human Development (ONDH) is to propose measures and actions that contribute to the establishment of human development programs implemented in Morocco, and to analyze and evaluate their impact.

It is a pluralistic and independent body that is in permanent contact with public authorities, the private sector, representatives of civil society (NGOs, political parties, trade unions) and international organizations working in the fields of human development, partnerships, exchange and cooperation.

In carrying out the actions carried out, the observatory:

- Contribute to the development of knowledge through the collection of quantitative and qualitative data related to its mission from public and private, and national and international legal entities;
- Contribute, through its information system, to inform, democratize and animate the debate on human development;
- Develop specific, synthetic and relevant indicators that allow the analysis and understanding of issues related to human development in Morocco;
- Prepare and disseminate an annual report on human development in Morocco, and summarize studies carried out at national and regional levels.

### Data and Topics Covered

The 11 topics covered are:

- Population and demography;
- Health: Medical supply and its dimension in relation with the population- The use of medical means provided to the population- Mortality and child and maternal health- International concerns and scourges to fight- Social welfare;
- Education and Knowledge: Literacy- Education and Training - Resources and Infrastructure, Education and Training - Admission, Completion, Progress, and Access to Knowledge;
- Housing: Household living conditions and housing supply;
- Access to services and basic infrastructure: Access to drinking water- Access to electricity- Access to sanitation - Access to roads and transport;
- Activity and Employment: Employment Status- Unbearable Work- Adequate Work and Income- Household-related Indicators- Workplace Stability and Safety- Fair Employment Equity- Healthy Work Environment- the Economic and Social Context of Work;
- Economy: Aggregates of National Accounts- Public Finance- Prices and Companies;
- Well-Being and Poverty: Income and Expenditure- Poverty Measures;
- Business climate;
- Governance, political rights and civil liberties;
- Synthetic indicators of human development.

The ONDH produces little data, apart from the spot surveys on the INDH and the annual survey of a panel of households. All the data it provides are collected from sources of national or international statistical production.

## The Geographical Information System of the ONDH

ONDH established its geographic information system (GIS) by geographic location projects and institutions, as well as the collection of statistical data relating to projects or human development institutions in Education, Vocational Training, Health, Higher Education, National Mutual Aid, Youth and Sports, Cooperatives, Mohammed V Foundation and INDH.

Today, the ONDH has geo-localized these institutions in the regions of Meknès-Tafilalet, Gharb Chrarda BniHsein, Rabat Sale Zemmour-Zaer, Grand Casablanca, Oriental, Taza Hoceima Taounate, Tangier-Tetouan(ONDH,2017).

More than that, the ONDH is a part of the new territorial division. In order to complete the geo-localization already undertaken according to the old division, it engaged the geo-localization of 16 Prefectures and Provinces belonging to the Regions of Fez Meknes, Drâa-Tafilalet, Grand Casablanca-Settat and Béni Mellal-Khénifra.

## Web Mapping Tool

The ultimate role of the "Territorial Information System" is to combine cartography and database management, to offer optimized representations of the space and to examine human development projects through taking into account the potential of territories and social interactions.

Figure 2 is an example of the projects map in the region of RABAT-SALE-KENITRA.

*Figure 2. Screenshot of the ONDH's GIS*
*Source: ALBACHARIA, 2017.*

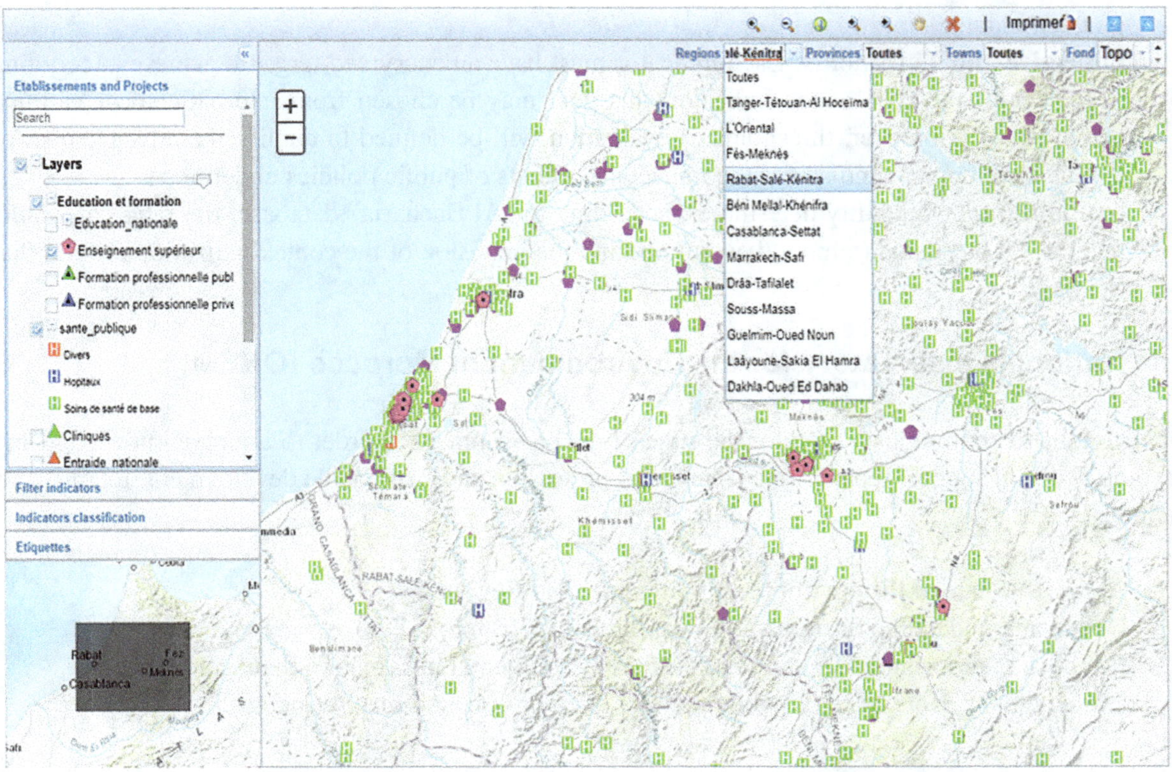

## Metadata

In each theme, choices have been made on the basis of a global context. The list of data/indicators, once confirmed by the ONDH, will result for each selected indicator in an exhaustive sheet containing the following elements:

- **Definition:** The definition of the indicator gives exactly what is measured, and how it is measured in order to know the exact framework of the results produced. It also determines the calculation rules used to obtain them and includes, for the composite indicators, the unitary content of their composition.
- **Functional Objective:** Each indicator has a functional objective of analysis that makes it possible to be included in perspectives and themes of human development.
- **Scope:** The scope of the indicator determines whether these are international standard indicators allowing comparison with other countries and to situate Morocco's position internationally, or indicators specific to the analysis of the country, its situation and the internal context, according to the priorities assigned in the conclusions of the fiftieth anniversary report.
- **Type of Indicator:** The type of indicator is used to define whether it is a key indicator of a domain, or a process indicator to explain or comment on a synthetic key indicator.
- **Source of the Indicator:** This information makes it possible to determine the reference frame from which the indicator was drawn. It lists the source (s) selected that provided the quantified results, and also the context of their acquisition and calculation.
- **Granularity:** It is the description of the existing criteria of disaggregation of the indicator allowing its contextualization for a better understanding. The focus will be on 3 major axes: time, geography (urban / rural context, international or local geographical location) and human component (gender and age).
- **Objective and Thresholds:** Each indicator must have an encrypted target to achieve. According to the scope and type of the indicator, this goal may be chosen from international or national perspectives. Moreover, thresholds of evaluation will be defined to qualify the advancement of human development and assess the impact and effects of public policies undertaken.
- **Availability:** Availability determines, according to "Al Bacharia" database, the range of results that can be presented (temporal evolution) and the precision of the contextualization criteria that can be used.

## The National Observatory for the Environment of Morocco (ONEM)

ONEM aims to continuously monitor the state of the environment in order to improve knowledge and develop decision support tools for environmental protection and sustainable development. Its missions include:

- Assess the state of the environment (nationally and Regionally);
- Define and update the Sustainable Development Indicators (SDIs);
- Establish appropriate tools and systems for environmental information management;
- Evaluate the performance of public policies in environmental management;

- Disseminate environmental data and share it as part of the national network or regional environmental networks.

The role of the National Observatory for the Environment of Morocco (ONEM) is to provide information on the environment and sustainable development. In this perspective, the approach is based on three fundamental principles:

- The development of a reliable, easily accessible and regularly updated multidisciplinary Information Database;
- The creation of a context for exchanging information and experiences around common goals;
- The promotion of an active involvement of different partners through carrying out joint projects and meeting shared objectives.

The ONEM currently has glossaries and dictionaries of data and indicators (with the scope "region" and "province" and sometimes even the scope "municipality"). They were developed within the framework of integrated environmental assessment studies. These glossaries and dictionaries have a 10-year traceable record, as from 2001, depending on the availability of data from the partners.

In a bid to decentralize the management of environmental information and to meet the needs of regional and local decision-makers, Morocco's economic regions need a powerful system to collect, analyze and manage environmental data. It is in this context that the Ministry in charge of the Environment is committed to the establishment of a Regional Information System on Environment and Sustainable Development (SIREDD). The overall aim is to have, at the regional level, a decision support and policy monitoring tool for environmental management and protection and sustainable development.

The information disseminated by the Department is public information collected from public administrations. However, once the databases and Regional Information Systems for the Environment and Sustainable Development (SIREDD) are finalized, they will be open and accessible to the general public with privileged rights to the partners of the regional networks for the collection and exchange of information. The data and indicators that the SIREDDs contain will be public data.

Similarly, the information that will be generated from these systems will also be made available to the public. The confidential data that partners provide will be managed within the framework of specific conventions. These conventions will be annexed to the global charter agreed with partners (4c, 2017).

The Department also writes a report on the State of the Environment in Morocco (every 4 years), regional reports of the Environment on the integrated assessment of the state of the environment (every 2 years) and a report on indicators of sustainable development in Morocco (4c, 2017).

## DIAGNOSIS AND INVENTORY OF SPATIAL DATA

### Introduction

The main purpose of this part is to provide a diagnosis of spatial data. This diagnosis identifies the existing information through the documentation and the data series held by each ministerial department, as well as the preparation of an inventory of the existing technical platform. The diagnosis will be dressed on both levels: nationally and regionally.

The national diagnosis is carried out by means of a documentary analysis, through reports and documents published by central administrations, as well as the web portals of information dissemination of the entities producing spatial information. A questionnaire is also implemented by the Directorate of Territory Planning to collect data. It contains two components: one deals with data and indicators, and the other focuses on computer tools and platforms.

Concerning the inventory and spatial data collection at the regional level, the first step is to define the data-providing bodies and entities within the region (local authorities, urban agencies, Regional Directorate of Rabat- Salé-Kenitra, which reports to the HCP, Regional Directorate of Agriculture, Regional Inspectorate of Urban Planning, Regional Direction of Equipment, etc.) and to study their predispositions to cooperate and exchange data via a specific questionnaire. This is the main difficulty of the project because it depends strongly on the degree of collaboration of the data providers.

## National Level

The purpose of this step is to carry out an inventory to identify and collect existing spatial data and information at the central government level:

The literary review is based on reports and documents of the central administrations, information available on public administrations' websites and data retrieved from teams in charge of the development of geographical information systems.

As a compliment to the literary review, direct interviews have been conducted with officials belonging to the Ministry of the urban planning, as well as public departments and agencies directly related to the use or production of spatial information.

In the context of these interviews, a questionnaire was developed to:

- Locate statistical data repositories and geo-referenced projects;
- Gather a global expression on the needs of the partners;
- Address issues to:
    - Identify system data;
    - Define the axes and indicators of the future system;
    - Identify system features;
    - Qualify users of the system;
    - Define the modes of exchange with the partners.

In order to optimize the exchange between the producers and users of spatial information, discussions are initiated with spatial-data supervisors so as to identify the themes and sources of their metadata. Moreover, these supervisors shed light on the cartographic processing, the development of synthetic indicators and the dissemination of information. Alongside these elements, the study aims to assess their information needs, their expectations of the future system and explore the partnership opportunities to institutionalize between data producers and the ONDT.

A list of twenty-five potential partners was selected. These partners are public departments and agencies directly related to spatial information.

The summary of the questionnaires analyses the structures that responded to it, through focusing on the information manipulated by the main producers and users of spatial information, the information to give to the future ONDT system and the partnership sought.

## Examples of the Questionnaire Results

1. Do you Have a Glassory of Spatial Data? (Figure 3)
2. What is the Main Source of Information do you use? (Figure 4)
3. Do you have a geographic information system within your department? (Figure 5)
4. What kind of software do you use? (Figure 6)
5. How often do you use GIS? (Figure 7)

*Figure 3. Answers about Data Glossary*
*Data source: Internal instigation*

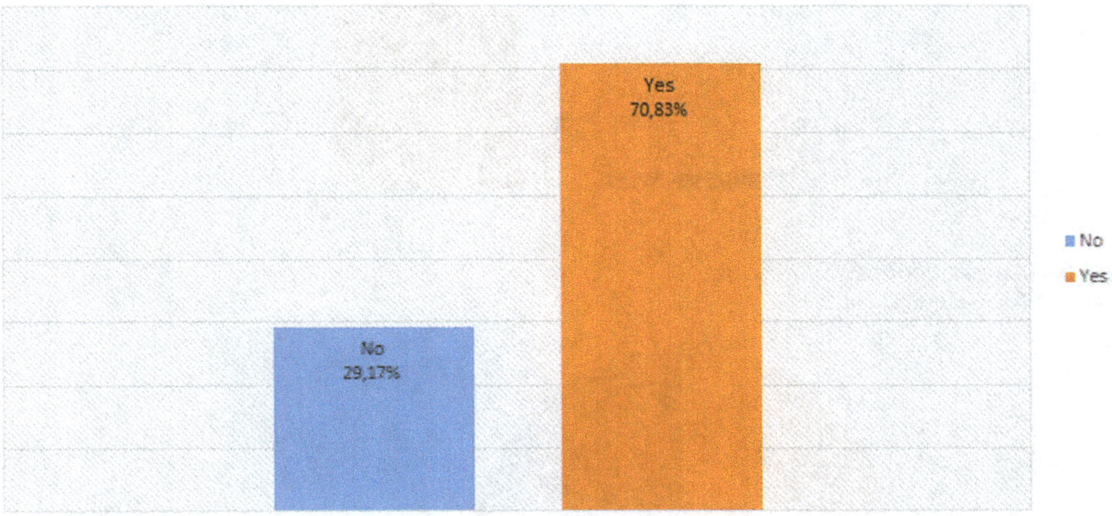


*Figure 4. Answers about Information Source*
*Data source: Internal instigation*

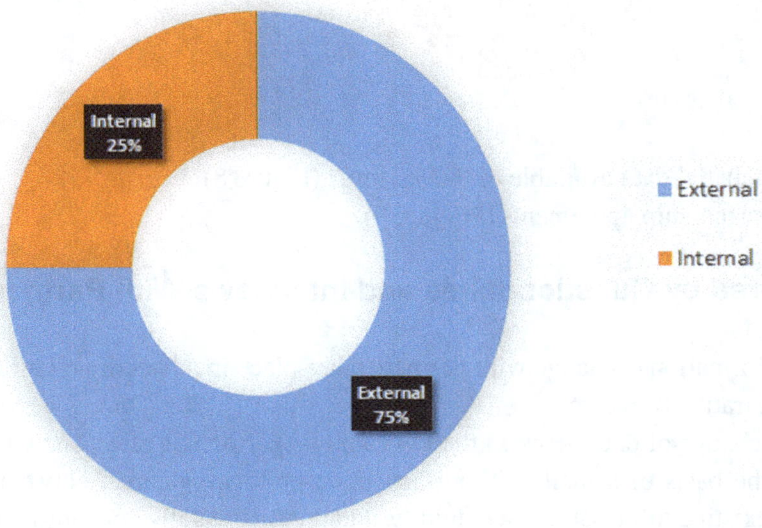

*Figure 5. Answers about GIS*
*Data source: Internal instigation*

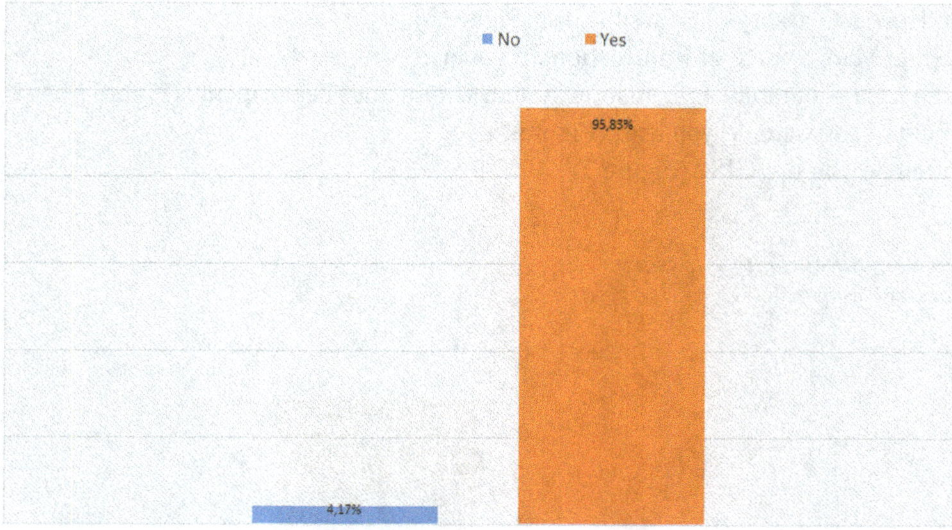


*Figure 6. Answers about GIS Software*
*Data source: Internal instigation*

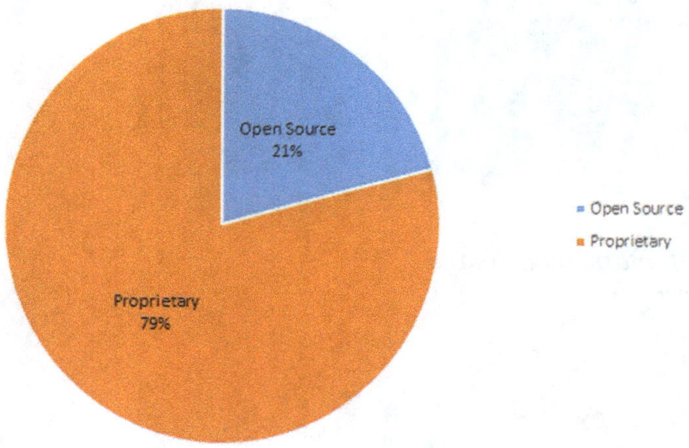


6. Can you make spatial data available to this system? (Figure 8)
7. Do you have partnership agreements? (Figure 9)

## Findings Gathered by Questionnaires and Interviews With Partners

- Most of the information exchange with partners takes place in a friendly relationship without any formal administrative framework.
- The official exchange of data between the DAT and its partners is rare. And when it is done, that would be on the basis of a mail, without any trade protocols defined between the two parties. Exchanges of existing information lack fluidity, and they are usually done through a dunning letter.

*Figure 7. Answers about Frequency of GIS use*
*Data source: Internal instigation*

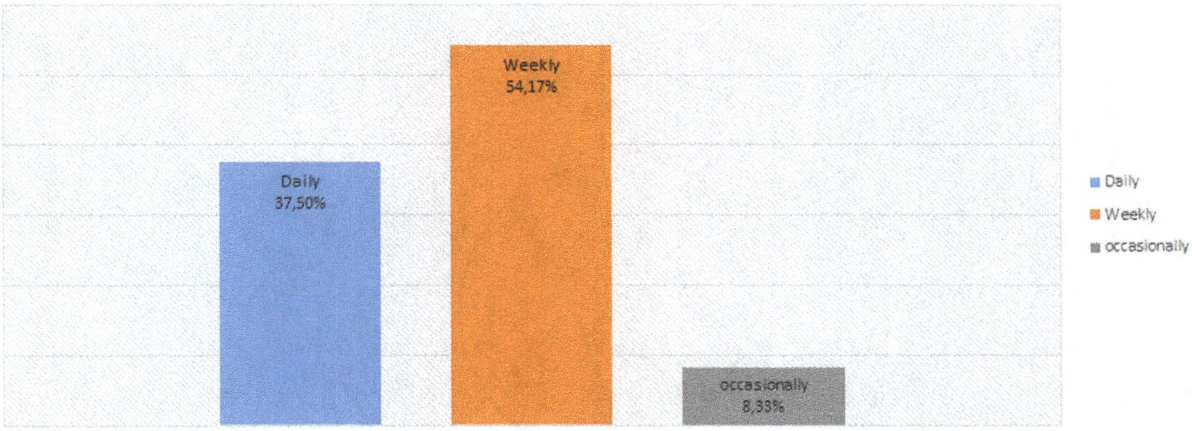


*Figure 8. Answers about Make Spatial data available*
*Data source: Internal instigation*

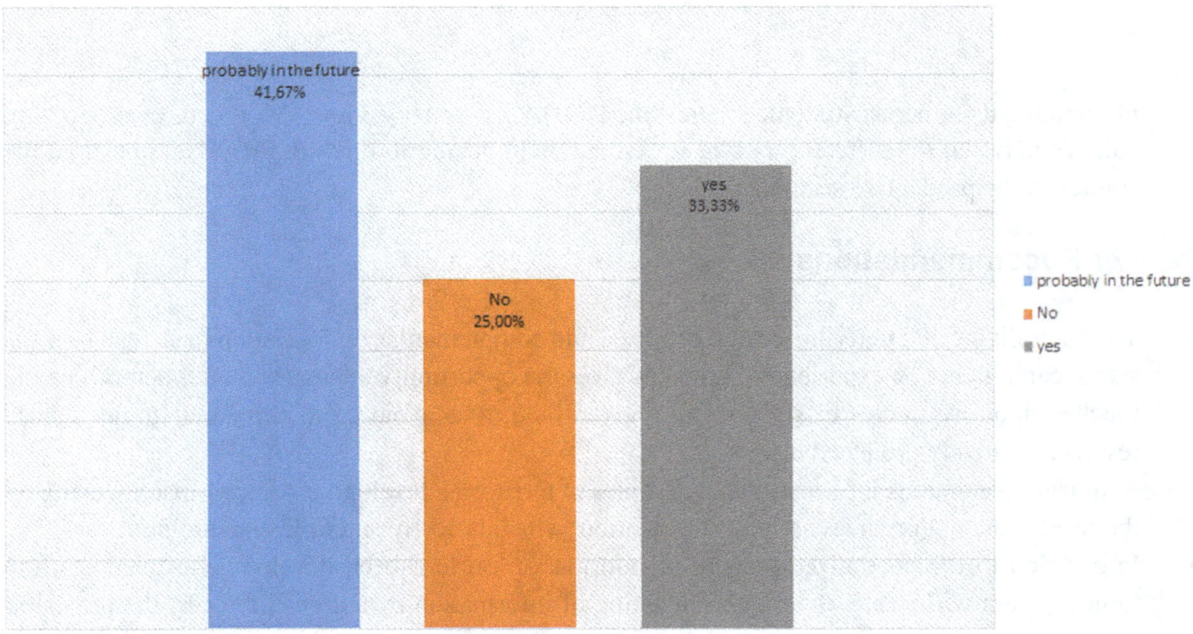


- The information exchanged consists mainly of flat files that do not favor the analysis and cross-checking of information. Some data are retrieved from the partners' web portals and are limited to the global data and indicators processed.
- Exchanges between the main producers should not be guided by administrative habits, but by the real needs expressed by the users.
- Proposed themes for possible partnerships are territorial planning and dynamics, sustainable development, social cohesion, energy efficiency, policy coherence, incorporating climate change

*Figure 9. Answers about Partnership Agreement*
*Data source: Internal instigation*

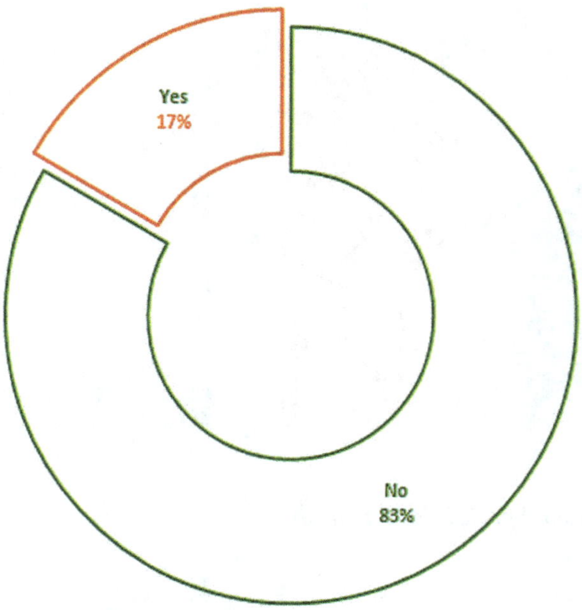


in planning documents, strategic studies, the ANDZOA strategy, Green Morocco, green growth, balance between the different uses of water, any field having to do with the urban question, the studies of the productive sectors.

## Partner Recommendations

- The centralizing and unifying system of varied and complementary information must also be a basis for capitalizing on experiences. It makes gives the opportunity to create an inter-actor dynamic touched upon by studies or statistics in a given field, thus avoiding to carry out studies whose results are already produced elsewhere.
- Acquiring continuous information on all topics of territorial development allows also focusing on the results and not on the search for information, which is costly and time-consuming.
- In addition to the centralized storage of information on territorial development dynamics, the future system will ensure the regular updating of information by automated feeds, the provision of information by the simple presentation of data and data export and generation of dashboards and maps.
- In an overall approach, the new territorial dynamics system will be a model in making information available, removing barriers for users in their acquisition of information.
- The system is meant to be a space of potentially useful features offered to users to help them go further in their research and exploitation of information

Real-time access to databases developed by the various partners is virtually non-existent, which hampers the deepening of the analysis to set up a platform for sharing spatial data. This is also compounded by the late release of data because of the time-consuming nature of the survey or census data. Exchanges between the main producers should not be guided by the administrative habits but by the real needs expressed by the users.

This information is judged to be of good quality and reliability, but its availability falls short of what is needed because of the delay caused while being disseminated by producers. Sometimes, the access to information is impossible, even in the event of official reminders. 90% of this data is obtained informally either in paper form or in CD-ROM / USB key. The data received on electronic media are in Excel, Text, Shapefile and MapInfo TAB format.

## Regional Level

To make an efficient and specific diagnosis for the study area, it is necessary to collect the needs and expectations of data providers and users. Afterwards, it is a question of studying and qualifying these needs and then classifying them according to the relevance to the need collected. And this before studying the various constraints related to these needs and challenges in terms of feasibility and deadline, while taking into account the priorities of key partners for the establishment of a regional spatial data infrastructure.

For this reason, a list of potential partners is presented for the success of this ambitious project. These partners are public departments and agencies directly or indirectly related to the production, use, processing, exploitation and dissemination of spatial data concerning the region. They will receive the questionnaire accompanied by a description of the project.

This interview with the questionnaire will:

- Present the project, identify their point of view on the contribution of such a project, and identify their needs as future partners.
- To know the information available to them and likely to feed the platform, and to define how to exchange this information.
- Identify the different agencies or administrations, sources of this information, and potential recipients of analyzes or publications produced.

*Table 1. Potential list of regional infrastructure stakeholders*

| Organization Name | Organization Name | Organization Name |
|---|---|---|
| Regional Council RABAT SALE KENITRA(RSK) | Regional Directorate of Agriculture Wilaya de RSK | Wilaya de RSK |
| Regional Agency for Project Executiondd | Urban agencies | Regional Investment Center |
| Council of the city of Rabat | Regional Direction of the Ministry of Health | Hydraulic Basin Agencies |
| Prefectural Council of TEMARA-SKHIRAT | Regional Delegation of the Ministry of Trade and Industry | Regional Directorate of the Ministry of the Environment |
| Distribution boards Water and Electricity | Regional Delegation of the Ministry of Tourism | Civil protection |
| Regional Direction of the Equipment | Regional Delegation of the Ministry of Crafts | |
| Regional Academy of Education and Training | Regional Inspection of Territorial Planning and Urban Planning | |

- Identify difficulties encountered in the acquisition, processing, exchange and quality or understanding of this information.
- Identify their expectations of the future project, including the consideration of their needs completing this project.
- Probe their interest and identify the factors are likely to attract them while completing this project.

## CONCLUSION AND PERSPECTIVES

The analysis of the current context, the diagnosis and the inventory in terms of spatial data were carried out in partnership with the Directorate of the Spatial Planning, which provided a set of results concerning the current state of spatial information use at central government level. This study is marked by the lack of a clear regulatory framework describing the effective management of spatial data.

Throughout this diagnosis, it is noticed that the study area has both advantages and disadvantages in terms of spatial data that have an impact on its development. In this region, administrations are home to hundreds of different spatial data as well as applications of geographic information systems. Moreover, the private sector has a number of datasets. But the way these data are managed constitutes a paradox for the region which needs to be urgently addressed in order to define a strategy for setting up a regional spatial data infrastructure.

This chapter presents Moroccan experiences that promote access and sharing of spatial information. First of all, the new law 31-13 on the right of access to information that was adopted by the Moroccan government's council on July 31, 2014. Then, the Moroccan E-Government program which aims at exploiting the technologies of information and communication to modernize the Moroccan public administration to guarantee the service of the citizen and the company by presenting two pilot projects that promote the sharing of spatial information. Finally, the presentation of ONDH, ONDT and ONEM observatories.

The following step will focus on the survey via the questionnaire mentioned in the final paragraph of this chapter in order to collect information on spatial data sharing within administrations of the RABAT-SALE-KENITRA region, as well as the development of a model of a SDI adapted to this problem.

## REFERENCES

ALBACHARIA. (2017). Retrieved from http://www.albacharia.ma/mapController?op=show

Craglia, M. (2010). *Building INSPIRE: The Spatial Data Infrastructure for Europe*. Retrieved from http://www.esri.com/news/arcnews/spring10articles/building-inspire.html

E-GOV Program. (2014). Retrieved from http://www.egov.ma/sites/default/files/Programme%20eGov%20 Morocco.pdf

HCP. (2016). *High Commission for Planning (Morocco)*. Retrieved from https://www.hcp.ma/region-rabat/attachment/775960/

Ibannain, F., & Donnay, J.P. (2012). *Prior analysis of a data infrastructure in Morocco*. Retrieved from http://orbi.ulg.ac.be/handle/2268/142431

M.C. (2011). *Moroccan Constitution.* Kingdom of Morocco, General Secretariat of the Government, Directorate of Official Printing. Retrieved from http://adala.justice.gov.ma/production/constitution/fr/Constitution.pdf

Michael, T., & Millward, L. (2014). *Accessing Information is Our Right" A Practical Guide to Promoting Access to Public Information in Morocco.* Retrieved from http://www.uneskit.org/sites/default/files/dai_illus_int.pdf

Plan Maroc Numeric. (2013). Retrieved from http://www.egov.ma/sites/default/files/Maroc%20Numeric%202013.pdf

Rajabifard, A., Feeney, M.-E. F., & Williamson, I. (2003). *Spatial data infrastructures: concept, nature and SDI hierarchy.* Academic Press.

## KEY TERMS AND DEFINITIONS

**ALBACHARIA:** Information system on human development. A device developed by the national observatory of human development, allows presenting valued information, organized in explicit and visual reading grids and illustrated by thematic maps.

**Data Collection Protocol:** The data collection protocol is a project launched by the Direction of the Spatial Planning, which aims to obtain precise, detailed, reliable, territorialized, and up-to-date information through mechanisms of sharing and exchange is made through sharing agreements between public institutions.

**ONDT:** National observatory of territorial dynamics that is a tool for decision making and the definition of territorialized public policies. This observation device sets itself the objective of being a place for reflection and cross-fertilization of different strategies and public actions in order to conceive a common, real, and prospective image of the territories.

**RDP:** The regional development program, as stipulated by the organic law (111-14), makes it possible to draw up the development axes of the region based on a participative approach that affects all the stakeholders for a period of 4 years.

**RSPS:** The regional spatial planning scheme is a planning document for the entire territory of the region which enables regional actors to design, for a period of 20 years, a strategic vision of the development of the region.

**SDIR:** The SDIR project (spatial data infrastructure of the RSK region) will be a powerful investment for the development, productivity and attractiveness of the territories of the Rabat-Salé-Kénitra region. The main objective of the development of this SDI is to provide an environment conducive to cooperation between stakeholders, users, and producers of spatial information, effectively and efficiently.

# Section 2

# The Geospatial Technology for Effective Natural Resources' Management

# Chapter 6
# Time Series Multispectral Images Processing for Crops and Forest Mapping:
## Two Moroccan Cases

**Loubna El Mansouri**
*Institut Agronomique et Vétérinaire Hassan II, Morocco*

**Said Lahssini**
*National School of Forestry Engineering, Morocco*

**Rachid Hadria**
*National Institute of Agriculture Research, Morocco*

**Nadia Eddaif**
*Mohammed V University, Morocco*

**Tarik Benabdelouahab**
*INRA, Morocco*

**Asmae Dakir**
*Mohammed V University, Morocco*

## ABSTRACT

*This chapter highlights time series image processing for accurate agriculture characterization through two Moroccan experiences. The first case aims at crop mapping. A new classification approach based on multiple classifiers combination (MCC) was developed and applied to multi-temporal enhanced vegetation index (EVI) bands. The whole process is performed in three stages: (1) Landsat data preparation and multi-temporal staked EVI image extraction, (2) MCC construction from six advanced and supervised classifiers, and (3) stacked EVI image classification using the build-up MCC. Some post-classification contextual rules were also added in order to optimize the crops classification and the final parcel shape. In the second case, a post-classification change detection process was implemented to detect changes in forest area. Many classification schemes with different vegetation and texture indices were investigated. The two experiences are cost-effective, reproducible, and transferable. Consequently, they can regularly be used to produce up-to-date land use maps.*

DOI: 10.4018/978-1-5225-5939-9.ch006

## INTRODUCTION

Mobilization of geospatial information contributes to the creation and sharing of a unique vision of territory. Earth observation with different sensors characteristics has been found to be a valuable tool for data acquisition for mapping, monitoring, evaluating and managing land, water and crop resources. Combined to geospatial information system (GIS) and global navigation spatial system (GNSS), they form together in a complementary way an effective geospatial technology tool for more land governance. These technologies can capture the present, go back in past and predict the future. As decision support tools, decision based on such technologies are helpful for ensuring sustainability and for dealing with global concerns such as sustainable and integrated development. In contrast to that, we note that if geo-information allows approaching set of themes related to territory management by bringing convincing elements in terms of decision support, it brings new scientific questionings as areas for future research. Precision agriculture and precision forestry are now and for the next decades the development trends in geo-information applied to agriculture monitoring and management (Beerech HV., Mrslatha BM., Thimmaraja Yadava G. and Navven Dandur, 2014; Oscar S. Dalmau,Teresa E., Alarcón and Francisco E. Oliva, 2017 and Paletto, Ferretti, Cantiani and De Meo, 2012).

Nowadays, agricultural domain is the most studied by the geospatial technics all over the world. Crops and forest are the two major studied themes. They remain for international communities a vital, environmental, economic and social question. Then, thematic crops and forests maps are produced according to process based mainly on supervised classification. Literature review reveals that more attention should be paid to classification parameters to ensure acceptable and correct derived thematic maps (Elmansouri 2013; 2014; Gumma, Thenkabail, Maunahan, Islam and Nelson, 2010; Waldhoff and Lussem, 2016).

Crop types, biophysical and vegetal parameters monitoring are the principal tasks needed for crop management. Further, many applications at various scales could be derived from them, such as: crop identification, vegetation health monitoring, identification of planting and harvesting date, yield estimation and modeling, crop rotation records, soil fertility, water consumption and irrigation scheduling.

In Morocco, crop plantation information is mostly collected through four ways: (1) farmer communications (FC), (2) spatially limited land survey (LLS), (3) visual photo-interpretation (PI) of a newly registered digital image and (4) automatic one-satellite image analysis (OSI). These procedures provide limited and subjective information with unguaranteed consistency, weak completeness and moderate accuracy. They also require a large team of skilled photo interpreters. Land survey could map accurately crop types but it is too time consumptive, high cost and labor-intensive which limits its use. Due to vegetation changes according to its phenological cycles over time and the variation of environmental conditions, image acquisition date is a critical point for the two last approaches (PI & OSI).

Recently consulted literature highlights that none of these methods can be used as a periodic process to monitor crop changes (Tingting and Chuang, 2008; 2010). Indeed, one image cannot depict the dynamic behavior of crops. Multi-date image analysis is more suitable. Time series of optical and radar remotely sensed imagery is shown to be a cost-effective data sources for automatic crops identification. They can be regularly used to produce an accurate and up-to-date crop map at different temporal and spatial resolutions, although the processing time is longer. The underlying hypothesis is that each crop species has both specific spectro-temporal signature in a time series of multispectral images and backscatter profile in radar data. This is logically true only if the taken time series images dates depict whole researched crops phonological cycle. In this way, associating spectral signatures with phenological stages which highly increase classification performance.

In this context, the used time series images depend on studied cropland detail's level, cost, processing time, data availability and study area conditions. Radar data suffers from high noise and low resolution while hyperspectral images are so expensive for a regular use. But note that in cloudiness and fog areas, data from Synthetic Aperture Radar are the only alternative for studying vegetation cover. Otherwise, Optical images from different types of sensors remains the most applied data in agriculture with the common condition of incorporating Near Infrared (NIR) Band. The coarse spatial resolution sensors such as AVHRR (Advance Very High Resolution Radiometer) or SPOT-VEGETATION have been successfully used to monitor agriculture land at daily frequency and continental scales. MODIS (Moderated Imaging Resolution Spectroradiometer) data classification has been extensively utilized to map single or limited number of cropland classes over a large area. However, high resolution images are more suitable to produce crop maps at finer scale but remain highly limited due to data availability.

Spectral vegetation indices (VI) have been widely used for evaluating, plant and crop biomass. Taking account on vegetation's reflective contrast within the near infrared and visible red wavelengths, the most used formulas are based on ratios or linear combination and exploit differences in the reflectance patterns of green vegetation and other objects. Enhanced vegetation index (EVI) has the particularity to reduce the adverse effect of environmental factors such as atmospheric conditions and soil background compared to classical Normalized Difference Vegetation Index (NDVI) which does not take into account the contribution of the atmosphere or the underlying ground to the canopy. Perpendicular vegetation index (PVI), Soil adjusted vegetation index (TSAVI) and Modified SAVI (MSAVI) are other vegetation indices developed and used in literature (Department of the interior U.S., 2016; Huete, Liu, Batchily and Van Leeuwen W, 1997).

From raw data, synthetic indices to classified maps an aggregation or decision methods are needed. Machine learning classifiers are prediction system defined by three main parameters: learning set, space attributes and decisions rules. Two classifiers that are different by at least one of these parameters do not give necessarily the same prediction and are thus considered different. Moreover, using more than one classifier in a single recognition task is shown to be an effective solution. Indeed, multiple classifier combination (MCC) approach has been introduced to remote sensing data classification and achieved better results compared to individual classifier (Benediktsson, Chanussot and Fauvel, 2002; 2007; Elmansouri, 2013; 2014; Geurts, Irrthum and Wehenkel, 2007; Marée; 2005, Kang and Doermann, 2005). The combination of classifiers is particularly recommended when classifiers show different behavior. Combining multiple classifiers could benefit from their complementarities and outperforms any single classifier. Moreover, in theory, the MCC prediction is characterized by minimal variance and seems to be less sensible to data noise. Another reason for using MCC arises from its faculty to diminish the impact of classifier's parameters adopted values on the final prediction.

Forest monitoring, through tracking state of the resources, is an effective tool for their multifunctional management leading to sustainable provision of ecosystem services. Besides resources assessment, multi-temporal remotely sensed images offer the possibility of locating and identifying changes in forest area through various change detection techniques. Currently, automated image-to-image comparison is the major method used to observe and evaluate change between multi-date images. Such case needs the compatibility of multi-temporal images, i.e- must be accurate and are matching (radiometric and geometric) (Elmansouri, 2014). Details about this technique will be given later in this paper.

Two cases study are presented in this chapter. They highlight both the enhancement of the process of crops mapping and land forest change detection through the use of sequential time series (in discrete long-time span) multispectral free fees images. The cases represent a wide diversity, either in studied territories or handled themes and specific objectives. The first experience concerns Gharb plain and deals with eight crop types mapping introducing an original combination system of supervised classifiers to enhance the accuracy of final land use map. The second experience aims at evaluating changes in forest area over a 27 years period. A post classification change detection process was implemented to detect changes. Many classification schemes with different vegetation and texture indices were investigated and quantitatively evaluated. Finally, change detection rules were developed in order to identify localization and change direction.

## CASE 1: TIME SERIES VEGETATION INDEX AND MULTIPLE CLASSIFIER COMBINATION FOR CULTIVATED CROPS MAPPING

### Background

#### Multiple Classifier Combination (MCC)

MCC consists in deriving the final classification decision by integrating the output of multiple learning machines according to a certain combination approach (Kang and Doermann, 2005). Two main strategies for combining classifiers exist: sequential and parallel. The first one consists on concatenation of organized connection of successive levels decisions. At each level, single classifier handles the decision supplied by the upstream classifier. This decision can be rejected or confirmed. The major inconvenience of the method is its big sensitivity to the classifiers order. On the other hand, in parallel scheme, classifiers operate independently from each other to produce their own decisions which will be posteriori merged to reach a consensus. Output information of member classifier could be a single class, a rank vector or a probability membership function to the in all classes (Benediktsson, Chanussot and Fauvel, 2007; Kang and Doermann, 2005). The parallel combination is more generic as it can be easily converted to rank or class. In these papers (Briem G. J., Benediktsson J. A., Sveinsson J.R., 2002; Elmansouri, 2014; Oscar S. Dalmau,Teresa E., Alarcón and Francisco E. Oliva., 2017 and Zhou, Hong-cha, 2006), the authors have used MCC for Land use mapping from satellite image.

In the case of competitive classifiers, the efficiency of an MCC is hardly dependent on complementary classifiers which can be verified only if they are different. The relationship between performance and complementarity becomes clear if we make the analogy with individual classifier which decides better with an additional attribute providing useful and complementary information to those supplied by the others. A machine learning classifier is a system of prediction which can be defined by three main parameters: learning set, space attributes and decisions rules. So, two classifiers which are different by at least one of these parameters do not give necessarily the same prediction and are thus considered different and can be then organized in a MCC.

When the output classifiers are probability functions, the final decision can be made simply through the use of summary statistics such as maximum, minimum, median, sum, product, average of probabilities or using some other more complex methods as nonlinear combination (Elmansouri, 2014; Kang and Doermann, 2005). Therefore, other elaborated ways are based on weighted combination rules on which each individual classifier is associated with a different weight that is, usually, linked to its performance.

## ADVANCED CLASSIFIERS

Crops can be identified based on their phenological profiles which could be approached from time series multispectral images. Most authors use supervised classification of staked time series vegetation index to extract crop types. Classical classifier cannot handle the high number of features in multi-temporal images analysis that is why advanced supervised learning classifiers are becoming more used in multi-date image classification. These include, but are not limited to, decision tree (DT) and tree-based ensemble methods. They are good non-parametric data reduction tools that are not based on any assumption related to data probability distribution. In DT algorithm, a learning set is successively split into binary homogenous subsets based on "if-then" rule tested at each internal node of the tree. The terminal node will be labeled by the majority class. At the end, the tree results in a number of class prediction rules that are used to create a predictive model. Classification accuracy from DT algorithm is often better compared to using maximum likelihood (ML) (Hazewinkel, 2001) or linear discriminate function (Benediktsson, Chanussot and Fauvel, 2007; Marée, 2005; Marée, Geurts, and Wehenkel, 2016).

The original classification single trees can overfit the training data. Ensemble methods overcome such concern. Indeed, tree-based ensemble methods, through randomization, generate a more stable prediction. These methods include Bagging, Random Forests (RF) and Extra-Trees. Bagging ("Bootstrap AGGregatING") builds from the primary learning set multiple training subsets by random replacement (bootstrapping). Each subset is inducted as explained above in the classical DT algorithm. Then, Bagging aggregates the predictions of all developed trees to produce final classification using a majority votes. RF combines Bagging with a random selection of subset of k attributes (predictors) at each node. In the Extra-Trees (for "EXTremely RAndomized Trees") method, each tree is built from the complete learning set but at each test node, the best split is determined by randomly selecting one attribute. A complete review can be found in (Marée, 2005; Wehenkel, Bastin, Phillips, and Geurts, 2017). Another advantage of these algorithms is their internal faculty for partitioning learning set into separate sets: training and validation subsets. This allows to automatically determining all their parameters importance through cross-validation.

## OBJECTIVE

In this case study, we focus on the practical use of free fees multispectral images for crop types mapping. As the accuracy of individual classifier is critical, a time series-MCC based approach is developed in order to enhance the accuracy of crops identification. The main objective here is to present a complete and validated procedure for regularly crops mapping. This study aims also to assess and to quantify real contributions of time series use and combining more than one classifier in the final obtained land use map accuracy.

## STUDY AREA

The study area is located in Gharb plain in the north of Morocco (Figure 1). It extends over an area of about 8805 km$^2$ and is centered at latitude 34° 15′ N and longitude 6° 35′ W (Datum WGS84). Crops within study area prevail on flat lands with an average altitude of 70 m above sea level. Further, good

*Figure 1. Geographical location of Gharb plain*

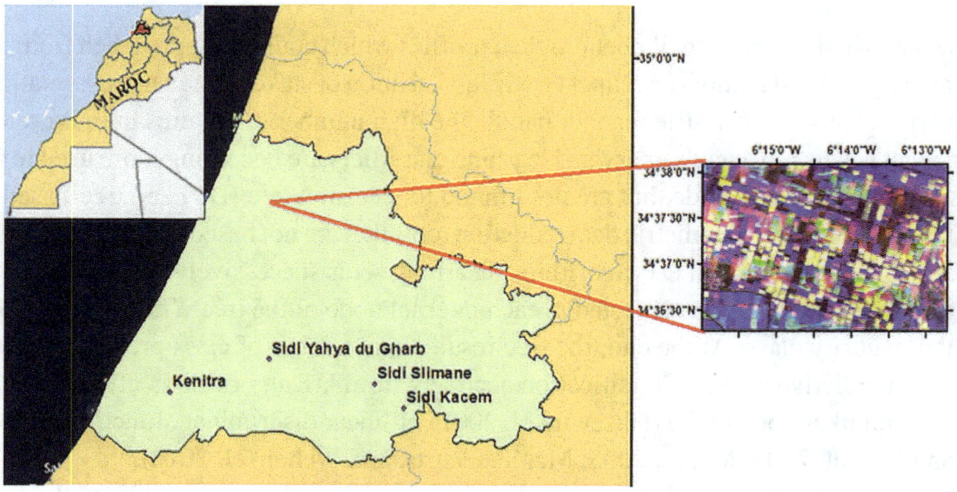


alluvium soils in addition to an efficient irrigation system promoted the development of intensive agriculture. Then, rice occupies low lying lands whereas the uplands are planted with fields crops such as cereals, maize, sugarcane and so on. Southern part of the region is dominated by "Maamora" forests.

## DATA SOURCES

### Collected Agricultural Parcels and Crop Classes

The ground truth data needed for calibration and validation of classification were collected in April 2011. During the field survey, plots corresponding to different land use classes (a part from water) were selected and digitalized to form a ground truth vectorial GIS layer. For each class a set of 400 to 3300 pixels was considered and divided separately for learning set (LS) and validation set (VS). Figure 2 presents the localization and the number of collected pixels for each class.

## IMAGE DATA

12 Landsat TM / ETM+ SLC-off cloud free images at 30m resolution for the year of 2010 and 2011 were used. Table 1shows the five optimal periods used for multi-temporal image acquisition according to agricultural calendar of main both winter and summer crop classes shown in Figure 2. A part from November and December, all others months are observed at least by one image. TM images are more desirables than ETM+ which suffer the loss of their scan line corrector and only captured about 80 percent of the data.

*Figure 2. Spatial distribution of ground truth data (centered at 34° 15′ N 6° 35′ W)*

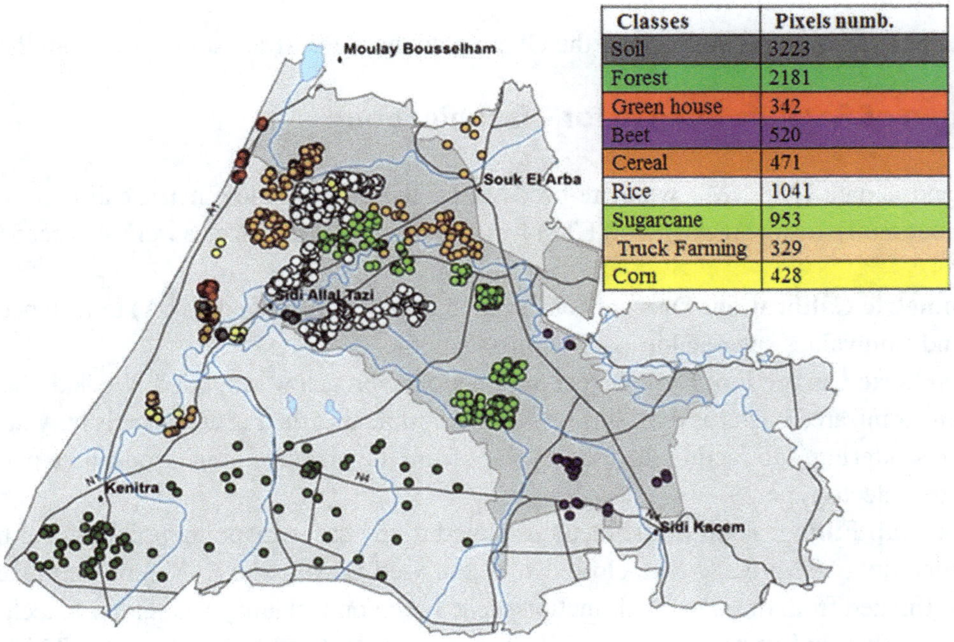


*Table 1. Optimal periods of images acquisition*

| Optimal Period | Sensor | Acquisition Date (*) |
|---|---|---|
| Period 1<br>05/10/ to 15/11 | ETM+ | 05-10-2010 |
| | ETM+ | 21-10-2010 |
| | ETM+ | 06-11-2010 |
| Period 2<br>15/02 to 15/03 | TM | 18-02-2011 |
| | TM | 06-03-2011 |
| | TM | 22-03-2011 |
| Period 3<br>15/04 to 15/06 | ETM+ | 15-04-2011 |
| | TM | 09-05-2011 |
| | ETM+ | 18-06-2011 |
| Period 4<br>15/07 to 15/08 | TM | 28-07-2011 |
| Period 5<br>15/09 to 15/10 | TM | 14-09-2011 |
| | TM | 16-10-2011 |

(*): The year 2010-2011 was chosen due to the availability of field data.

## SHAPE DATA

Shape data consists on a limit polygon of the Gharb plain used to extract study area from the images.

### Preparation of Satellite Images for Classification

The L1T Landsat data TM/ETM+ were pre-processed at the levels of radiometric calibration and atmospheric correction. To restore the SLC-off ETM+ malfunctions ETM+ images were also been Gap filled.

- **Radiometric Calibration:** DNs are converted to Top of Atmosphere (ToA) reflectance using the bias and gain values specified in Landsat metadata file.
- **Atmospheric Correction:** Dark Object Subtraction approach was applied. We had identified water on near infrared bands as dark surfaces and assuming that their reflectance is only due to atmospheric scattering, subtracting then their values from the image to omit atmosphere part in captor detected reflected energy.
- **ETM+ Gap Filling:** Fault in the methods suited to the crops mapping problem and because of unavailability of TM image with close date to the used ETM+, the ETM+ missed pixel's values were estimated from the same malfunction scene based on Delaunay triangulation technique.
- **Temporal Profile Extraction:** First EVI bands were calculated for all corrected TM/ETM+ images as indicated in equation (1). They were stacked by after and the study area was extracted according to the Gharb plain limit. The temporal profile for each pixel was constructed. It presents the spectral-temporal attributes to be conducted into classification process.

$$EVI = 2.5*((\text{Near infrared} - \text{Red}) / ((\text{Near infrared} + 6*\text{Red} - 7.5*\text{Blue} + 1)) \quad (1)$$

## MCC CLASSIFICATION

Based on land survey and available images we have identified 10 classes with the eight major crop types: Water, bare-soil/buildup area, forest, rice, corn, cereal, beet, sugarcane, vegetables and greenhouse. From six supervised learning algorithms, an unbiased MCC has been constructed and applied on staked multi-temporal EVI images to predict classes and map crop types. The MCC building details will be presented in the following section.

## MCC BUILDING

In parallel MCC scheme, a subset of independent classifiers is selected. Their decisions are then aggregated to predict finale decision. Their implementation consists on defining: (1) input classifiers, (2) selection of criteria for maintaining, (3) combination rules and (4) criteria for assessing classifier performance. Omission (f-measure) and commission errors are widely used as quality index as they assess the global accuracy and classification errors.

Literature recognizes that classical classifier cannot handle the high number of features in phenology-based analysis. For this reason, advanced supervised learning classifiers become more used. Including, but are not limited to, SVM (Vapnick, 1995) and tree-based ensemble methods, such non-parametric methods are a good data reduction tools that are easy to adapt for new dataset and able to detect multivariate interacting effects between temporal features. The main idea beyond SVM classification is to fit the optimal hyperplane separating the two classes. Kernel functions could construct the optimal hyperplane for the complex data which cannot be separated with linear hyperplanes. As described before, tree-based ensemble methods, based on "if-then" rules, splits data and predictors to multiple subsets.

To build our MCC, six classifiers have been used: the DT, the KNN (K Nearest Neighborhood), the SVM, the Bagging, the RF (Random Forest) and the Extra Trees (Extremely Randomized Trees). The first ones are both implemented to put forward their ineffectiveness in studied classification problem. These six classifiers are independent from each other. SVM, induction of trees based methods and KNN are of different theoretical foundations. On the other hand, the three tree-based algorithms methods introduce randomness into the process in a different way and at different levels which vary from one algorithm to the others and ensure their diversity.

The three selected classifiers in the previous step will be combined according to five linear operators: the maximum, the minimum, the product, the average and the weighted average. Individual classifier performance per class will be used as weight for the last combination. The most performing of tested combinations is the one that will be retained for the classification of the stacked EVI image to map crop types

To ensure an unbiased MCC, the three aforesaid operations: parameters optimization for individual classifier, classifier selection and the best MCC selection must be executed on independent datasets. Consequently, for statistical reliability, LS should be big enough to be divided into 3 independent sets of data. To deal with a reduced LS size, cross validation was adopted. Then, as presented in figure 3, LS was successively divided into three levels of cross validation. At each level (CV_1, CV_2, CV_3), data set is splitted into training dataset (LSi) and validation dataset (VSi). CV_1 level serves to optimize individual classifiers parameters (number of developed trees: T and stop splitting criterion: Nmin for based-trees methods; Kernel type, ϒ in kernel function and penalty parameter: C for SVM), CV_2 level is used to select the three best classifiers and their weights, whereas the CV_3 level serves to determine the best CMCC to recommend. In this way the final prediction is obtained without bias.

Table 2 presents the performance of individual classifiers achieved at CV_2 level based on their optimal parameters fixed at CV_1 level. As expected, classical DT, KNN and Bagging are clearly less successful than RF, Extra Trees and SVM that have similar performances. Classification result is en-

*Figure 3. Three level cross-validation scheme to construct MCC*

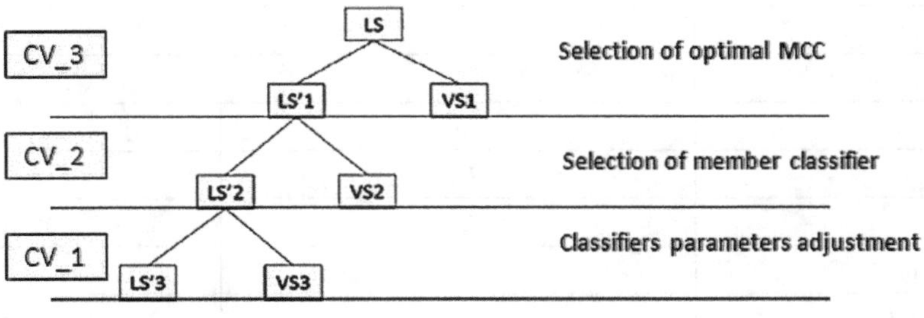

*Table 2. HE performance of individual classification based on their optimal parameters*

| Classifier | f-Score % | σf-Score |
|---|---|---|
| Classical DT (nmin=3) | 53.73 | 5.18 |
| KNN (k=5) | 59.34 | 1.04 |
| Bagging (T=100; nmin=1) | 62.55 | 3.19 |
| RF (T=200; nmin=1) | 71.14 | 1.94 |
| Extra Trees (T=500; nmin=1) | 71.62 | 1.61 |
| SVM (radial, C= 200, ϒ= 0.1) | 72.15 | 1.55 |

hanced at least by 9% with tree-based ensemble method compared to original DT. These results prove the performance of advanced algorithms and confirm the conclusion met in the literature. The SVM is very close to RF and Extra Trees and can be slightly better. Extra Trees requires more trees to increase and stabilize the accuracy of its prediction.

SVM, RF and Extra Trees were maintained to build 5 MCC. Table 2 is the f-measure score and its standard deviation of individual classifier and the 5 MCC obtained at CV_3 level.

From Table 3, some comments can be drawn. First, it can be found that the combination by the min and the product operators suffer from high variance and decrease classification accuracy compared to all individual classifiers. In the fact, this conclusion is already mentioned in a lot of papers, but our motivation behind their use is that the performances of the classifiers are rather good to return the results of such reliable statistical rules. Second, the maximum function leads to a light improvement with regard to the best classifier (SVM). Third, average function and the weighted average behave better than the best classifier and enhance significantly the classification accuracy and its variance. An enhancement of about 5.5% is achieved in accuracy score and 1% in standard deviation score. We observed that the improvement concerns mainly the cases when the 3 classifiers make uncorrelated wrong predictions for the same pixel. When comparing these two average ways, one can observe that no technique is clearly the best because the introduced weights rely on classifiers accuracies, which are quasi-similar. Therefore, the weighted average converges to the simple average. For the rest of the process, the predictions from simple average will be adopted. It will be indicated by MCC in the remaining sections.

*Table 3. HE performance of individual classifiers and the applied MCC*

| Classifier | Moyenne f-Measure % | σf-Measure |
|---|---|---|
| Max | 71.6 | 2.77 |
| Min | 68.9 | 3.9 |
| Product | 69.2 | 3.12 |
| Arithmetic average | 76.4 | 0.89 |
| Weighted average | 76.3 | 1.23 |
| RF (T=200; nmin=1) | 70.4 | 1.89 |
| Extra Trees (T=500; nmin=1) | 70.9 | 1.91 |
| SVM (radial, C= 200, ϒ= 0.1) | 71 | 1.53 |

## POST CLASSIFICATION ENHANCEMENT

Traditional per-pixel classification leads to “salt and paper’ effects in thematic maps. For this reason, a majority filter is applied to reduce noises. On the other hand, due to the complexity of crops environment, spectral confusion is common among crop cover classes. Thus, ancillary data are used to improve the classification result based on established expert rules. For example, based on some existing agriculture data and on prior crops knowledge as irrigated perimeter, crops parcels adjacency, crop parcels homogeneity, crops mixture in the same parcel and parcels regularity limits, a series of contextual rules were defined in order to optimize the crops classification result and to refine the final parcels shape.

## RESULTS AND DISCUSSION

The general crop-related LUCL map produced by MCC is presented in figure 4. Visually the geographical distribution of the major land use such as forest and beet estimated in this map coincided well with the landforms. Also, it was found that most area along “Sebou” river was occupied for rice which is true. Overall accuracy, f-measure, kappa coefficient and its standard deviation obtained based on validation set points were respectively 75%, 73%, 69% and 2%. These relatively moderate values are usually obtained in remote sensing-based crop mapping literature. In this research, they are principally due to the low classification rate achieved for vegetables and sugarcane classes. Collected Training set for the first class suffers from a phenological variability due to irregular agricultural practices. Sugarcane has the same spectro-temporal profile as forest which generates symmetric confusion. Best classification scores are obtained for rice and maize. Their phenology behavior is well traced from the used images dates.

*Figure 4. General crop-related LUCL map*

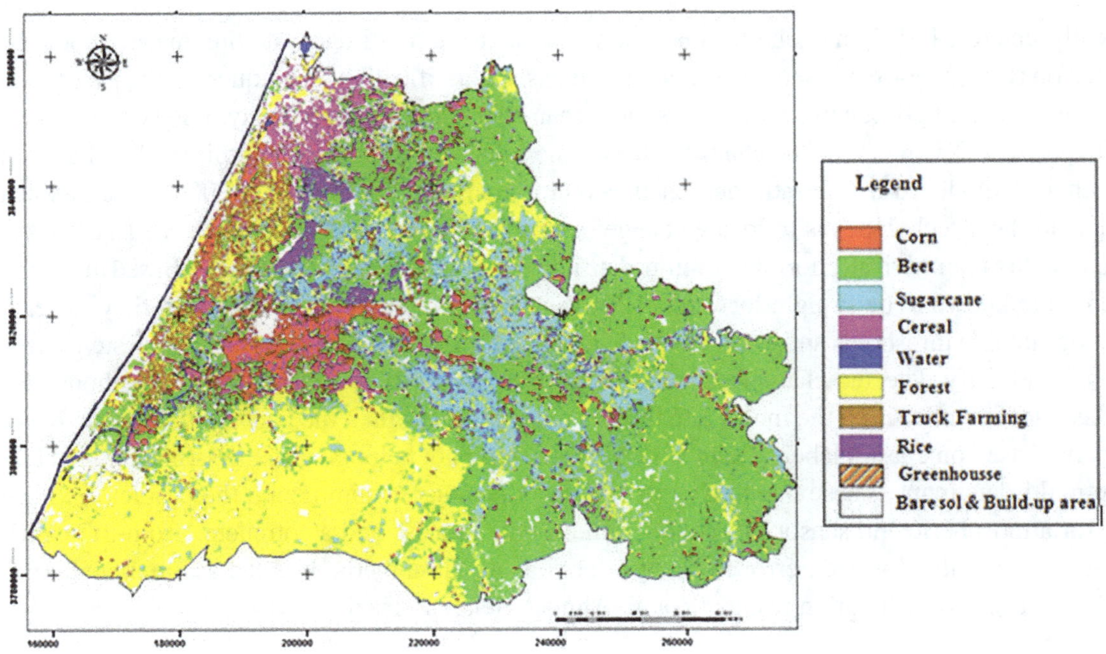

## PARTIAL CONCLUSION

Remote sensing has been proved a powerful, reliable and effective technology to map crop types at different spatial and temporal scales. Based on experiment and discussions below, three conclusions can be drawn. The first one concerns the MCC. An area of about 29.000 ha was correctly mapped thanks to MCC compared to the single best classifier. This proves that combining multiple algorithms can play an important role in image classification enhancement and crops identification by making full use of the abundant and detailed information in time series and integrating the benefits of different classifiers. But note that for MCC use, there are still many issues for further investigation, for example, selection of member classifier criteria and determination of combination strategy. Secondly, about 41.000 ha were correctly mapped when using Multi-date EVI bands compared to single EVI band. Finally, this study demonstrated that time series Landsat data sets has sufficient spectral and spatial resolutions to discriminate the major crop types. The temporal profile based classification yielded a crop map with satisfactory accuracy and consistency. Which was difficult to achieve based only on one-image analysis. However, this approach is rather inefficient for crop mapping at a large scale and for minor classes because of low spatial resolution, the limited availability of cloud free imagery and SLC ETM+ problem.

Tests based on Sentinel 2 images will be a perspective. Indeed, using more than one sensor is another way to enhance derived crops map. On the other hand, the evaluation of any other vegetation indices incorporating NIR band and water indices including MIR band should be done as future work to better describe the space of attributes, using an object-based classification as an addition to MCC.

## CASE 2: POST-CLASSIFICATION CHANGE DETECTION TO MONITOR ARGAN WOODLANDS CHANGE USING MULTI-TEMPORAL LANDSAT DATA

### Background

Basically, changes based on image-to-image comparison using multi-date satellite imagery could be organized on two categories: based mask change and based classification techniques. In the first category, a change indicator between two dates is first built than a threshold or statistically based decision criteria lead to generate “change” vs. “no change” areas. This category includes, but not limited to, band image differencing, modified image ratio and change vector analysis (Coppin et al., 2004). The elementary premise in these techniques is to locate change comparing images reflectance values. That’s why, it’s mandatory to radiometrically normalize multi-date images before comparison process. Based mask change methods are known to be straight forward, effective and easy to implement. However, they suffer from selecting suitable thresholds and cannot provide details of nature change. A classification step is usually needed to identify detected change nature (Elmansouri, 2014). Based classification methods involve the classification of each of the multi-date images independently, then the generated thematic maps are compared based on classes labels. Changed pixels are those not belonging to the same thematic classes in the multi-date maps. These methods present many advantages: minimizing the problem of normalizing for atmospheric and sensor difference in multi-date process and a complete “from-to” matrix of change can be built. However, error in change derive from these methods is the accumulated errors of the all multi-date classification results (Rokni, Ahmad, Selamat, Hazini, 2014).

Regardless of the used method, the result of change detection from image-to-image comparison can be affected by many factors: clouds, fog and shadow present in the images, acquisition time, geometric viewing configuration, atmospheric and illumination conditions, landscape and topography of studied areas, phenological phase of vegetation and expert's skill and experience (Coppin, 2004; Elmansouri, 2014).

For deforestation studies, it is recommended to transform multispectral images to vegetation indices before implementing change detection. This reinforce vegetation component. Furthermore, summer images are useful as they avoid spring vegetation. Many studies have addressed such concerns, a complete review on change detection in forest area could be found in (Kangas and Maltamo, 2006; Lu, Mausel, Brondizio and Moran, 2004). Currently, LiDAR technology appears more suitable for forest application. Outputs measurement are cloud of [$x$, $y$, $z$] information from which it is possible to restitute the ground surface and the canopy surface (Kershaw et al., 2016). However, this technology is far from economic use.

Argan woodlands in south-western Moroccan has been classified as Unesco MaB natural heritage. Studies tried to analyze its spatial distribution (Moukrim et al., 2018) or to evaluate the exploitation trends and spatial changes of this heritage. However, in the major part, these studies do not provide similar statistics and seems to overestimate the degradation trends. Furthermore, there is no consensus on the current extent of the Argan tree or on its degradation status (Aouragh, 2012; El Abboudi, 2015). These confusions are due to the diversity of the used change detection methods, ranging from visual interpretation to fully automated classification technique. Literature review demonstrates that success of classification results depends on image processing process. Extraction of useful attributes of remotely sensed data and selection of a suitable classifier are significant especially for improving classification accuracy. Nonparametric classifiers and knowledge-based classification have increasingly become important methods for supervised image classification. More investigation, however, is necessary to optimize the image processing chain to enhance classification accuracy (Eddaif, 2015; El Abboudi, 2015).

## OBBJECTIVE

The purpose of this research is to study the potential use of image-to-image comparison in change detection process to evaluate changes in Argan forest area over a 25 years period and at five temporal points: 1988-2000-2005-2010 and 2015. The core post classification change analysis was based on landsat sensors. Many classification designs with different classification algorithms and vegetation / texture indices were investigated. A reclassification GIS tool based on arithmetic combination was used in GIS environment to compare the classifications outcomes. Then, 'from–to' change information, as well as the area and the type of landscape transformations, are provided in map and histogram format.

## STUDY AREA

Study concerned Mesguina forest located in the south-western part of Morocco (30° 25 '11'' N& 9° 28' 02 ''W). It covers about 49625 hectares (HCEFLCD, 2017). Mesguina presents a great ecological diversity and consists on different types of argan facies with high socio-economical added value. In order to identify and understand Mesguina forest changes within the last 25 years, three different sites according to trees density, ground topography and build-up area presence have been chosen. The 3 sites are shown in figure 5.

## IMAGE DATA

Five Landsat images from different Landsat sensors (presented in Table 1) and covering the same scene from 1988 to 2015 were used. These sensors present similar specifications which allow a meaningful comparison of the data sets. On the other hand, choosing, as possible anniversary dates of satellite images, is necessary to minimize discrepancies in reflectance caused by seasonal vegetation fluxes. Table 4 specify the used images.

*Figure 5. Study area localization (centered at 30° 25 '11'' N& 9° 28' 02 ''W)*

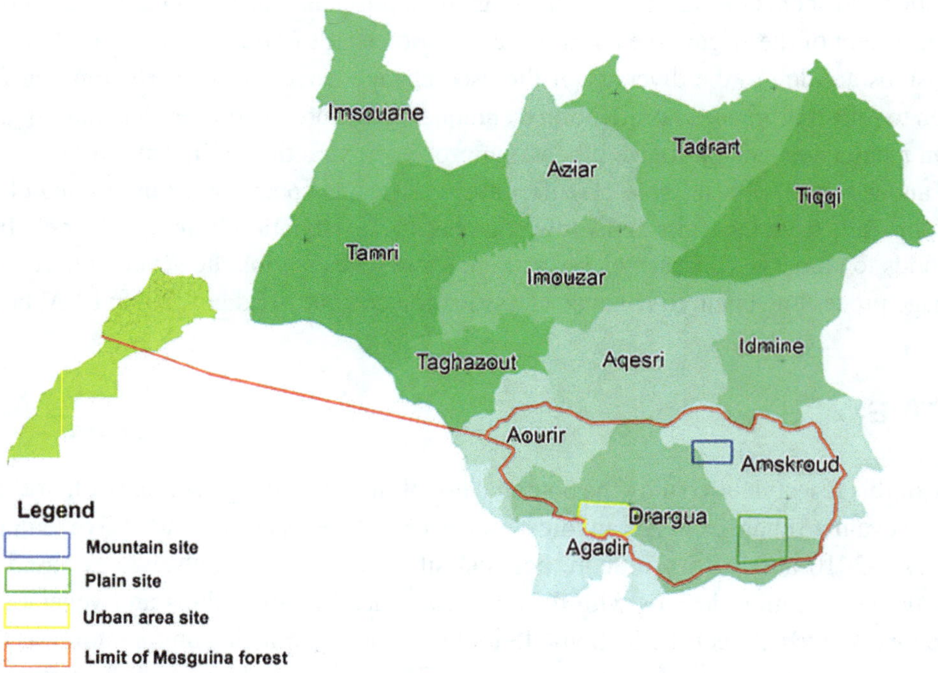


*Table 4. Image specifications*

| Acquisition Date | Sensor | Spatial Resolution |
|---|---|---|
| 28/03/1988 | Landsat TM L1T | 30 m for all spectral bands |
| 20/05/2010 | | |
| 25/01/2000 | Landsat ETM+ | 30 m for multispectral bands, 15m for the panchromatic band. |
| 12/04/2005 | | |
| 31/03/2015 | Landsat OLI_TIRS | |

## IMAGES PRE-PROCESSING

All Landsat imagery are provided in a UTM projection and a WGS84 datum and ellipsoid respectively. Thus, a geometric correction was not necessary. The only geometric process consists on clipping images to study area limits. Data from ETM+ have been processed in order to restore missing lines data.

As, change detection process will be applied on classified results obtained from the same sensors family, there is no need to calibrate radiometric raw sensor data. Also, atmospheric correction was not performed. Literature demonstrates the very small effect of an atmospheric correction for post classification change detection accuracy. That's why we save bands with their original digital numbers.

Multispectral transformation as false color composite and vegetation indices were calculated. Digitized training and validation sites and their corresponding thematic class were collected directly from images for six classes: agriculture, dense forest, sparse forest, build-up areas, bare soil and water. According to existing maps of some subsets of the studied areas, agriculture and forest sub classes were grouped to vegetation class. Build-up area was also assigned to roads and constructions. No fieldwork was carried out.

## METHODOLOGY

The whole proposed approach is developed in three steps: optimization of the classification scheme, post classification process and change detection rules.

### Optimization of the Classification Scheme

Aiming at generating reliable change maps of Argan woodlands successively between 1988, 2000, 2005, 2010 and 2015, based on post-classification comparison, there is a need to optimize the adopted supervised classification technic. Indeed, three classification's basic elements must quantitatively be adjusted: classifier parameters, image spatial resolution and pixels attributes.

Twelve classification schemes were tested. To ensure an unbiased optimization, all schemes are tuned based on the same training and validation independent sets and evaluated based on kappa metric (Congalton, 1991). In the first two schemes, the classical Maximum Likelihood (ML) algorithm and the advanced SVM were applied to original spectral bands to find the best classifier, which will be retained for all the coming tests. Then, original spectral bands spatial resolution was enhanced using the high pass filter (HPF) method (Gangkofner et al., 2008). Resulting pansharpened bands with 15-meter resolution are quantitatively validated. Their spatial validation consists on assessing detected edges from them relatively to ones derived from panchromatic one. While spectral validation is based on correlation coefficient with the original spectral bands. Scheme 3 allows quantifying and assessing pansharpened bands classification's contribution compared to the original spectral bands one. The other schemes utilize different attributes combination. Several spectral and textural commonly used indices in similar forest literature context are tested. Textural indices are derived from the 1 pixel distance and 3*3 co-occurrence matrix (Haralick, 1973). Table 5 presents their motivation, use and their computing formula.

The first scheme when ML applied to original spectral bands is taken as a reference scheme. The classification schemes were applied to the five used images. For each image, adopted classification result is obtained from the most accurate scheme according to the achieved kappa values.

*Table 5. List of calculated indices*

| Category | Index | Formula (*) | To Discriminate |
|---|---|---|---|
| Spectral | NDVI (Normalized Difference Vegetation Index) | (PIR-R)/(R+PIR) | Vegetation / no vegetation |
| | TSAVI (Transformed Soil-Adjusted Vegetation Index) | $\frac{\left[a\left(PIR-aR-b\right)\right]}{\left[R+aPIR-ab+X\left(1+a^2\right)\right]}$. | Sparse / Dense vegetation |
| | IBS (brightness soil index) | $\sqrt{(R^2+PIR^2)}$ | Bare soil |
| | SI (Shadow index) | $\sum_{k=b1}^{n} band\ /n$ | Dark area |
| Textural | Entropy | [Haralick, 1973] | Smooth surfaces / rough surfaces |
| | Mean | | |
| | Homogeneity | | |
| | Second moment | | |
| | Dissimilarity | | |

(*): R, PIR are spectral bands (R: red, NIR: near infrared); a and b are respectively the slope and the intercept of soil linear regression line; X: mitigating factor for soil brightness.

## POST CLASSIFICATION PROCESSING

In order to produce acceptable land use land cover (LULC) maps, a post classification process was necessary. It consists on these three operations:

- Build-up area subclasses and agriculture one are merged respectively to agglomeration class and to agriculture land class in all classification outputs,
- In order to exclude scattered pixels, caused by the used pixel-by-pixel classification approach, and classified as forest in an almost homogenous agriculture area, a majority filter was applied based on logical operators,
- All unclassified pixels in classifications results were masked out to and were not considered in the following change detection steps.

## CHANGE DETECTION

In this paper, we were interested on both identifying and quantifying changes. It was approached through the use of classification results that are less noise sensitive than direct comparison carried out by images' arithmetic operations. For each pixel, their affectations to defined land class were compared for each timestamp (i.e., its class at date T1 compared to its newly affected class in date T2).

Change location and identification of its direction (from – to) have been implemented in GIS environment according to five steps approach:

- Assigning values ranging from 1-6 to the output of the both classifications to be compared (at T1 and T2 date T2),
- For T1 date, multiplying the assigned value (ranging from 1 to 6) of the classification results by 10,
- Generating change map through algebraic addition of the reclassified outputs maps at T1 and T2 timestamps,
- Reclassifying the resulting change map through the affectation of 0 to pixels having nn values,
- Interpretation of the reclassified change maps. 0 means no change, while other values indicate change and its direction.

## RESULTS, VALIDATION, AND DISCUSSION

### Classification

Details about implemented classification schemes components and their achieved kappa score are given in Figure 6. It shows obtained kappa scores for the all implemented classifications schemes. We can note that: 1) Classification performance depends on the discriminatory power of classifier; more than 6% is achieved with SVM compared to ML. 2) Finer spatial resolution improves kappa value for about 4%. 3) Combined use of original bands and spectral vegetation indices enhance classification quality which

*Figure 6. Obtained kappa score for tested classification schemes*

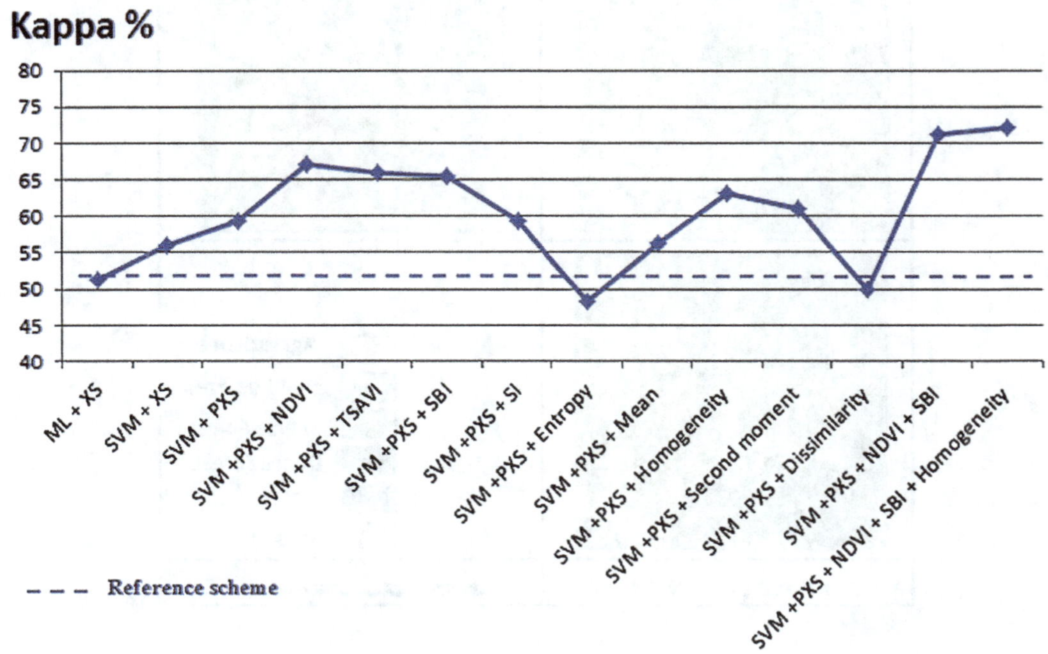

is very expected in forest area. 4) Blindly use of textural bands could decrease classification accuracy and 5) Reliability level increases from 52% to 72% when using simultaneously: pansharpened bands, SVM algorithm, vegetation indices (NDVI and SBI) and textural homogeneity band. This last scheme is maintained for the five images classification process. An example of resulted LULC maps for plain area is presented in Figure 7.

Based on visual inspection for derived maps, we can assume that all LULC are homogenous. Build-up areas, particularly, the road network is well extracted which is probably due to enhanced spatial resolution thanks to pan-sharpened bands. In the plain area and over all investigated years, sparse forest is the most present class which is particularly validated from Google earth scene for 2015. A new region of high density forest appears since 2000 and then will expand. This could be explained by the regional reforestation programs during the last decade and to natural vegetation dynamics. Such conclusion confirms other finding (HCEFLCD, 2017).

*Figure 7. LULC Maps of plain area*

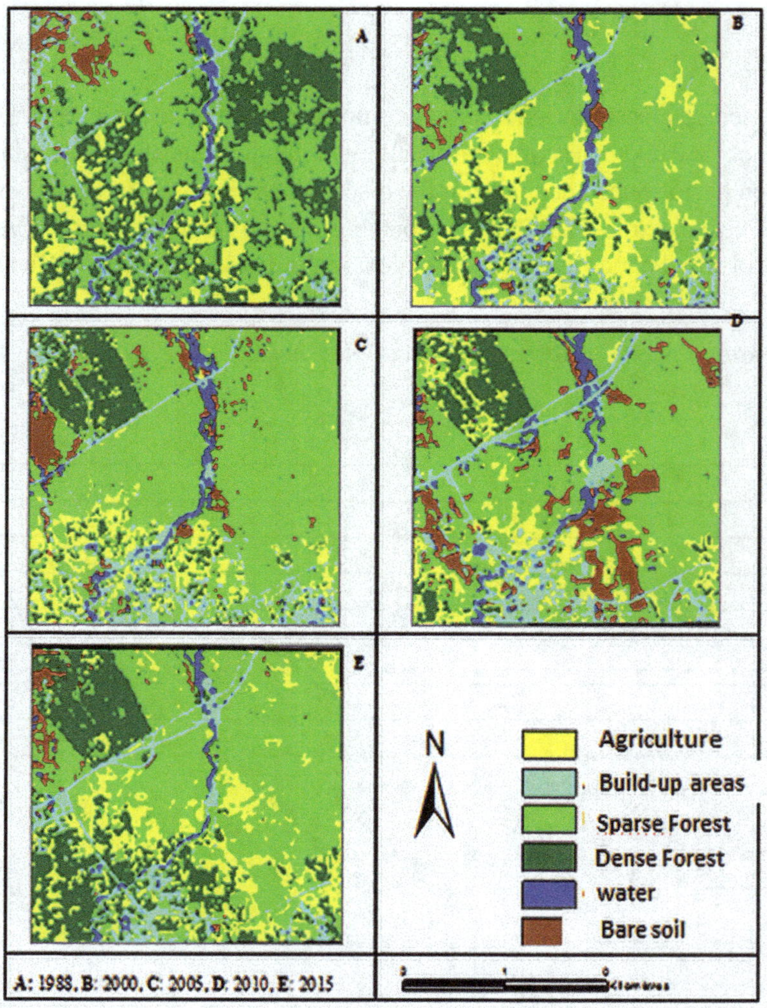

On the other hand, much confusion between forest and agriculture classes exists. This is could be explained by the agroforestry feature of such species and to the fact that one image analysis is not suitable to identify species vegetation. A time series images analysis is needed in this case.

## Change Detection

According to the change detection maps presented in Figure 8 and statistic results showed in Figure 9, we deduce that the plain area is the most affected by change (an annual rate of 15.40 ha per year) followed by urban area (8 ha per year) and the mountainous lands (2ha per year). The peaks of change in the plain area are recorded in the period between 1988 and 2000 and between 2005 and 2010. A strong decline is raised between 2005 and 2010 in the urban area. The Mountainous site is almost stable with a variation between 2000 and 2005.

*Figure 8. Change Maps of plain area*

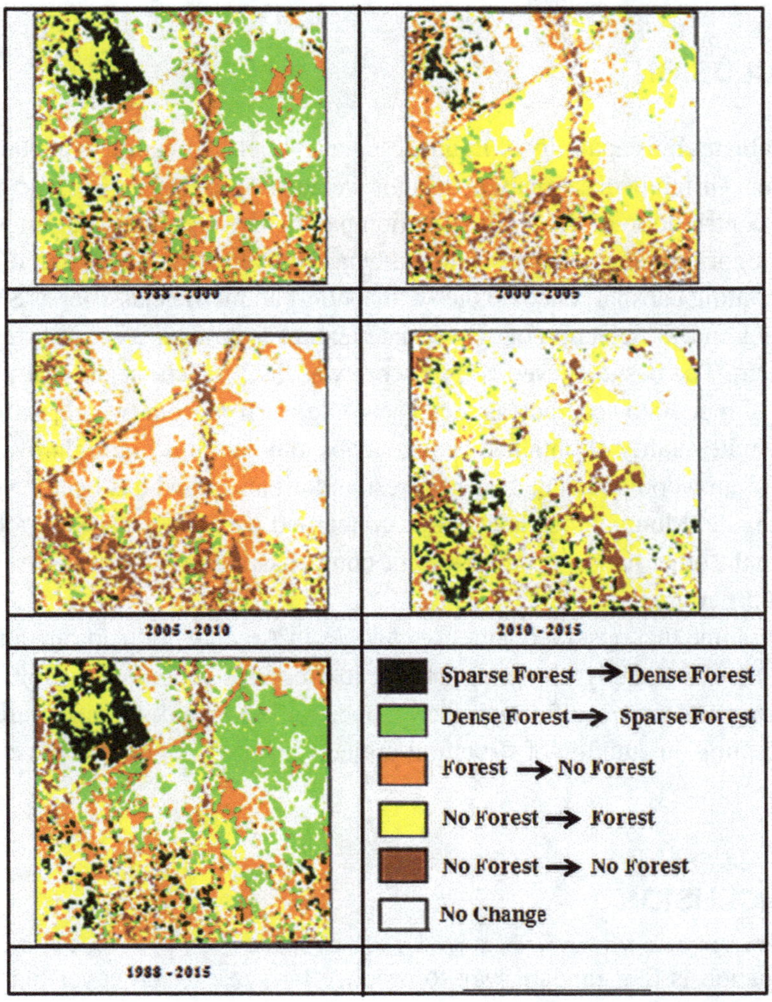

*Figure 9. Area Argan woodlands lost between 1988 and 2015*

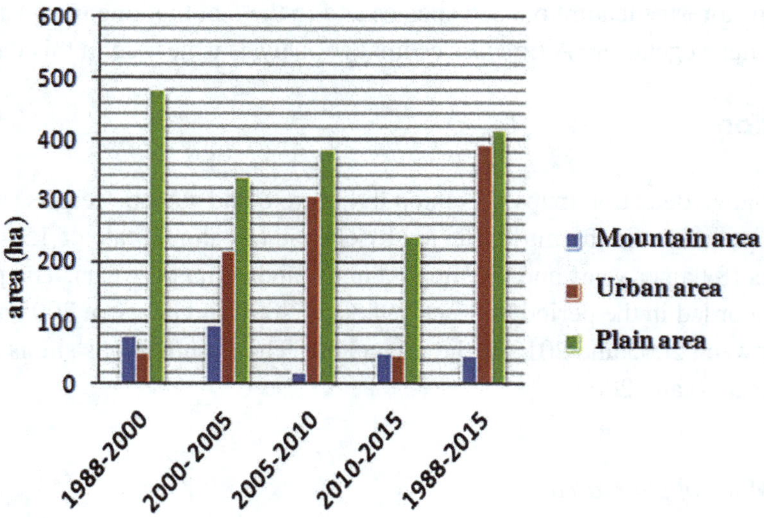


## PARTIAL CONCLUSION

This experiment evaluates the capability of Landsat supervised classification to capture spatio-temporal changes in forest area. Different classification designs were tested to find the most accurate one to use in forest context. The contribution of three classification parameters was assessed: images spatial resolution, classification algorithm and selected features. Quantitative results show that a better accuracy is achieved when combining pansharpened images with a non-parametric classifier as SVM and including texture and vegetation index. Selection of appropriate texture features is a critical step in texture based classification problem. The best achieved Kappa score was 0.72 which means that the derived change detection maps are of moderate accuracy as they accumulate errors from two classification results.

Despite the moderate quality of obtained change maps, our findings highlight overall trends in land use change in the Argan woodland. The loss of forest and urban sprawl has been revealed intensive in plain area and urban one. Mountainous area is less concerned with deforestation problem thanks to its hard topography. On the other hand, we can see some compensation in lost forest area which is probably due to reforestation program.

Finally, we can assume that observed land use changes in Argan woodlands are attributed to a combination of natural factors (mainly climatic), anthropic pressures (population growth exacerbated by migration, overgrazing and the need for mobilizing lands for urban and infrastructure development ...) and political one (the implementation of structural projects that do not promote the conservation of the Argan tree).

## GENERAL CONCLUSION

Supervised classification is the standard way to produce land use / land cover maps and to monitor land use changes. In the first case study of this chapter, a crops map was generated using an optimal classification design. The second case study deals with post-classification change detection analysis to determine the nature and the extent of land use changes over 27 years in forest area. The findings of the two researches demonstrate that designing a suitable image processing procedure is primordial for a successful classification of remotely sensed data into a thematic map. According to our results, we can assume that in vegetation context, the following considerations are especially significant for improving thematic map accuracy: (1) using time series vegetation index allows separating forest from agriculture. Also it can depict vegetation variability species if the plants does not have similar seasonal trends in vegetation index values, (2) complementary use of spectral and textural features of remotely sensed data reduce confusion between some plants with similar spectral behavior (3) advanced non-parametric classifiers release better the multivariate interacting effects between features, (4) effective combination of multiple independent classifiers increases the rate of correct classes prediction, (5) integrating prior knowledge and existing data in post-classification step improves the geometric and semantic final quality of thematic map and (6) high spatial resolution or pansharpened images use is recommended to map smaller land parcels and isolated parcels even if it requires more processing time. Note that besides these six mentioned considerations, there are other general factors impacting classification and should be fixed according to user's need, scale of the study area and economic condition. These factors includes: sensor characteristics, image preprocessing, training samples, validation samples, classification approach and accuracy assessment. Analyst's skills are also important to be considered.

Finally, in post-classification change detection analysis scheme, generated change maps accumulate errors from classification results at the considered timestamps. This may lead to unsatisfactory change maps. Consequently, a comparison between mask-based change method and post-classification is highly encouraged in forest area.

At operational level, once developed, the MCC is able to produce acceptable, quick and low-cost crops map. The whole procedure could be applied to different data with only few adjustments according to local agriculture conditions. The cost of the entire procedure is controlled to a very low level, which will enable its usage in routine annual crop mapping in Morocco. However more attention should be paid to local agriculture practices. They have to be formulated as contextual features and added to the classification system. Therefore, on one side more details knowledge on the seasonal trends of crops and intercropping should be integrated, and on the other, crops historic data should be considered to optimize post-classification process and field visits rate.

## ACKNOWLEDGMENT

We gratefully acknowledge the use of facilities provided by Royal Center of Remote Sensing (CRTS). We would like to thank Mrs. Leila Rasmy.

## REFERENCES

Achard, F., Eva, H. D., Stibig, H.-J., Mayaux, P., Gallego, J., Richards, T., & Malingreau, J.-P. (2002). Determination of deforestation rates of the world's humid tropical forests. *Science*, *297*(5583), 999–1002. doi:10.1126cience.1070656 PMID:12169731

Aouragh, M. (2012). *Dynamique du « paysage-arganeraie » dans le Sud-ouest marocain. Apport des données de télédétection et perspectives de les intégrer dans un SIG* (Thèse de Doctorat). Géographie, l'université Paris IV Sorbonne.

Barandela,, R.M., & Valdovinos, J.S. (2003). New applications of ensembles of classifiers. *Pattern Analysis & Applications*, *6*, 245–256. doi:10.100710044-003-0192-z

Beerech, H.V., Mrslatha, B.M., Thimmaraja Yadava, G., & Dandur, N. (2014). An approcach for identification and classification of crops using multispectral images. *IJERT, 3*(5).

Benediktsson, J. A., Chanussot, J., & Fauvel, M. (2007). Multiple classifiers in remote sensing: from basics to recent developments. *7th International Workshop on Multiple Classifier Systems (MCS)*, 501-512. 10.1007/978-3-540-72523-7_50

Bharathkumar, L., & Mohammed-Aslam, M. (2015). A Crop Pattern Mapping of Tumkur Taluk Using NDVI Technique" A Remote Sensing and GIS Approach. *Aquat. Procedia*, *4*, 1397–1404. doi:10.1016/j.aqpro.2015.02.181

Briem, G. J., Benediktsson, J. A., & Sveinsson, J. R. (2002). Multiple Classifiers Applied to Multisource Remote Sensing Data. *IEEE Transactions on Geoscience and Remote Sensing*, *40*(10), 2291–2300. doi:10.1109/TGRS.2002.802476

Congalton, R. (1991). Areview of Assessing the Accuracy of Classifications of Remotly Sensed Data. *Remote Sensing of Environment*, *37*(1), 35–46. doi:10.1016/0034-4257(91)90048-B

Coppin, P., Jonckheere, I., Nackaerts, K., Muys, B., & Lambin, E. (2004). Review ArticleDigital change detection methods in ecosystem monitoring: A review. *International Journal of Remote Sensing*, *25*(9), 1565–1596. doi:10.1080/0143116031000101675

Dalmau, Alarcón, & Oliva. (2017). Crop Classification in Satellite Images through Probabilistic Segmentation Based on Multiple Sources. *Sensors (Basel)*, *17*(6). PMID:28608825

Department of the Interior U.S. (2016). *Geological survey product guidelandsat surface reflectance-derived Spectral indices version 3.3 December 2016*. Author.

Diatta, M. (2007). *Application de la télédétection et des Systèmes d'Information Géographique à la gestion des forêts classées du SENEGAL: cas de la forêt de Thiès. Mémoire de DESS, RECTAS*. Ilé-Ifè, Nigéria: OAU Campus.

Eddaif, N. (2015). *Suivi spatio-temporel des surfaces de l'arganier par imagerie satellitaire optique à moyenne résolution spatiale: Application à la forêt Mseguina-Sud Ouest du Maroc*. Mémoire de Master: Sciences et Technologie de l'Espace, Facultés des sciences de Rabat.

El Abboudi, A. (2015). *Cartographie de la densité du couvert arboré de l'arganier de la province d'Ait Baha*. Résumé des communications, Congrès international de l'arganier, 3éme édition.

El Mansouri, L. (2013). *Object-based approach and Tree-based ensemble classifications for mapping Building changes*. IARIA.

El Mansouri, L. (2014). Une approche basée objet combinée avec les classificateurs avancés (SVM, RF, Extra Trees) pour la détection des changements des bâtiments. La revue internationale de Géomatique, Février 2014.

Florian, S. (2009). *Potential of a Post-Classification Change Detection Analysis to Identify Land Use and Land Cover Changes. Bachelor's Degree, Physical Geography and Ecosystem Analysis*. Lund University.

Fouad M. (2015). *Analyse basée télédétection pour la révision de la carte de répartition de l'écosystème à arganeraie et l'étude diachronique de sa dynamique spatio-temporelle*. Résumé des communications, Congrès international de l'arganier, 3éme édition.

Gangkofner, U. G., Pradhan, P. S., & Holcomb, D. W. (2008). Optimizing the High-Pass Filter Addition Technique for Image Fusion. *Photogrammetric Engineering and Remote Sensing*, *74*(9), 1107–1118. doi:10.14358/PERS.74.9.1107

Geurts, P., Irrthum, A., & Wehenkel, L. (2009). Supervised learning with decision tree-based methods in computational and systems biology. *Molecular BioSystems*, *5*, 1593–1605.

Gumma, M. K., Thenkabail, P. S., Maunahan, A., Islam, S., & Nelson, A. (2010). Mapping seasonal rice cropland extent and area in the high cropping intensity environment of Bangladesh using MODIS 500m data for the year 2010. *ISPRS Journal of Photogrammetry and Remote Sensing*, *91*, 98–113.

Haralick, K. (1973). *Textural Features for image Classification.* Institute of Electrical and Electronics Engineers.

Haut Commissariat au Plan. (2008). *Monographie régionale GHARB – CHRADA – BENI HSSEN*. Author.

Hazewinkel, M. (2001). Encyclopedia of Mathematics. Springer.

HCEFLCD. (2017). *Mesguina Forest management plan* (Draft document). Author.

Huete, A. R., Liu, H., Batchily, K., & Van Leeuwen, W. (1997). A comparison of vegetation indices over a global set of TM images for EOS-MODIS. *Remote Sensing of Environment*, *59*(3), 440–451. doi:10.1016/S0034-4257(96)00112-5

Kang, H. J., & Doermann, D. (2005). Selection of Classifiers for the Construction of Multiple Classifier Systems. *Proceedings of the Eight International Conference on Document Analysis and Recognition*, 263-268.

Kangas, A., & Maltamo, M. (2006). *Forest inventory methodology and applications*. Dordrecht: Springer. doi:10.1007/1-4020-4381-3

Kershaw, J. A., Ducey, M. J., Beers, T. W., & Husch, B. (2016). *Forest mensuration* (5th ed.). Chichester, UK: John Wiley & Sons. doi:10.1002/9781118902028

Lu, D., Mausel, P., Brondizio, E., & Moran, E. (2004). Change detection techniques. *International Journal of Remote Sensing*, *25*(12), 2365–2401. doi:10.1080/0143116031000139863

Maree R. (2005). *Decision trees for automatic images classification* (PhD Thesese). Ulg University.

Marée, R., Geurts, P., & Wehenkel, L. (2016). Towards Generic Image Classification using Tree-based Learning: An Extensive Empirical Study. *Pattern Recognition Letters*, *74*, 17–23. doi:10.1016/j.patrec.2016.01.006

Moukrim, S., Lahssini, S., Mharzi Alaoui, H., Rifai, N., Arahou, M., & Rhazi, L. (2018). Modélisation de la distribution spatiale des espèces endémiques pour leur conservation: Cas de l'Argania spinosa (L.) Skeels. *Revue d'Écologie*, *73*(2), 153–166.

Paletto, A., Ferretti, F., Cantiani, P., & De Meo, I. (2012). Multi-functional approach in forest landscape management planning: An application in Southern Italy. *Forest Systems*, *21*(1), 68–80. doi:10.5424/fs/2112211-11066

Rokni, K., Ahmad, A., Selamat, A., & Hazini, S. (2014). Water Feature Extraction and Change Detection Using Multitemporal Landsat Imagery. *Remote Sensing*, *6*(5), 4173–4189. doi:10.3390/rs6054173

Tingting, L., & Chuang, L. (2008). Crop classification by support vector machine with intelligently selected training data for an operational application. *International Journal of Remote Sensing*, *29*(8), 2227–2240. doi:10.1080/01431160701395203

Tingting, L., & Chuang, L. (2010). Study on extraction of crop information using time-series MODIS data in the Chao Phraya Basin of Thailand. *Advances in Space Research*, *45*(6), 775–784. doi:10.1016/j.asr.2009.11.013

Vapnik, V. N. (1995). *The nature of statistical learning theory*. Springer-Verlag New York, Inc. doi:10.1007/978-1-4757-2440-0

Waldhoff, G., & Lussem, U. (2016). *Enhanced Land Use Classification 2015 of the Rur Catchment – Update*. CRC/TR32 Database (TR32DB). .10.5880/TR32DB.19

Wehenkel, L., Bastin, C., Phillips, C., & Geurts, P. (2017). Tree Ensemble Methods and Parcelling to Identify Brain Areas Related to Alzheimer's Disease. In *2017 International Workshop on Pattern Recognition in Neuroimaging (PRNI), Proceedings*. IEEE.

Zhou, W., & Hong-cha, M. A. (2006). Combining methods and diversity measures in multiples classifier systems for thematic classification of landsat TM image. *Journal of Heilongjiang Institute of Technilogy*, *20*(4), 14–21.

# Chapter 7
# Geospatial Technics, Modelling, Meteorological, and Ground Data for Crop Management in Semi-Arid Zones of Morocco

**Rachid Hadria**
*National Institute for Agricultural Research, Morocco*

**Loubna El Mansouri**
*Institut Agronomique et Vétérinaire Hassan II, Morocco*

**Tarik Benabdelouhab**
*National Institute for Agricultural Research, Morocco*

**Pietro Ceccato**
*Columbia University, USA*

## ABSTRACT

*Recent geospatial technologies offer an infinite number of opportunities in the domain of land management and governance. This is due principally to the fact that new satellites allow the collection and monitoring of important information about soil, crops, weather, and climate. In this context, this chapter presents a review of studies conducted in three semi-arid plains in Morocco during the last two decades on the combined use of geospatial technologies, modelling tools, meteorological, and ground data to manage irrigated lands in semi-arid zones. The three studied regions are located in three different sites well distributed over the country. This chapter presents a simple and clear summary of what has been performed using geospatial technologies by Moroccan researchers to improve agricultural management in irrigated areas in Morocco. Such work could help policymakers in developing efficient land governance policies by taking into account the latest scientific results.*

DOI: 10.4018/978-1-5225-5939-9.ch007

## INTRODUCTION

Land governance consists in better managing earth's surface and its various resources such as crops, water, air, and biodiversity. This governance is increasingly important especially in the actual context marked by a population increase, extension of urban areas and climate changes. These factors affect mainly agricultural sector in underdeveloped countries where, unfortunately, agriculture represents the main activity of the population. Morocco is a good example of these countries since agriculture accounts for 15% to 20% of gross domestic product (GDP) of the country, depending on the amount and the distribution of rainfall during each year (Balaghi, Jlibene, Tychon, & Eerens, 2013). The effect of rainfall fluctuations on agricultural production explains 75% of the year-to-year variability in national GDP. Agriculture sector is the largest employer in the country and employs about 40% to 50% of the nation's workforce. In term of area, about 9,895,000 hectares or 22.1% of the total land area is arable (excluding Moroccan Sahara). About 43% of arable land is devoted to cereals and in a normal year, Morocco produces about two-thirds of the grains (mostly wheat, barley, and maize) needed for domestic consumption. Irrigated land accounts for about 15% of cropped area and produces over 50% of the value of production and irrigation accounts for 87% of water use.

However, water resources constraint the development of irrigated agriculture in Morocco. Water is exploited beyond its renewable limits, and the expansion of irrigated areas is not an adequate solution. Agriculture, which currently accounts for 87% of water use, suffers from increasing competition from urban and industrial demands. In irrigated agriculture, productivity is undermined by farmers' uncertainty about water supply, water scarcity being a key factor in lower-than-potential agricultural revenues. Introduction of more efficient irrigation techniques can help sustain production while conserving water resources, especially in the climate change context.

Realizing the gravity of the situation, Morocco has invested significant efforts since 1961 in the mobilization of water resources. The country currently has more than 100 dams with a total storage capacity of approximately 16 billion $m^3$. To reinforce this infrastructure, 30 other dams will be built by 2030. Regarding the development of agriculture, many irrigation areas have been created in several regions of Morocco. Result, about 1.5 million hectares are now irrigated from dams. However, the context of climate changes imposes to develop new good land governance approaches to better manage these irrigated areas and to preserve the limited water resources of the country.

Recent developed geospatial technologies offer new opportunities in the domain of crop management. This is due mainly to the fact that new satellites allow the collection and monitoring of important information about soil, crops, weather and climate. Thanks to new cloud computing services, it is now possible to analyse large amount of data at high spatial and temporal resolutions. Remote sensing data have been widely used around the world to monitor crops and lands at large spatial scale ((de Wit, Boogaard, & van Diepen, 2004) ; (Dinku, Chidzambwa, Ceccato, Connor, & Ropelewski, 2008); (Du et al., 2013)).

In Morocco, important efforts were deployed by Moroccan researchers during the last two decades to improve water management in irrigated areas by combining remote sensing data, crop modelling, meteorological data and field measurements. This chapter reviews three example of such studies conducted in three regions of Morocco. Firstly, a brief synthesis of adopted methodologies and used data are presented. The main results obtained are then summarised in this chapter.

The authors expected that this modest contribution will present a simple and clear summary of what has been done by Moroccan researchers to improve agricultural management in three regions based on the recent geospatial technologies. Such work could also help policy makers in developing efficient land governance policies by taking into account the latest scientific results obtained within the Moroccan context.

## PRESENTATION OF THE THREE STUDIED REGIONS

The three studies were conducted in three different sites well distributed over the agricultural area of Morocco; as presented in *Figure 1*. These sites are Tansift Alhaouz plain in Marrakech region (South), Tadla plain around Beni Mellal city (Centre) and Trifa plain in the extreme North-Eastern of Morocco. The common irrigated crops in the three zones include mainly wheat, olives threes, citrus, vegetables and alfalfa.

### Case Study of Tensift Alhaouz Region

The Tansift Alhaouz plain is located in the centre of Morocco (centered on to the longitude 7° 36′ W and latitude 31° 40′ N (Datum WGS84)), 40 km east from the city of Marrakech. This Plain is limited in

*Figure 1. Geographical location of the three considered study sites*

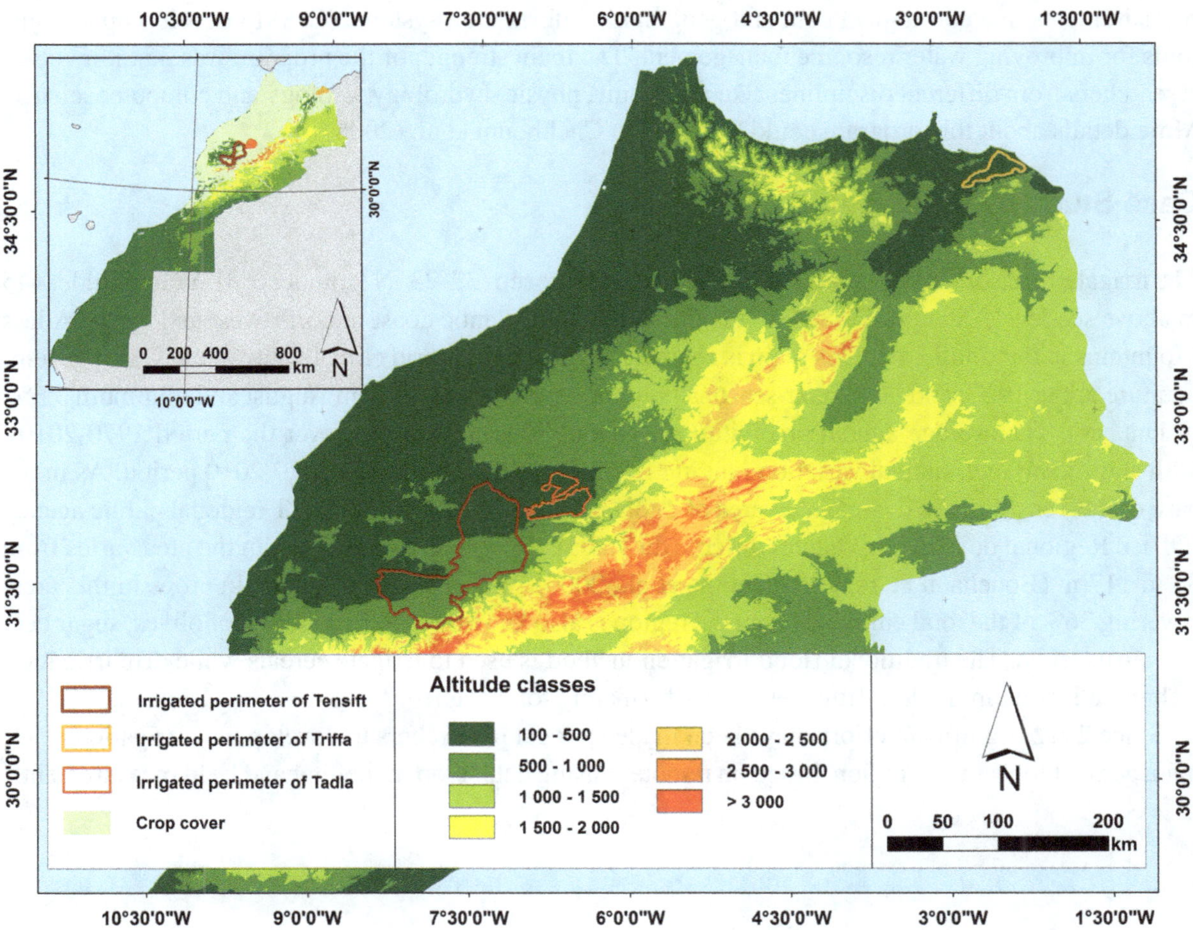

the south by high Atlas Mountains and, in the North by Jbialet hills. As the general climate of Morocco, the climate of Tansift plain is semi-arid and it is characterised by low and irregular rainfall. The average annual rainfall varies around 240mm/year while the evaporative demand is around 1500 mm/year (B. Duchemin et al., 2006). The study conducted by Chaponnière et al. (2005) showed that Atlas Mountains represent the main plain's water bank ensuring irrigation water of many irrigated areas. Indeed, the average annual rainfall varies in these mountains around 600mm/year. Agriculture water management is assured mainly by the ORMVAH regional public agency (Office Régional de Mise en Valeur Agricole du Haouz. This agency is in charge of dam water distribution, in collaboration with farmer associations. The water is distributed according to a predetermined schedule during irrigation period, with the same quantity in each irrigation event. The number of irrigations and the total schedule is predetermined at the beginning of the growing season regardless of the actual climate and agricultural practices. Therefore, improper quantities of water are usually delivered; thus, there is an urgent need to improve the management of irrigation water in Tensift plain.

The Tansift Alhaouz plain can be considered as the most important open laboratory in Morocco to study the issue of improving agricultural water management by combing geospatial technics, modelling and ground data of semi-arid zones. In 2002, an important program, entitled SUDMED program (Chehbouni et al., 2008), has been designed to address the issue of improving the understanding of the hydrological functioning of the Tensift basin. The main objectives of this program were to: 1) Test existing models or develop new ones to better understand the hydrological responses to future scenarios about climate change and human pressure, and, 2) Use remote sensing data combined with models, field measurements, meteorological data and geographic information systems (GIS) to develop operational tools for improving water resource management. The main strength of the program was that it involved researchers from different disciplines like agronomy, physic, hydrology, geology and computer science. More details about this program could be found in Chehbouni et al. (2008).

## Case Study of Tadla Plain

The irrigated area of Tadla plain is located in central Morocco (32°23´N latitude; 6°31´W longitude; 445 m above sea level, (Datum WGS84)). It is limited by the Atlantic coast in north-west and by the Atlass Mountains in the south-east. This plain is characterized by a semi-arid climate with annual average temperature about 19°C, and large inter-seasonal variation (maximum 38°C in August and minimum 3.5°C in January). The average annual precipitation is about 300 mm (average over the period 1970-2010), with significant inter-annual variation ranging from 130 to 600 mm in the 1970-2010 period. As in the case of the Tensift plain, the irrigated area of Tadla is managed by the ORMVAT regional public agency (Office Régional de Mise en Valeur Agricole du Tadla). The groundwater depth in the area varies from 31 to 117 m (Bouchaou et al., 2009; Najine et al., 2006). Wheat is one of the main crops in this area, covering 36% of the total cultivated area the region is also known by the production of olives, sugar beet and citrus fruits. The traditional flood irrigation method is used to irrigate cereals while drip irrigation technique is used in modern fruit tree orchards (mainly for citrus).

Since 2012, significant efforts have been made by local researchers to develop new irrigation water management tools in this region, based on remote sensing data, crop modelling and field measurements.

## Case Study of Trifa Plain

Trifa plain is located at the North-eastern part of Morocco, centered at longitude 2° 22′ W and latitude 34° 58′ N (Datum WGS84) (see *Figure 1*) and covers about 36000 hectares. It is limited in the east by the Algerian Massif of Mzirda (613 m), on the south by the Beni-Snassen Massif (1.532 m), on the north by the Mediterranean Sea and on the west by the plain of Boughriba. The climate of the region is of semi-arid Mediterranean type. The average annual rainfall is 350 mm with variations between 250 and 425 mm depending on the year. Rainfall period is generally limited between January and April. Average air temperatures are about 4°C to 6 °C during the winter and 27°C to 30 °C in summer. The water resources of the region are mainly surface water supplied by the Moulouya River. These water resources are rather fragile and are very influenced by the inter annual rainfall distribution. Successive drought events occurred during the last decades, and the over exploitation of groundwater resources decreased the piezometric level. In addition, recent studies have shown that groundwater chemical quality of the Trifa plain is being degraded due to the excessive use of fertilizers and pesticides (Fekkoul, 2013).

The irrigated perimeter of Trifa, managed by the Office Régional de Mise en Valeur Agricole de la Moulouya (ORMVAM) regional public agency, is the most important in the region both in terms of agro-economic importance and in terms of irrigation water consumption. Citrus represents the main sector in the irrigated perimeter of Trifa plain covering about 19600 hectares in 2017 (ORMVAM, 2017). In addition to citrus, Trifa plain produces also vegetable and olive.

Moroccan researchers have focused on the development of new approaches for good management of the limited resources of the region and the geospatial technics have been mainly used to develop irrigated crop maps.

# MAIN MATERIALS AND METHODS USED IN THE THREE STUDIES

Geospatial technics, modelling tools, meteorological and ground data have been widely used in the three studies presented in this chapter. A summary of the main materials and methods used in the three studies is presented below. More details can be found in the references cited in the chapter.

## Remote Sensing Data

Remote sensing data are the most promising tool for monitoring irrigated areas at regional scale. In the three considered studies reported in this chapter, radar and optical satellite images have been used at low, medium and high spatial and temporal resolutions. Table 1 summarises the main remote sensing images used. Before exploring these images, they were firstly pre-processed following complex procedures (co-registration using a cross-correlation algorithms; geographic corrections through ground control points and atmospheric corrections). Images of vegetation index and others biophysical and meteorological variables have been then deduced from processed images (see table 1).

In the three sites, these satellite images have been used to initialize, force and control models (Hadria et al., 2007). They have also been used to retrieve some biophysical variables such as leaf area index and

*Table 1. Summary the main remote sensing images used in the framework of the three studies presented in this chapter*

| Image | Study Site | Retrieved Variable | Reference |
|---|---|---|---|
| Landsat 7 ETM+<br>(spatial resolution: 60m in IR and 15m for panchromatic band)<br>Temporal resolution:<br>16 day | Alhaouz plain | • Normalised difference vegetation index<br>• Leaf area index | B. Duchemin et al. (2006) |
| FORMOSAT-2<br>(spatial resolution:<br>high resolution of 2m for panchromatic data and 8m for multispectral data)<br>Temporal resolution:<br>1 day | Alhaouz plain | • Green leaf area index<br>• Land cover | Hadria et al. (2010)<br>(B. Duchemin et al., 2008) |
| ENVISAT/ASAR -Radar data-<br>spatial resolution:<br>30m<br>Temporal resolution:<br>35 days repeat cycle | Alhaouz plain | Soil moisture | Hadria, Duchemin et al. (2010) |
| ASTER<br>(spatial resolution: 15m for VNIR, 30 for SWIR and 90m for TIR)<br>Temporal resolution:<br>16 days | Alhaouz plain | Actual evapotranspiration | Er-Raki et al. (2008) |
| SPOT-5<br>(spatial resolution: panchromatic: 5m or 2.5m; multispectral: 10m; SWIR: 20m)<br>Temporal resolution:<br>2-3 days, depending on latitude | Tadla plain | • soil water content<br>• vegetation water content | Benabdelouahab et al. (2015); Benabdelouahab, Balaghi, Hadria, Lionboui, and Tychon (2016) |
| Sentinel-2A<br>(spatial resolution: four bands at 10 m, six bands at 20 m and three bands at 60 m spatial resolution)<br>Temporal resolution:<br>5 days | Trifa plain | Citrus map | Moutaib and Oujamaa (2017)<br>Lahmer (2017) |
| NOAA-AVHRR<br>spatial resolution:<br>1.09km)<br>Temporal resolution:<br>Two images/day | All sites | air temperature | Hadria et al. (2018) |
| TRMM<br>Spatial resolution:<br>0.25°<br>Temporal resolution:<br>3 hours | All sites | Rainfall | Ouatiki et al. (2017), Ezzine et al. (2014) |

land cover (Duchemin et al., 2006; Duchemin et al., 2008), green leaf area index (Hadria et al., 2010), soil moisture ((Hadria et al., 2010); (Benabdelouahab et al., 2015)), vegetation water content (Benabdelouahab, Balaghi, Hadria, Lionboui, & Tychon, 2016), meteorological variables such as rainfall (Ezzine, Bouziane, & Ouazar, 2014; Ouatiki et al., 2017) and air temperature (Hadria et al., 2018).

## Modelling Tools

Models have the capability to predict crop development, grain yield, water consumption as influenced by the climatic conditions, soil characteristics, and agricultural practices. Modelling tools are becoming more and more efficient to manage water resources, to estimate grain yield, as well as to provide water and nutrients balance (Panda, Behera, & Kashyap, 2003);(Bandyopadhyay & Mallick, 2003)). In the literature, there are many crop models and some are designed for particular crops, e.g. ARCWHEAT for wheat, (Weir, Bragg, Porter, & Rayner, 1984) and CERES-maize for maize (Jones, Kiniry, & Dyke, 1986); while others are generic models, e.g. EPIC (R. Williams, A. Jones, R. Kiniry, & A. Spanel, 1989) and STICS (Brisson et al., 1998).

Modelling tools have been used in three regions. These tools included simple models like FAO model (Allen, Pereira, Raes, & Smith, 1998) to estimate crop evapotranspiration and complex models such as STICS crop model (Brisson et al., 1998) and Acquacrop model (Steduto, Hsiao, Raes, & Fereres, 2009) to simulate the whole process related to crop development from sowing to harvesting. In addition to existing models, new ones have been developed. Among these models, Benoît Duchemin, Maisongrande, Boulet, and Benhadj (2008) developed the SAFY model *(for Simple Algorithm For Yield Estimate)*. This simple model was developed to simulate the main processes of annual crop development and growth with a limiting number of input parameters in order to facilitate the spatialization of the model.

Simonneaux et al. (2009) proposed a new software called Satellite Monitoring of irrigation (SAMIR) to spatialize water balance of irrigated crops. This software is based on the combining use of FAO model to estimate crop evapotranspiration (Allen et al., 1998) and remote sensing data.

## Micro-Meteorology Tools

In this chapter, micro-meteorology tools refer to micro-meteorological measurements based on Bowen ratio (Bowen, 1926), eddy covariance system (Burba, 2013) and scintillometers (Andreas, 1989) technics. Its refer also to eco-physiological techniques such as sap flow (Valancogne & Nasr, 1989) and isotope (Ehleringer, Roden, & Dawson, 2000) methods. A review of these methods had been presented by Er-Raki et al. (2013). These tools have been used mainly to assess crop evapotranspiration at local or regional spatial scales of olives, citrus and wheat in the Tansfit Alhouz plain in the framework of Sudmed project (Chehbouni et al., 2008). Mentioned technics have been tested under drip and flooding irrigation.

# MAIN RESULTS OBTAINED FROM THE THREE STUDIES

From the research performed in the three regions, interesting results have been obtained and more than thirty scientific papers have been published in peer-reviewed scientific journals between 2003 and 2017. In this section, a summary of the main results obtained in the framework of the three studies is presented.

## Estimation of Biophysical and Meteorological Variables From Remote Sensing Data

To better understand the behaviour of Moroccan crop systems and the processes that affect the water and energy balances in irrigated semi-arid areas under local conditions of the country (social, climatic, economic), important field's experiments have been conducted in the three study areas. A full description of these field's experiments have been presented by B. Duchemin et al. (2006) for Tensift plain, by Benabdelouahab, Balaghi, Hadria, Lionboui, Djaby, et al. (2016) for Tadla plain and by Moutaib and Oujamaa (2017) for Trifa plain.

Field measurements are required to parametrize models or to develop new local approaches for monitoring water and energy balances. Field surveys are also required to develop and update land cover maps in each study site. In the literature, leaf area index which quantifies the area of green leaves per unit area, is an important variable used to simulate biomass production, grain yield production, and water and nitrogen balances. It is also the key variable used to monitor crop covers from remote sensing data (see for example: (Asrar, Fuchs, Kanemasu, & Hatfield, 1984); (Baret, Guyot, & Major, 1989); (Richardson, Wiegand, Wanjura, Dusek, & Steiner, 1992)). Direct correlations have been found between leaf area index and the normalised difference vegetation index, defined as the difference between near-infrared and red reflections divided by their sum (Rouse, Haas, Schell, Deering, & Harlan, 1974). A particular attention have been devoted to these two variables in case of Tensif plain and multiple methods have been tested to propose local correlations between the two variables (see for example: (B. Duchemin et al., 2006); (Hadria et al., 2006), (Khabba et al., 2009)).

Soil moisture and vegetation water content are two key indicators in the monitoring of water stress of irrigated crops in arid and semi arid zones. The monitoring of these variables helps to better manage irrigation especially in irrigated areas. Tradionnaly, soil moisture and vegetation water observations are based on in situ measurements (field sampling, laboratory analysis). This conventional method is destructive, fastidious in terms of time and efforts and gives only local information. Retrieving surface soil moisture and vegetation water content based on satellite remote sensing data is more appropriate, less expensive and gives information at regional scale. This issue have been extensively studied in the studies presented in this chapter ((Hadria et al., 2010); (Benabdelouahab et al., 2015), (Benabdelouahab, Balaghi, Hadria, Lionboui, & Tychon, 2016)). The authors evaluated the potential of different spectral indices, calculated from remote sensing data, to retrieve soil moisture and vegetation water content values of irrigated wheat. The results obtained showed the existence of high correlations between measured soil moisture and vegetation water content with spectral indices. Thus, the use of remote sensing data could be used to improve irrigation and crop management of wheat at regional levels.

Meteorological variables are essential for understanding the physics of many land surface processes. In crop management, these variables are required as inputs of all crop models. These variables are measured by weather station. However, the low density of meteorological stations is the main problem encountered when conducting agrometeorological studies over large geographic areas. Data collection, their analysis, and the maintenance of instruments constitute significant factors affecting the accuracy and quality of measured data at weather station level. To overcome these problems, remote sensing represents an efficient tool to estimate some key meteorological variables like rainfall and air temperature. Indeed, numerours studies in the literature explored this issue (see for example: (Shen & Leptoukh, 2011), (Fabiola Flores & Mario Lillo, 2010), (Hachem, Duguay, & Allard, 2012), (Hengl, Heuvelink, Perčec Tadić, & Pebesma, 2012), (Vogt, Viau, & Paquet, 1997), (Hsu, Gao, Sorooshian, & Gupta, 1997), (Joyce, Janowiak, Arkin, & Xie, 2004), (Su, Hong, & Lettenmaier, 2008), (Gu et al., 2010), (Arvor D., Dubreuil V., Ronchail J., & S., 2008), (Li et al., 2009), (Vancutsem, Ceccato, Dinku, & Connor, 2010), (Moffitt, Hossain, Adler, Yilmaz, & Pierce, 2011), (Zhang & Jia, 2013), (Du et al., 2013), (Yaduvanshi, Srivastava, & Pandey, 2015).

In Morocco, many research studies have been undertaken on using the Tropical Rainfall Measurement Mission (TRMM) products. Milewski, Elkadiri, and Durham (2015) have evaluated four TRMM 3B42 products against rain gauges to estimate average annual precipitation. The work done by Tramblay, Thiemig, Dezetter, and Hanich (2016) in northern Morocco shows the potential use of these data in hydrological modelling. Ezzine et al. (2014) confirmed the potentiality of TRMM data to estimate monthly average rainfall precipitations in Morocco. The authors also introduced the new standardized water index derived from the normalized difference water index (NDWI) short time-series that can be used for the monitoring of drought at the scale of the country. Recently, Ouatiki et al. (2017) confirmed the accuracy of the TRMM 3B42 Version 7 product to estimate monthly and annual precipitation over the Oum Er Rbia Watershed in Morocco. Finally, Hadria et al. (2018) proposed simple linear correlations to retrieve the three components of air temperature (maximum, minimum and average) from surface temperature retrieved from satellite images acquired by the Advanced Very High Resolution Radiometer sensor, on board the National Oceanic and Atmospheric Administration (NOAA) satellites.

## Testing, Improving, and Validation of Models

In the three studied plains, wheat is one of the main dominant crops, along with olive and orange orchards. Water supplies are scarce and the availability of water is one of the main factors that control plant growth and grain yield. Because of the shortage of rainfall and its irregular distribution in Morocco, irrigation is a common practice to avoid water stress during the occurrence of drought events. In irrigated area of Morocco, irrigation is generally scheduled according to a predetermined program at the beginning of growing season regardless of the actual climate and agricultural practices. The number of irrigations and the total amount of applied water depend on the dams' water level and thus, improper quantities of water are usually delivered (Hadria et al., 2007). In this context, a special attention has been devoted to test, improve and validate modelling tool in order to better manage irrigation water in the three studies presented in this chapter. Two approaches have been used: the first approach is based on simple models, while the second one was based on complex models.

## Approach Based on Simple Models

This approach consisted to explore tools that are simple to use and do not require a lot of input data. Er-Raki et al. (2007) tested firstly the method based on the FAO-56 dual crop coefficient approach, described by (Allen et al., 1998), to estimate actual evapotranspiration of wheat in Tensift Alhaouz plain. The last approach describes the relationship between maximal evapotranspiration and reference evapotranspiration by separating the standard crop coefficient into a basal crop coefficient and a soil water evaporation coefficient. The authors showed that the values of the basal crop coefficient and the fraction of soil surface covered by vegetation given in the FAO-56 tables are not appropriate for wheat in Moroccan conditions. New values of these paramatrers have been proposed based on the real measurements of crop evapotraspiration measured by Eddy Correlation systems.The same authors proposed also simple relationships between the crop basal coefficient and the fraction of soil surface covered by vegetation from one hand, and the Normalized difference vegettaion Index, on the other hand.

In the same context, Er-Raki et al. (2008) showed that the default values of crop coefficients proposed by (Allen et al., 1998) overestimate the actual evapotranspiration of olive orchards in Morocco by about 18% when compared against eddy covariance measurements. The authors proposed thus a simple method to improve FAO-56 apporach to better estimate actual evapotranspiration of olives threes in Tensift Alhaouz plain. This improvement has been done by assimilating actual evaportanspiration, derived from thermal infrared remote sensing data into the FAO-56 single source model. The study realised by (Er-Raki, Chehbouni, Boulet, & Williams, 2010) aimed at evaluating the potential of the dual approach of FAO-56 (Allen, 2000) for partitioning evapotranspiration into soil (evaporation) and plant (transpiration) components for olive orchards under semi arid conditions. In this study, sap flow technics have been used to measure the exact amount of water absorbed by the threes (transipration). The results obtained in this study showed that the dual approach of FAO-56 gives accurate estimate of plant transpiration and soil evaporation. The associated RMSE of plant transpiration and soil evaporation were 0.59 mm and 0.73mm per day, respectively.

## Approach Based on Complex Models

Although the simple methods are more attractive from the point of view of their applicablity, the main limitation of such apporaches is related to their accuracy. To overcome this problem, crop models can simluate accurately and simultaneously multiple processes related to a given crop like dailly evaporation, transpiration, grain yield, carbon emission. These models have been tested in Tanssift and Talda plains to simulate wheat cylce under semi arid conditions of Morocco.

In Tansift Alhaouz Plain, Hadria et al. (2007) calibrated and validated the shoot growth module and grain yield of the STICS crop model (Brisson et al., 1998) by using field data. The authors proposed to use this model to simulate wheat final grain yield under semi arid conditions of Morocco. They suggested also to apply the model to schedule irrigation of wheat. In this application, the model was used to manage the time of irrigation according to a threshold for water deficit fixed as input of the model.

Almost smilar exercice was performed in Tadla plain by Benabdelouahab, Balaghi, Hadria, Lionboui, Djaby, et al. (2016) by using Acquacrop model (Steduto et al., 2009). The authors proposed a local parametrization of the model under the local pedoclimatic condtions of Tadla plain. The application of the model to schedule irrigaton of wheat confirmed the capacity of crop models te be used as decisional tools to better manage irrigation in irrigated area of Morocco.

Although, their low spatial accuracy, the numerous input paramaters of crop models required and shortage of input data at the appropriate space-time scales represent their major limitation for operational use at regional scale ((Guérif & Duke, 1998); (Moulin, 1999)). To solve this problem, many studies in the litterature porposed alternative methods to combine crop models with remote sensing data. These methods are based on driving, calibration or assimilation techniques ((Olioso, Chauki, Courault, & Wigneron, 1999); (Jacquemoud, Bacour, Poilvé, & Frangi, 2000);(Verhoef & Bach, 2003); (Demarty et al., 2004)). In Morocco, Hadria et al. (2006) compared calibrarting and driving technics to monitor irrigated wheat in Tensift Alhaouz plain by using STICS crop model and remote sensing data. In case of calibration method, the model has been calibrated by adjusting its input parameters to estimate accurately the leaf area index. In case of driving technic, leaf area index, retrived from remote sensing data, has been used as an input variabe into STICS crop model. The authors concluded that calibrating method is more accurate than driving technic. The accuracy of the last technic is directly related to the availablity of of remote sensing data (satelite time revisit frequency, availability of cloud free images).

## CONCLUSION AND RECOMMANDATIONS

Moroccan researchers deployed important efforts during the last two decades to improve water management in irrigated areas by using new geospatial technics. They mainly focused on combining remote sensing data, crop modelling, meteorological data and field measurements. This chapter reviews three example of studies conducted in three regions of Morocco: Tansift Alhaouz in Marrakech region (South), Tadla around Beni Mellal city (Centre) and Trifa in the extreme North-Eastern of Morocco. Important financial and human resources were deployed and interesting results were obtained in the framework of these studies. These results have been published in more than thirty scientific papers in peer-reviewed scientific journals. The studies obtained proved the great interest of new geospatial technics to monitor irrigated areas in semi-arid regions.

Despite the innovative character of the tools and methods developed and the good results obtained, the outputs of the research had low impacts on land governance policies and famers in the studied zones and at country level. This problem could be explained by many reasons. It is obvious that all proposed tools and results were based on high scientific knowledge level. For example, the use of crop models need a strong scientific background and an important input data. In addition, almost all scientific published papers cited in this chapter have been published in English scientific journals. The access to that information is mainly restricted to English speaking scientists and practitioners and not to the farmers.

The same reasons cited above are valid for policies makers. The latter do not have the necessary resources (human and/or material), time and needed skills to apply the results of scientific research. Policy makers often expect simple application tools that answer specific request raised by them. Researchers are thus strongly invited to take into account both famers and policy makers request to access easy to understand information. The ideal would be to involve them in joint projects to answer questions of great interest to the country. In addition to the publication of scientific results, the organization of training sessions in favour of the decision-makers and educated farmers on these results would be necessary for their good diffusion.

## REFERENCES

Allen, R. G. (2000). Using the FAO-56 dual crop coefficient method over an irrigated region as part of an evapotranspiration intercomparison study. *Journal of Hydrology (Amsterdam)*, *229*(1), 27–41. doi:10.1016/S0022-1694(99)00194-8

Allen, R. G., Pereira, L. S., Raes, D., & Smith, M. (1998). Crop evapotranspiration —guidelines for computing crop water requirements. FAO Irrigation and drainage paper 56. Food and Agriculture Organization, Rome. Rome.

Andreas, E. L. (1989). Two-wavelength method of measuring path-averaged turbulent surface heat fluxes. *Journal of Atmospheric and Oceanic Technology*, *6*(2), 280–292. doi:10.1175/1520-0426(1989)006<0280:TWMOMP>2.0.CO;2

Arvor, D., Dubreuil, V., & Ronchail, J., & S., P. (2008). Apport des données TRMM 3B42 à l'étude des précipitations au Mato Grosso. *Climatologie*, *5*, 49–69.

Asrar, G., Fuchs, M., Kanemasu, E. T., & Hatfield, J. L. (1984). Estimating Absorbed Photosynthetic Radiation and Leaf Area Index from Spectral Reflectance in Wheat1. *Agronomy Journal*, *76*(2), 300–306. doi:10.2134/agronj1984.00021962007600020029x

Balaghi, R., Jlibene, M., Tychon, B., & Eerens, H. (2013). Agrometeorological cereal yield forecasting in Morocco (2013 ed.). National Institute of Agronomic Research (INRA): Devision of information and communication (INRA).

Bandyopadhyay, P. K., & Mallick, S. (2003). Actual evapotranspiration and crop coefficients of wheat (Triticum aestivum) under varying moisture levels of humid tropical canal command area. *Agricultural Water Management*, *59*(1), 33–47. doi:10.1016/S0378-3774(02)00112-9

Baret, F., Guyot, G., & Major, D. J. (1989). Crop biomass evaluation using radiometric measurements. *Photogrammetria*, *43*(5), 241–256. doi:10.1016/0031-8663(89)90001-X

Benabdelouahab, T., Balaghi, R., Hadria, R., Lionboui, H., Djaby, B., & Tychon, B. (2016). Testing Aquacrop to Simulate Durum Wheat Yield and Schedule Irrigation in a Semi-Arid Irrigated Perimeter in Morocco. *Irrigation and Drainage*, *65*(5), 631–643. doi:10.1002/ird.1977

Benabdelouahab, T., Balaghi, R., Hadria, R., Lionboui, H., Minet, J., & Tychon, B. (2015). Monitoring surface water content using visible and short-wave infrared SPOT-5 data of wheat plots in irrigated semi-arid regions. *International Journal of Remote Sensing*, *36*(15), 4018–4036. doi:10.1080/01431161.2015.1072650

Benabdelouahab, T., Balaghi, R., Hadria, R., Lionboui, H., & Tychon, B. (2016). Assessment of vegetation water content in wheat using near and shortwave infrared SPOT-5 Data in an irrigated area. *Revue des Sciences de l'Eau*, *29*(2), 97–107. doi:10.7202/1036542ar

Bouchaou, L., Michelot, J. L., Qurtobi, M., Zine, N., Gaye, C. B., Aggarwal, P. K., ... Vengosh, A. (2009). Origin and residence time of groundwater in the Tadla basin (Morocco) using multiple isotopic and geochemical tools. *Journal of Hydrology (Amsterdam)*, *379*(3–4), 323–338. doi:10.1016/j.jhydrol.2009.10.019

Bowen, I. S. (1926). The ratio of heat losses by conduction and by evaporation from any water surface. *Physical Review*, *27*(6), 779–787. doi:10.1103/PhysRev.27.779

Brisson, N., Mary, B., Ripoche, D., Jeuffroy, M. H., Ruget, F., Nicoullaud, B., ... Delécolle, R. (1998). STICS: A generic model for the simulation of crops and their water and nitrogen balances. I. Theory and parameterization applied to wheat and corn. *Agronomie*, *18*(5-6), 311–346. doi:10.1051/agro:19980501

Burba, G. (2013). Eddy Covariance Method for Scientific, Industrial, Agricultural and Regulatory Applications: a Field Book on Measuring Ecosystem Gas Exchange and Areal Emission Rates. Lincoln, NE: Academic Press.

Chaponnière, A., Maisongrande, P., Duchemin, B., Hanich, L., Boulet, G., Escadafal, R., & Elouaddat, S. (2005). A combined high and low spatial resolution approach for mapping snow covered areas in the Atlas mountains. *International Journal of Remote Sensing*, *26*(13), 2755–2777. doi:10.1080/01431160500117758

Chehbouni, A., Escadafal, R., Duchemin, B., Boulet, G., Simonneaux, V., Dedieu, G., ... Sobrino, J. A. (2008). An integrated modelling and remote sensing approach for hydrological study in arid and semi-arid regions: The SUDMED programme. *International Journal of Remote Sensing*, *29*(17-18), 5161–5181. doi:10.1080/01431160802036417

de Wit, A. J. W., Boogaard, H. L., & van Diepen, C. A. (2004). Using NOAA–AVHRR estimates of land surface temperature for regional agrometeorogical modelling. *International Journal of Applied Earth Observation and Geoinformation*, *5*(3), 187–204. doi:10.1016/j.jag.2004.03.003

Demarty, J., Ottlé, C., Braud, I., Olioso, A., Frangi, J. P., Bastidas, L. A., & Gupta, H. V. (2004). Using a multiobjective approach to retrieve information on surface properties used in a SVAT model. *Journal of Hydrology (Amsterdam)*, *287*(1), 214–236. doi:10.1016/j.jhydrol.2003.10.003

Dinku, T., Chidzambwa, S., Ceccato, P., Connor, S. J., & Ropelewski, C. F. (2008). Validation of high-resolution satellite rainfall products over complex terrain. *International Journal of Remote Sensing*, *29*(14), 4097–4110. doi:10.1080/01431160701772526

Du, L., Tian, Q., Yu, T., Meng, Q., Jancso, T., Udvardy, P., & Huang, Y. (2013). A comprehensive drought monitoring method integrating MODIS and TRMM data. *International Journal of Applied Earth Observation and Geoinformation*, *23*, 245–253. doi:10.1016/j.jag.2012.09.010

Duchemin, B., Hadria, R., Erraki, S., Boulet, G., Maisongrande, P., Chehbouni, A., ... Simonneaux, V. (2006). Monitoring wheat phenology and irrigation in Central Morocco: On the use of relationships between evapotranspiration, crops coefficients, leaf area index and remotely-sensed vegetation indices. *Agricultural Water Management*, *79*(1), 1–27. doi:10.1016/j.agwat.2005.02.013

Duchemin, B., Hagolle, O., Mougenot, B., Benhadj, I., Hadria, R., Simonneaux, V., ... Chehbouni, A. G. (2008). Agrometerological study of semi-arid areas: An experiment for analysing the potential of time series of FORMOSAT-2 images (Tensift-Marrakech plain). *International Journal of Remote Sensing*, *29*(17-18), 5291–5299. doi:10.1080/01431160802036482

Duchemin, B., Maisongrande, P., Boulet, G., & Benhadj, I. (2008). A simple algorithm for yield estimates: Evaluation for semi-arid irrigated winter wheat monitored with green leaf area index. *Environmental Modelling & Software*, *23*(7), 876–892. doi:10.1016/j.envsoft.2007.10.003

Ehleringer, J. R., Roden, J., & Dawson, T. E. (2000). Assessing ecosystem-level water relations trough stable isotope ratio analysis. In Methods in Ecosystem Science. Springer.

Er-Raki, S., Chehbouni, A., Boulet, G., & Williams, D. G. (2010). Using the dual approach of FAO-56 for partitioning ET into soil and plant components for olive orchards in a semi-arid region. *Agricultural Water Management*, *97*(11), 1769–1778. doi:10.1016/j.agwat.2010.06.009

Er-Raki, S., Chehbouni, A., Guemouria, N., Duchemin, B., Ezzahar, J., & Hadria, R. (2007). Combining FAO-56 model and ground-based remote sensing to estimate water consumptions of wheat crops in a semi-arid region. *Agricultural Water Management*, *87*(1), 41–54. doi:10.1016/j.agwat.2006.02.004

Er-Raki, S., Chehbouni, A., Hoedjes, J., Ezzahar, J., Duchemin, B., & Jacob, F. (2008). Improvement of FAO-56 method for olive orchards through sequential assimilation of thermal infrared-based estimates of ET. *Agricultural Water Management*, *95*(3), 309–321. doi:10.1016/j.agwat.2007.10.013

Er-Raki, S., Ezzahar, J., Khabba, S., Jarlan, L., Kharrou, M. H., & Chehbouni, G. (2013). *Micrometeorology tools for measuring evapotranspiration from the leaf to the region. In Evapotranspiration: processes, sources and environmental implications* (pp. 1–22). New York: Nova Science.

Ezzine, H., Bouziane, A., & Ouazar, D. (2014). Seasonal comparisons of meteorological and agricultural drought indices in Morocco using open short time-series data. *International Journal of Applied Earth Observation and Geoinformation, 26*, 36–48. doi:10.1016/j.jag.2013.05.005

Fabiola Flores, P., & Mario Lillo, S. (2010). Simple Air Temperature Estimation Method from MODIS Satellite Images on a Regional Scale. *Chilean Journal of Agricultural Research, 70*, 436–445.

Fekkoul, H. (2013). Groundwater contamination by nitrates, salinity and pesticides: Case of the unconfined aquifer of triffa plain (Eastern Morocco). *Revue Marocaine des Sciences Agronomiques et Vétérinaires, 2*, 12–36.

Gu, H.-h., Yu, Z.-b., Yang, C.-g., Ju, Q., Lu, B.-h., & Liang, C. (2010). Hydrological assessment of TRMM rainfall data over Yangtze River Basin. *Water Science and Engineering, 3*(4), 418–430. doi:10.3882/j.issn.1674-2370.2010.04.005

Guérif, M., & Duke, C. (1998). Calibration of the SUCROS emergence and early growth module for sugar beet using optical remote sensing data assimilation. *European Journal of Agronomy, 9*(2), 127–136. doi:10.1016/S1161-0301(98)00031-8

Hachem, S., Duguay, C. R., & Allard, M. (2012). Comparison of MODIS-derived land surface temperatures with ground surface and air temperature measurements in continuous permafrost terrain. *The Cryosphere, 6*(1), 51–69. doi:10.5194/tc-6-51-2012

Hadria, R., Benabdelouahab, T., Mahyou, H., Balaghi, R., Bydekerke, L., El Hairech, T., & Ceccato, P. (2018). Relationships between the three components of air temperature and remotely sensed land surface temperature of agricultural areas in Morocco. *International Journal of Remote Sensing, 39*(2), 356–373. doi:10.1080/01431161.2017.1385108

Hadria, R., Duchemin, B., Jarlan, L., Dedieu, G., Baup, F., Khabba, S., ... Le Toan, T. (2010). Potentiality of optical and radar satellite data at high spatio-temporal resolutions for the monitoring of irrigated wheat crops in Morocco. *International Journal of Applied Earth Observation and Geoinformation, 12*(1), S32–S37. doi:10.1016/j.jag.2009.09.003

Hadria, R., Duchemin, B., Lahrouni, A., Khabba, S., Er-raki, S., Dedieu, G., ... Olioso, A. (2006). Monitoring of irrigated wheat in a semi-arid climate using crop modelling and remote sensing data: Impact of satellite revisit time frequency. *International Journal of Remote Sensing, 27*(6), 1093–1117. doi:10.1080/01431160500382980

Hadria, R., Khabba, S., Lahrouni, A., Duchemin, B., Chehbouni, A., Carriou, J., & Ouzine, L. (2007). Calibration and Validation of the Shoot Growth Module of STICS Crop Model: Application to Manage Irrigation Water in the Marrakesch/Al Haouz Plain. *Arabian Journal for Science and Engineering, 32*, 87–101.

Hengl, T., Heuvelink, G. B. M., Perčec Tadić, M., & Pebesma, E. (2012). Spatio-temporal prediction of daily temperatures using time-series of MODIS LST images. *Theoretical and Applied Climatology*, *107*(1-2), 265–277. doi:10.100700704-011-0464-2

Hsu, K., Gao, X., Sorooshian, S., & Gupta, H. V. (1997). Precipitation Estimation from Remotely Sensed Information Using Artificial Neural Networks. *Journal of Applied Meteorology*, *36*(9), 1176–1190. doi:10.1175/1520-0450(1997)036<1176:PEFRSI>2.0.CO;2

Jacquemoud, S., Bacour, C., Poilvé, H., & Frangi, J. P. (2000). Comparison of Four Radiative Transfer Models to Simulate Plant Canopies Reflectance: Direct and Inverse Mode. *Remote Sensing of Environment*, *74*(3), 471–481. doi:10.1016/S0034-4257(00)00139-5

Jones, C. A., Kiniry, J. R., & Dyke, P. T. (1986). *CERES-Maize: a simulation model of maize growth and development*. College Station, TX: Texas A&M University Press.

Joyce, R. J., Janowiak, J. E., Arkin, P. A., & Xie, P. (2004). CMORPH: A Method that Produces Global Precipitation Estimates from Passive Microwave and Infrared Data at High Spatial and Temporal Resolution. *Journal of Hydrometeorology*, *5*(3), 487–503. doi:10.1175/1525-7541(2004)005<0487:CAMTPG>2.0.CO;2

Khabba, S., Duchemin, B., Hadria, R., Ezzahar, J., Chehbouni, A., Lahrouni, A., & Hanich, L. (2009). Evaluation of digital hemispherical photography and plant canopy analyser for measuring Vegetation area index of orange orchards. *Journal of Agronomy*, *8*(2), 67–72. doi:10.3923/ja.2009.67.72

Lahmer, I. (2017). Extraction stratifiée par âge des agrumes à partir d'une série temporelle des images Sentinel 2 dans la plaine de triffa, région de l'oriental du Maroc. In Rapport de projet de fin d'étude. Institut Agronomique et Veterinaire Hassan II.

Li, L., Hong, Y., Wang, J., Adler, R. F., Policelli, F. S., Habib, S., ... Okello, L. (2009). Evaluation of the real-time TRMM-based multi-satellite precipitation analysis for an operational flood prediction system in Nzoia Basin, Lake Victoria, Africa. *Natural Hazards*, *50*(1), 109–123. doi:10.100711069-008-9324-5

Milewski, A., Elkadiri, R., & Durham, M. (2015). Assessment and Comparison of TMPA Satellite Precipitation Products in Varying Climatic and Topographic Regimes in Morocco. *Remote Sensing*, *7*(5), 5697–5717. doi:10.3390/rs70505697

Moffitt, C. B., Hossain, F., Adler, R. F., Yilmaz, K. K., & Pierce, H. F. (2011). Validation of a TRMM-based global Flood Detection System in Bangladesh. *International Journal of Applied Earth Observation and Geoinformation*, *13*(2), 165–177. doi:10.1016/j.jag.2010.11.003

Moulin, S. (1999). Impacts of model parameter uncertainties on crop reflectance estimates: A regional case study on wheat. *International Journal of Remote Sensing*, *20*(1), 213–218. doi:10.1080/014311699213730

Moutaib, O., & Oujamaa, A. (2017). Cartoghrahie agrumes dans la plaine de Tiffa. In Rapport de projet de fin d'étude. Institut Agronomique et Veterinaire Hassan II.

Najine, A., Jaffal, M., Khammari, K. E., Aïfa, T., Khattach, D., Himi, M., ... Aqil, H. (2006). Contribution de la gravimétrie à l'étude de la structure du bassin de Tadla (Maroc): Implications hydrogéologiques. *Comptes Rendus Geoscience*, *338*(10), 676–682. doi:10.1016/j.crte.2006.04.015

Olioso, A., Chauki, H., Courault, D., & Wigneron, J.-P. (1999). Estimation of Evapotranspiration and Photosynthesis by Assimilation of Remote Sensing Data into SVAT Models. *Remote Sensing of Environment*, *68*(3), 341–356. doi:10.1016/S0034-4257(98)00121-7

ORMVAM. (2017). Potentialités agricoles de la région de l'Oriental. Berkane: Office Régional de Mise en Valeur Agricole de la Moulouya.

Ouatiki, H., Boudhar, A., Tramblay, Y., Jarlan, L., Benabdelouhab, T., Hanich, L., ... Chehbouni, A. (2017). Evaluation of TRMM 3B42 V7 Rainfall Product over the Oum Er Rbia Watershed in Morocco. *Climate (Basel)*, *5*(1), 1. doi:10.3390/cli5010001

Panda, R. K., Behera, S. K., & Kashyap, P. S. (2003). Effective management of irrigation water for wheat under stressed conditions. *Agricultural Water Management*, *63*(1), 37–56. doi:10.1016/S0378-3774(03)00099-4

Richardson, A. J., Wiegand, C. L., Wanjura, D. F., Dusek, D., & Steiner, J. L. (1992). Multisite analyses of spectral-biophysical data for Sorghum. *Remote Sensing of Environment*, *41*(1), 71–82. doi:10.1016/0034-4257(92)90062-O

Rouse, J. W., Haas, R. H., Schell, J. A., Deering, D. W., & Harlan, J. C. (1974). Monitoring the vernal advancement and retrogradation (green wave effect) of natural vegetation. Texas A&M Univ.

Shen, S., & Leptoukh, G. G. (2011). Estimation of surface air temperature over central and eastern Eurasia from MODIS land surface temperature. *Environmental Research Letters*, *6*(4), 045206. doi:10.1088/1748-9326/6/4/045206

Simonneaux, V., Lepage, M., Helson, D., Metral, J., Thomas, S., Duchemin, B., ... Chehbouni, A. (2009). Estimation spatialiséee de l'évapotranspiration des cultures irriguées par télédétection: Application à la gestion de l'irrigation dans la plaine du Haouz (Marrakech, Maroc). *Sécheresse*, *20*(1), 123–130.

Steduto, P., Hsiao, T. C., Raes, D., & Fereres, E. (2009). AquaCrop—The FAO Crop Model to Simulate Yield Response to Water: I. Concepts and Underlying Principles All rights reserved. No part of this periodical may be reproduced or transmitted in any form or by any means, electronic or mechanical, including photocopying, recording, or any information storage and retrieval system, without permission in writing from the publisher. *Agronomy Journal*, *101*(3), 426–437. doi:10.2134/agronj2008.0139s

Su, F., Hong, Y., & Lettenmaier, D. P. (2008). Evaluation of TRMM Multisatellite Precipitation Analysis (TMPA) and Its Utility in Hydrologic Prediction in the La Plata Basin. *Journal of Hydrometeorology*, *9*(4), 622–640. doi:10.1175/2007JHM944.1

Tramblay, Y., Thiemig, V., Dezetter, A., & Hanich, L. (2016). Evaluation of satellite-based rainfall products for hydrological modelling in Morocco. *Hydrological Sciences Journal*, *61*(14), 2509–2519. doi:10.1080/02626667.2016.1154149

Valancogne, C., & Nasr, Z. (1989). Une méthode de mesure du débit de sève brute dans de petits arbres par bilan de chaleur. *Agronomie*, *9*(6), 609–617. doi:10.1051/agro:19890609

Vancutsem, C., Ceccato, P., Dinku, T., & Connor, S. J. (2010). Evaluation of MODIS land surface temperature data to estimate air temperature in different ecosystems over Africa. *Remote Sensing of Environment*, *114*(2), 449–465. doi:10.1016/j.rse.2009.10.002

Verhoef, W., & Bach, H. (2003). Remote sensing data assimilation using coupled radiative transfer models. *Physics and Chemistry of the Earth Parts A/B/C*, *28*(1), 3–13. doi:10.1016/S1474-7065(03)00003-2

Vogt, J. V., Viau, A. A., & Paquet, F. (1997). Mapping regional air temperature fields using satellite-derived surface skin temperatures. *International Journal of Climatology*, *17*(14), 1559–1579. doi:10.1002/(SICI)1097-0088(19971130)17:14<1559::AID-JOC211>3.0.CO;2-5

Weir, A. H., Bragg, P. L., Porter, J. R., & Rayner, J. H. (1984). A winter wheat crop simulation model without water or nutrient limitations. *The Journal of Agricultural Science*, *102*(2), 371–382. doi:10.1017/S0021859600042702

Williams, R. (1989). The EPIC Crop Growth Model. *Transactions of the ASAE. American Society of Agricultural Engineers*, *32*(2), 497. doi:10.13031/2013.31032

Yaduvanshi, A., Srivastava, P. K., & Pandey, A. C. (2015). Integrating TRMM and MODIS satellite with socio-economic vulnerability for monitoring drought risk over a tropical region of India. *Physics and Chemistry of the Earth Parts A/B/C*, *83*, 14–27. doi:10.1016/j.pce.2015.01.006

Zhang, A., & Jia, G. (2013). Monitoring meteorological drought in semiarid regions using multi-sensor microwave remote sensing data. *Remote Sensing of Environment*, *134*, 12–23. doi:10.1016/j.rse.2013.02.023

# Chapter 8
# Support Irrigation Water Management of Cereals Using Optical Remote Sensing and Modeling in a Semi-Arid Region

**Tarik Benabdelouahab**
*INRA, Morocco*

**Hayat Lionboui**
*INRA, Morocco*

**Rachid Hadria**
*INRA, Morocco*

**Riad Balaghi**
*INRA, Morocco*

**Abdelghani Boudhar**
*Sultan Moulay Silmane University, Morocco*

**Bernard Tychon**
*University of Liège, Belgium*

## ABSTRACT

*Irrigated agriculture is an important strategic sector for Morocco, contributing to food security and employment. Nowadays, irrigation scheme managers shall ensure that water is optimally used. The main objective was to support the irrigation monitoring and management of wheat in the irrigated perimeter using optical remote sensing and crop modeling. The potential of spectral indices derived from SPOT-5 images was explored for quantifying and mapping surface water content changes at large scale. Indices were computed using the reflectance in red, near infrared, and shortwave infrared bands. A field crop model (AquaCrop) was adjusted and tested to simulate the grain yield and the temporal evolution of soil moisture status. This research aimed at providing a scientific and technical approach to assist policymakers and stakeholders to improve monitoring irrigation and mitigating wheat water stress at field and irrigation perimeter levels in semi-arid areas. The approach could lead to operational management tools for an efficient irrigation at field and regional levels.*

DOI: 10.4018/978-1-5225-5939-9.ch008

## GENERAL CONTEXT

Irrigated areas produce more than half of all foodstuffs in the world and therefore contribute significantly to food security. This activity, however, consumes about 72% of available water resources (Geerts & Raes, 2009; Seckler, Barker, & Amarasinghe, 1999). In Morocco, water availability is one of the main limiting factors in achieving good yields. Irrigated agriculture occupies only 15% of the cultivated area (about 1.5 million ha) in the country, but accounts for about 45% of the agricultural Gross Domestic Product and 75% of agricultural exports, depending on the season. This contribution is greater during dry seasons when production in rainfed areas is severely affected (MAPM, 2012). The challenge for stakeholders and managers in the irrigated perimeter is to increase production, control water management and rationalize irrigation. In order to save water and help farmers meet this challenge, they are given technical supervision and coaching, as well as subsidies for irrigation equipment, and legislation governing the mobilization and rational use of water resources has been enacted (Conseil Supérieur de l'Eau et du Climat, Law No. 10-95) (Hayat Lionboui, Benabdelouahab, Hasib, & Boulli, 2016).

## WATER RESOURCE MANAGEMENT IN THE STUDY AREA

The study area is located in central Morocco, between the Atlantic coast in the north-west and the Atlas Mountains in the south-east (32°23´ north latitude; 6°31´west longitude; 445 m above sea level), within the irrigation perimeter of the Tadla region. The area is characterized by a semi-arid climate: the annual average temperature is about 19°C, with large inter-seasonal variation. The average cropping season precipitation is about 300 mm (average over the 1970-2010 period), with significant inter-annual variation ranging from 130 to 600 mm.

The area is characterized by a semi-arid climate; the annual average temperature is about 19°C, with large inter-seasonal variation (maximum = 38°C in August and minimum = 3.5°C in January). The average annual precipitation is about 300 mm (average over the 1970-2010 period), with significant inter-annual variation (from 130 mm to 600 mm). The irrigated area covers 100,000 hectares (ha) and is characterized by a flat topography. Durum wheat is one of the main crops in this perimeter (12% of total cultivated area).

Created in the 1940s, the Tadla irrigated perimeter was among the first large irrigation schemes in the country. It is on a plain in central Morocco (32°23´ N latitude; 6°31´ W longitude; 445 m above sea level) that covers about 100,000 hectares (ha) and is characterized by a flat topography. The plain has a semi-arid climate, with about 300 mm average annual precipitation over the 1970-2010 period and a high inter-annual variation, ranging from 130 to 600 mm over the same period.

The Tadla irrigated perimeter is divided into two sub-schemes by the Oum-Er-Rbia river, which flows from the Middle-Atlas Mountains (east) to the Atlantic Ocean (west) (Figure 1).

The irrigation water used in the Tadla perimeter comes mainly from surface water (87.1% of the total amount of irrigation water consumed in 2009/2010). Two dams, Ahmed-Al-Hansali (capacity of 750 million $m^3$) and Bin-El-Ouidane (1.5 billion $m^3$), supply irrigation water to the Tadla perimeter, in addition to groundwater pumping (Hayat Lionboui, Benabdelouahab, Elame, Hasib, & Boulli, 2016).

The groundwater depth in the area varies from 31 to 117 m (Bouchaou et al., 2009; Najine et al., 2006). The over-exploitation of groundwater has led to reduced piezometric levels (FAO, 2011). The proportion of groundwater, however, has increased in recent years due to frequent droughts. Groundwater

*Figure 1. Location of the Tadla irrigated perimeter*

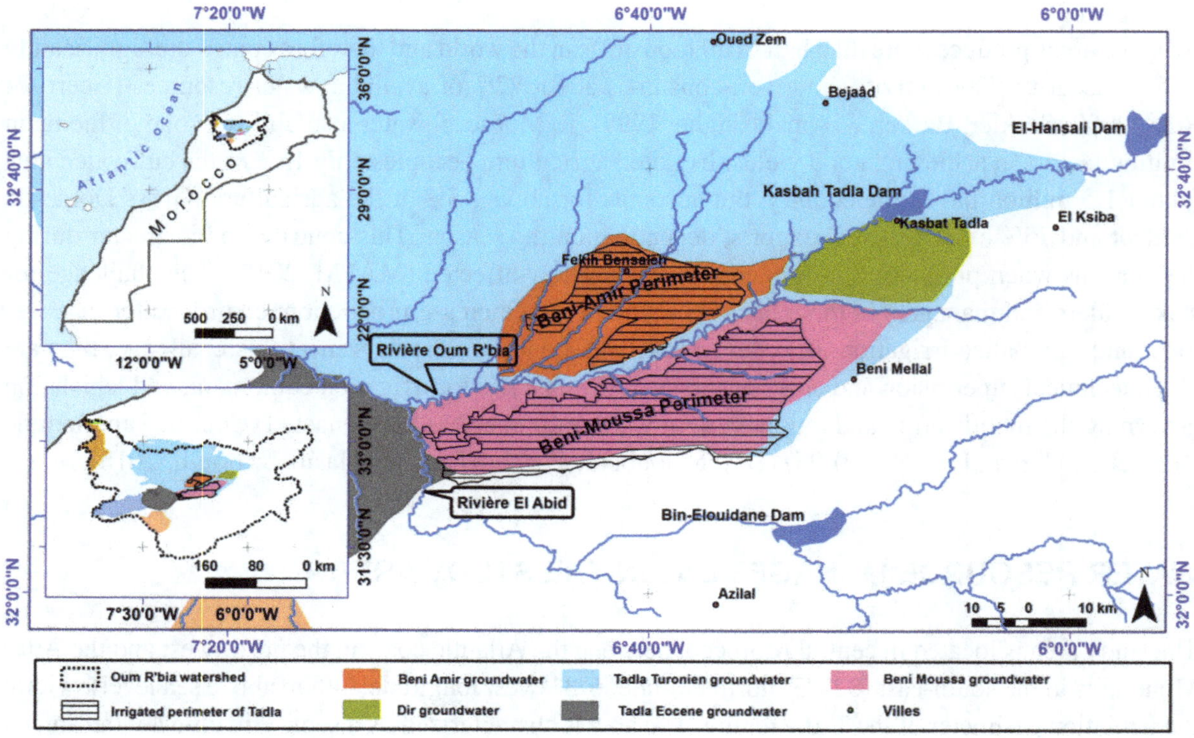


used for agricultural purposes in the Tadla comes from Beni-Moussa and Beni Amir groundwater and the Turonian deep water table. The total amount groundwater used by the Tadla perimeter was 12.9% (120 million m$^3$ per year) of the total amount of irrigated water used in the 2009-2010 cropping season.

Irrigation decision-making is based on the amount of water reserves in dams at the beginning of the cropping season and the estimated need for irrigation water. In drought years, restrictions on the water allocation to the Tadla perimeter are set by the Agence du Bassin Hydraulique d'Oum Er Rbia (ABHOER), which is responsible for the assessment and management of water resources in the Oum Er Rbia watershed area.

The Tadla irrigated perimeter is managed by the Regional Office for Agricultural Development of Tadla (ORMVAT), a public sector institution within the Ministry of Agriculture that is responsible for organizing the distribution of the water in the perimeter to more than 27,000 farmers.

The Département de la Gestion du Réseau d'irrigation et de drainage (DGRID), which is responsible for managing the irrigation and drainage network, establishes the provisional distribution program in the irrigated perimeter. This program is transmitted to the district-level agency, Arrondissement de Gestion du Réseau (AGR), for the release of water.

The irrigation water is supplied via a canal system consisting of a network of main canals (212 km), primary and secondary canals (742 km) and tertiary canals (2,166 km). Distribution agents and valve guards are responsible for implementing and controlling the distribution of the irrigation water. This system involves many field staff and agents at the local level, but lacks the ability to monitor and optimize control of the entire irrigated area.

Since the 1980s, several studies of the Tadla perimeter have been carried out by national and regional stakeholders in partnership with international institutions and development bodies. They have focused on identifying periods of likely water scarcity and improving irrigation management across the perimeter, but have not provided a spatio-temporal approach for the monitoring and control of the main production parameters. Such an approach, however, is essential for decision-making on large-scale schemes, such as Tadla. Our research sought to contribute to improving the spatio-temporal monitoring and control of irrigation in the Tadla irrigated perimeter.

## WHEAT PRODUCTION

Wheat is the main crop in the Tadla irrigated perimeter, covering more than 36% (40,000 ha) of the total irrigated area (ORMVAT). Despite the large amounts of irrigation water used, wheat yields remain low, with high inter-seasonal variations due to fluctuating water availability and poor management practices. The average yield in the 1994-2010 period was 32 quintals/ha, with a coefficient of variation of 9.84% (ORMVAT, 2009). Water is one of the main factors limiting wheat production in the Tadla perimeter, and good management of irrigation water on a large-scale is required in order to address this challenge.

The volume of irrigation water used by wheat in the 1994-2002 period in Tadla rose to 136 million m3/year, on average. This is equivalent to 18% of all irrigation water used across the irrigated perimeter (ORMVAT, 2009).

The wheat-growing cycle in the region runs from November-December to June. It is usually sown between mid-November and mid-January, depending when the first significant precipitation occurs, and is harvested between May and June, depending on temperature conditions. During this period, wheat is irrigated following traditional flood irrigation practice, from two to five times, depending on rainfall availability in the autumn and the volume of water accumulated in dams during winter and spring.

Traditional flood irrigation practice (Robta) involves compartmentalizing land into elementary basins with an average size of 50-60 $m^2$ and supplying water to these basins, one by one. This technique results in irrigation efficiency losses of about 50% (ORMVAT, 2002).

The spatio-temporal monitoring of wheat development, irrigation supplies and surface water content could be an interesting basis for improving irrigation scheduling and preventing water stress from adversely affecting yield (Duchemin et al., 2006).

## CHAPTER OUTLINE

The objective of the research described here was to improve irrigation management and surface water content monitoring for wheat throughout a large perimeter, based on remote sensing and crop modeling in the semi-arid area of Morocco. The research was aimed primarily at decision-makers and managers of large-scale irrigated perimeters. Two studies using optical satellite imagery and modeling were presented, in this document, based on scientific papers published in peer-reviewed international journals (Tarik Benabdelouahab et al., 2016; T. Benabdelouahab et al., 2015).

The first case of study aims to evaluate the potential of two spectral indices to assess surface water content in wheat fields in order to detect irrigation water supplies and wheat stress management in the irrigated perimeter, to improve irrigation monitoring at field and regional levels. Optical imagery could also be used for water stress warning by detecting plots with under-threshold water content that are probably suffering from water stress. This situation generally arises from the limited availability of water in the soil. In order to intervene in time, there needs to be a short gap between image acquisition and image processing and high acquisition frequency of satellite images.

This limitation of remote sensing data can be overcome by integrating the modeling tools that will allows to simulate the missing daily evolution of crop development and soil water content to bridge the gap between image acquisition. This research comes to satisfy a need of stakeholders, involved in managing irrigated areas to access to decision making tools to help them schedule irrigation and assess its impact on yield.

Hence the interest of the second case of study which addresses the management of irrigation water in plots throughout the cropping season using a soil-plant-atmosphere model (i.e., the AquaCrop model developed by the Land and Water Division of Food and Agriculture Organization of the United Nations, FAO).

The used model provides an appropriate tool for detecting water stress and simulating the effects of water stress on crop productivity in order to help decision-makers to improve irrigation management and to optimize water-use efficiency in arid and semi-arid conditions.

The crop model AquaCrop and spectral indices derived from remote sensing data could be a key factor in optimizing growing conditions throughout the growing season in order to maintain good grain yields and improve water-use efficiency, while avoiding excess water or water deficit situations.

## CASE 1: MONITORING SURFACE WATER CONTENT USING VISIBLE AND SHORTWAVE INFRARED SPOT-5 DATA OF WHEAT PLOTS IN IRRIGATED SEMI-ARID REGIONS

Remotely sensed reflectance has been used to estimate soil and vegetation water content for various crops and to monitor water irrigation per surface unit (Ben-Gal et al., 2010; Ceccato, Flasse, & Grégoire, 2002; Cheng et al., 2012; Hadria et al., 2010; Penuelas, Pinol, Ogaya, & Filella, 1997; Tian, Tong, Pu, Guo, & Zhao, 2001; Trombetti, Riaño, Rubio, Cheng, & Ustin, 2008), drawing on the high temporal and spatial resolution of satellite images. Several indices based on wavelengths ranging between 400 and 2,500 nm have been developed to describe land-surface moisture conditions (Kogan, 2000). Estimation of surface water content values from remote sensing data is usually based on reflectance in the red (R; 610-680 nm), Near Infrared (NIR; 780-890 nm) and Shortwave Infrared (SWIR; 1,580-1,750 nm) regions of the spectrum (Lobell, Asner, Ortiz-Monasterio, & Benning, 2003; Moreno et al., 2014; Muller & Décamps, 2000; Skidmore, Dickerson, & Shimmelpfennig, 1975).

Many indices for the simultaneous estimation of vegetation water content and soil moisture have been proposed for different land surfaces, from bare soils to vegetated areas, among which are the visible and shortwave infrared drought index (VSDI) (Ning, Yang, Qiming, & Lu, 2013), the modified shortwave infrared perpendicular water stress index (MSPSI) (Feng, Chen, Dong, Wang, & Meng, 2013), the modified perpendicular drought index (MPDI) (Ghulam, Qin, Teyip, & Li, 2007), the normalized difference water index (NDWIRog) (Rogers & Kearney, 2004) and the moisture stress index (MSI) (Hunt Jr & Rock, 1989). In this study we explore the potential of $NDWI_{Rog}$ and MSI for comparing, quantifying and mapping the surface water content of wheat plots, from bare soil to completely covered soil.

An operational index for simultaneously measuring surface water content of bare soil, mixed bare and covered soil has become crucial for irrigation management, especially in arid and semi-arid regions. This is required, especially for large irrigated areas and throughout the cropping season, when the vegetation cover is continuously changing. This index could lead to an operational tool for monitoring surface water content and managing irrigation, at least for the study area.

## Field Data

The experiments were conducted during the 2012-2013 and 2013-2014 wheat cropping seasons to assess changes in soil moisture and vegetation. Dates and amounts of irrigation water supply and physiological crop data were collected.

Soil moisture was measured weekly for all 30 plots during the two cropping seasons, starting from sowing until grain filling, at 0-10 cm depth, with three random replications per plot (Figure 2). Soil moisture was measured using gravimetric methodology (dried in an oven at 105°C for 24 hours). Vegetation water content was also measured weekly, starting from tillering until wheat grain filling (January to May 2013). Soil and vegetation water content were quantified on a gravimetric basis (i.e., grams water/grams soil or biomass), expressed as a percentage (%).

*Figure 2. Location of the irrigated area (upper left inset shows a map of Morocco; the study area is indicated by diagonal lines and the experimental plots are in grey).*

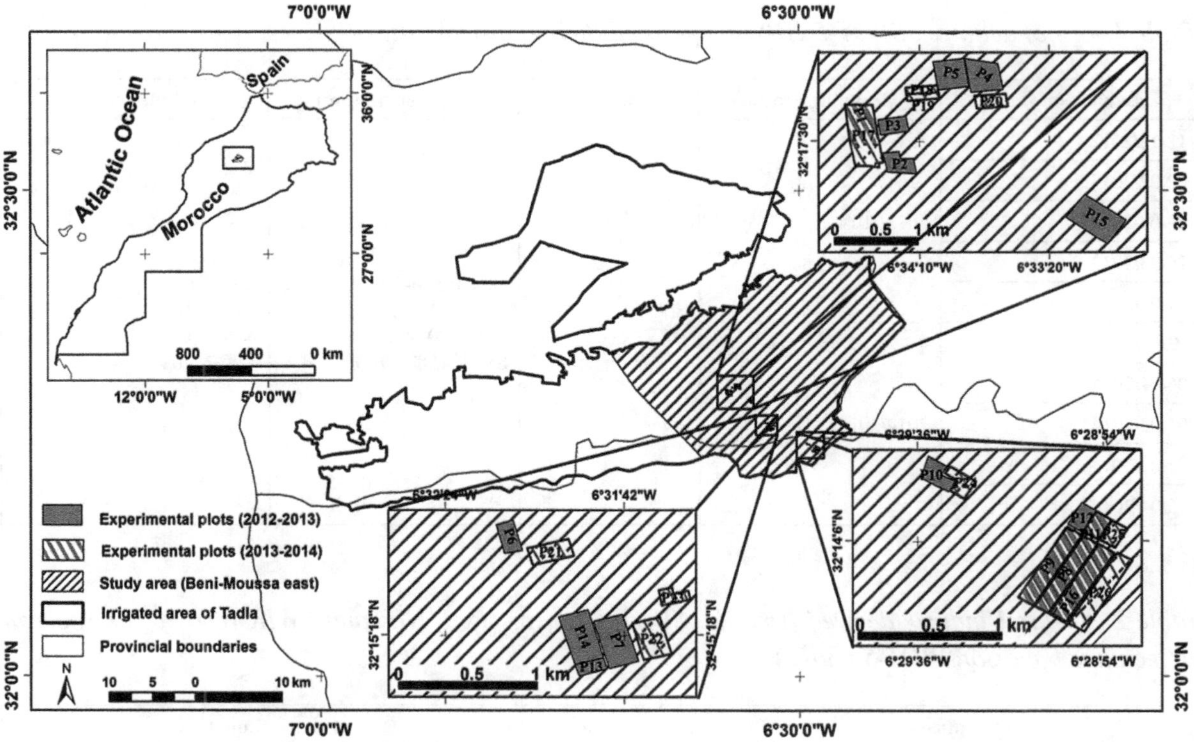

## Satellite Images and their Processing

Ten SPOT-5 HRV satellite images were acquired between December (at wheat emergence) and April (at grain filling) for the 2012-2013 and 2013-2014 cropping seasons (Table 1). The processing level of the acquired images was 1B, which included radiometric and geometric corrections. Atmospheric corrections were performed from radiance images.

The first step of images post-processing involved computing two spectral indices, the $NDWI_{Rog}$ (Lasaponara & Masini, 2012; Rogers & Kearney, 2004) and the MSI (Ceccato, Flasse, Tarantola, Jacquemoud, & Grégoire, 2001; Ceccato, Gobron, Flasse, Pinty, & Tarantola, 2002; Hunt Jr & Rock, 1989) (Table 2), using the spectral reflectance in the Red, NIR and SWIR bands for each SPOT-5 HRV image.

The second step involved delineating the region of interest (ROI) used as a mask of wheat sub-plots. The average values of the $NDWI_{Rog}$ and MSI spectral indices were then computed for each corresponding sub-plot (7×7 pixels) where field measurements were conducted (Figure 3).

## Soil Moisture Assessment During the Main Growth Stages of Wheat

Following the strong ability of $NDWI_{Rog}$ and MSI to estimate soil moisture separately for bare soil and full vegetation cover, we tested the capacity of these spectral indices to estimate this parameter throughout wheat cropping season, apart from the senescent stage which was not studied, since no irrigation is applied during this stage of wheat development.

*Table 1. List of acquired SPOT-5 HRV images and their characteristics*

| Acquisition Date | Cropping Season | Sensor | Wavelength (nm) | Resolution (meters) |
|---|---|---|---|---|
| 12/12/2012 | 2013/2014 | SPOT-5 HRV | Green: 500-590<br>Red: 610-680<br>NIR: 780-890<br>SWIR:1580-1750 | Red: 10<br>Green: 10<br>NIR: 10<br>SWIR: 20 |
| 02/02/2013 | | | | |
| 21/03/2013 | | | | |
| 26/03/2013 | | | | |
| 11/04/2013 | | | | |
| 02/12/2013 | 2014/2015 | | | |
| 06/01/2014 | | | | |
| 01/02/2014 | | | | |
| 26/03/2014 | | | | |
| 15/04/2014 | | | | |

*Table 2. Spectral indices derived from the SPOT-5 sensor (Red, NIR and SWIR refer to the spectral reflectance bands of SPOT-5 image)*

| Indices | Abbreviation | Equation |
|---|---|---|
| Normalised Difference Water Index | NDWI Rog | (Red – SWIR)/(Red + SWIR) |
| Moisture Stress Index | MSI | (SWIR/NIR) |

*Figure 3. Schematic diagram illustrating field data and satellite image processing*

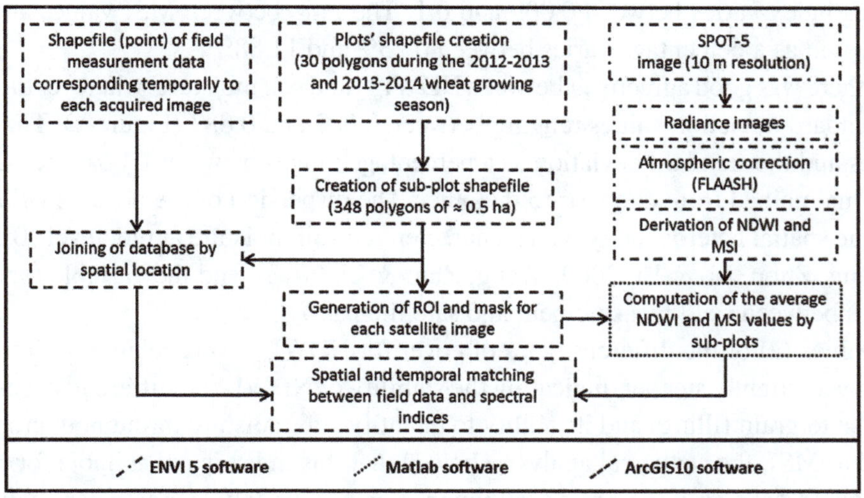


Figure 4 shows the comparison between observed soil moisture values and those derived using the spectral indices for the 10 acquisition dates. The statistical indicators $R^2$ and RMSE were 0.86 ($p<0.01$) and 1.62% for NDWIRog, respectively.

The point clouds for MSI, representing different kinds of cover, show opposite trends according to the main growth stages of wheat. This indicated that there was no unique linear relationship between MSI

*Figure 4. Relationship between soil moisture and the $NDWI_{Rog}$ values derived from all the acquired images (cropping season 2012/2013 in blue and cropping season 2013/2014 in Red: squares = bare soil; cross = covered soil; circles = mixed cover).*

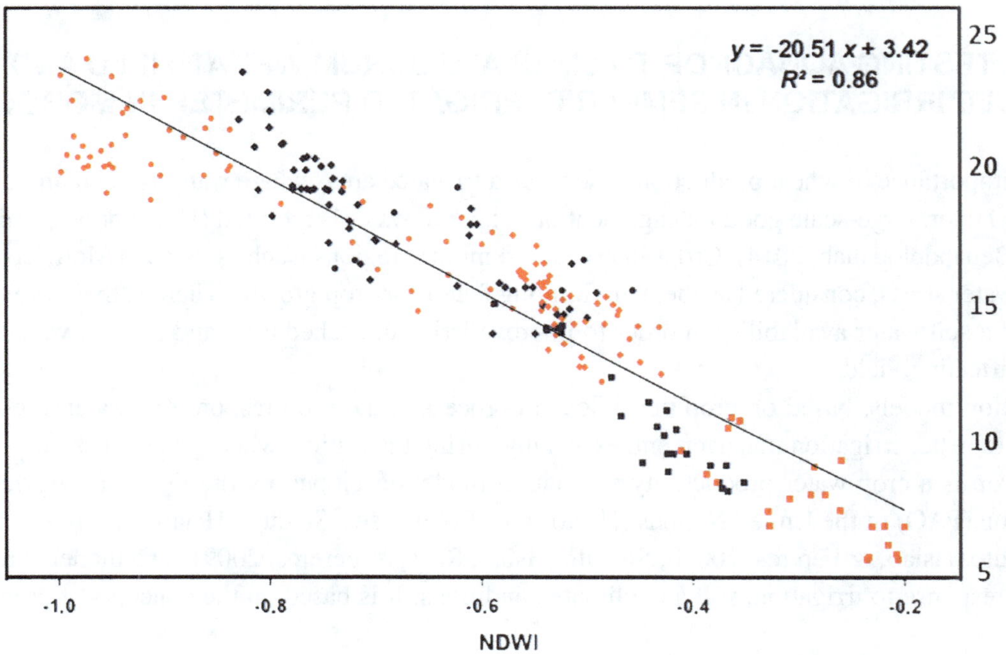

and surface soil moisture, for the entire wheat crop cycle (apart from the senescent stage). The standard deviation of this index varies between 0.009 and 0.1. The ratio between MSI values and the standard deviation expressed as a percentage varies between 0.23% and 11.88%.

In contrast, there was good agreement between $NDWI_{Rog}$ and soil moisture, whatever the wheat growth stage with a standard deviation values ranging between 0.007 and 0.087 (Figure 4). The ratio between $NDWI_{Rog}$ values and the standard deviation as a percentage ranges between 1.12% and 12.87%. The relationship was maintained from one year to the other. The dispersion of the observed cloud points was mainly due to the spatial heterogeneity which characterized soil moisture at plot level (Bi, Li, Liu, Guo, & Li, 2009; Song, Zhou, Li, & Li, 2009; Wang, Zhu, & Li, 2013), and the variable time lags ranging from 0 to 3 days between field measurement and satellite pass.

The slope values (a) of the different types of cover for $NDWI_{Rog}$ were relatively similar. For mixed cover, the slope was slightly steeper, indicating the stability of $NDWI_{Rog}$ at different stages of crop cover (from emergence to grain filling) and its ability to quantify soil moisture throughout crop growth.

With regard to MSI, the statistical analysis showed that this index is not suitable for estimating soil moisture throughout crop growing period, although it can accurately estimate bare soil moisture and vegetation water content separately. The NIR reflectance of covered soil is significantly higher than that of bare soil (Ning et al., 2013). As MSI uses NIR as the reference band, the values of this index are much higher for bare soil than for covered soil, which means that MSI cannot be used to compare dissimilar land-cover types.

Feng et al. (2013) simulated soil moisture using the MSPSI model and obtained an $R^2$ of 0.66. They also obtained R2 values of 0.54, 0.48 and 0.60 for the PDI, MPDI and SPSI models, respectively, for both bare and covered surfaces.

Ning et al. (2013) proposed using the VSDI for monitoring soil and vegetation moisture simultaneously over different land-cover types. This index is based on exploiting the SWIR and red bands. In a comparison between VSDI and the fractional water index over different land-cover types, they obtained an $R^2$ of 0.54.

## CASE 2: TESTING AQUACROP TO SIMULATE DURUM WHEAT YIELD AND SCHEDULE IRRIGATION IN SEMI-ARID IRRIGATED PERIMETER IN MOROCCO

Given the importance of wheat production in semi-arid irrigated areas where water is the main limiting production factor, large-scale good management of irrigation water is required (H. Lionboui, Fadlaoui, Elame, & Benabdelouahab, 2014). Crops in arid and semi-arid regions, such as those in Morocco, regularly face water stress, considered as the main limiting factor for crop growth. There is therefore a need for assessing soil water availability in order to improve irrigation scheduling and prevent water stress adversely affecting yield.

Simulation models, based on crop physiological processes and crop response to water stress, can contribute to better irrigation management, especially during the critical wheat growth period.

AquaCrop is a crop water productivity simulation model developed by the Food and Agriculture Organization (FAO) of the United Nations (Hsiao et al., 2009; Raes, Steduto, Hsiao, & Fereres, 2009a; Raes, Steduto, Hsiao, & Fereres, 2009b; Steduto, Hsiao, Raes, & Fereres, 2009). The model simulates crop yield response to irrigation, soil and climate conditions. It is based on the concepts of crop yield response to water (Doorenbos & Kassam, 1979; Doorenbos & Pruitt, 1977) and is suitable for areas where the water is a limiting factor for agricultural production. Aimed at balancing precision and simplicity, it uses a small number of explicit and mostly intuitive parameters and input variables that require simple methods for their estimation (Hsiao et al., 2009; Steduto et al., 2009).

Designing and implementing operational tools that can provide accurate estimates of crop water needs and the impact on production, and can quantify crop water consumption and production, would facilitate the monitoring of irrigation efficiency and crop water use. The main objective of this study was to test the ability of AquaCrop to simulate durum wheat (Triticum turgidum) biomass production, grain yield and the soil water content profile (0-100 cm layer) in an irrigated area, on the one hand, and on the other hand to study the capacity of the model to manage the irrigation water and optimize a timely application of irrigation supply to increase water use efficiency.

## Field Experiments

The experimental sites are located across the eastern Beni-Moussa municipality in the 2009-2010, 2010-2011 and 2011-2012 cropping seasons. Data were collected from 15 fields of durum wheat, located at Tadla's Regional Agricultural Research or belonging to farmers, thus providing a valid representation of the soil-plant relationship in the study area. The field data related to Marzak and Karim cultivars, which are widely cultivated in the study region.

*Figure 5. Location of the Tadla irrigated perimeter (upper left window represents Morocco map; in the upper right window, the experimental plots in black diamond)*

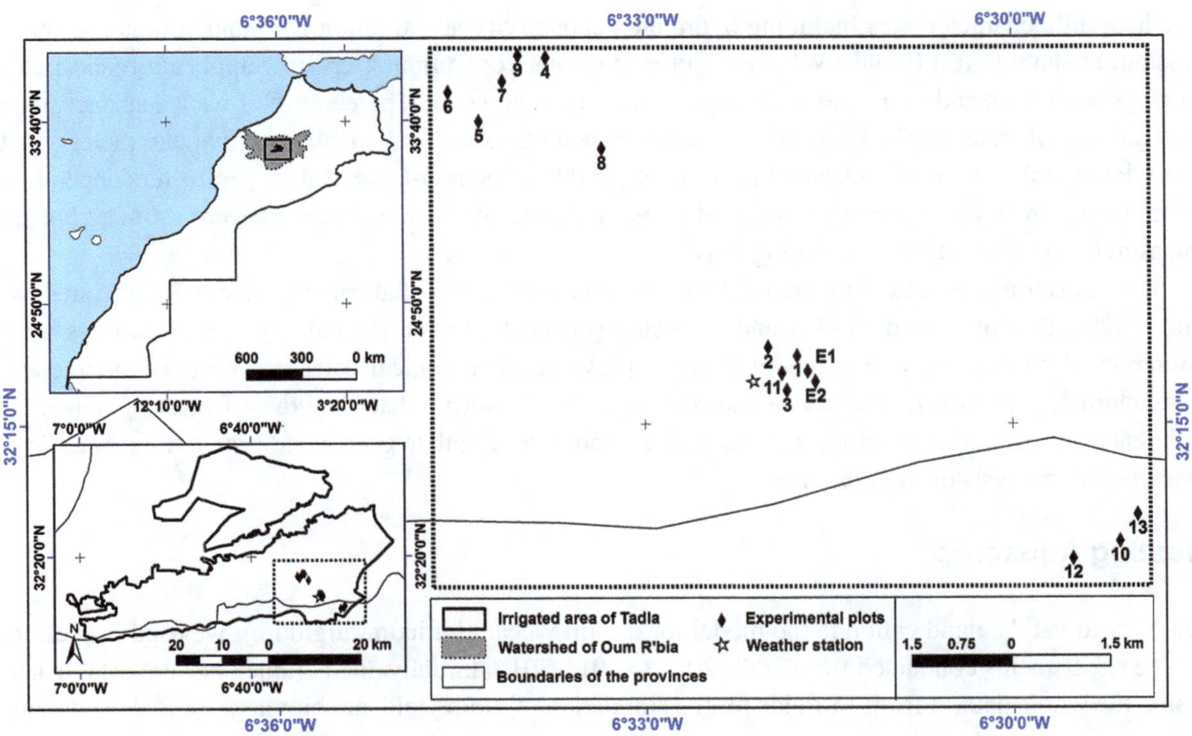

All the studied plots were harvested 10-15 days after physiological maturity, and the grain yield was measured. Aboveground biomass quadrats of 1 $m^2$ were cut at ground level with three replicates per plot. The collected samples were placed in the oven at a temperature of 65 °C for 48 hours to get the dry aboveground biomass (Iqbal et al., 2014).

We quantified the water provided at plot level for each irrigation supply. Soil moisture was measured using gravimetric methodology in two of the fifteen fields, E1 and E2. The measurements were dried in an oven at 105°C for 24 h and made every 10-15 days and at depths of 0-80 cm, during the 2009-2010 growing season for E1 and during the 2010-2011 growing season for E2. The measurements were performed in three replications by sampling.

## Aquacrop: Presentation and Parameterization

AquaCrop is a FAO crop model (Steduto et al., 2009) that simulates crop and soil response to water stress under various climatic, soil and crop management conditions. Crop yield is estimated as the product of dry biomass by the harvest index (HI). At the flowering stage, HI increases linearly as a function of time after a latent phase, up to near physiological maturity. Transpiration is calculated and then translated into biomass using biomass water productivity, evaporative demand and air CO2 concentration data (Allen, Pereira, Raes, & Smith, 1998; Steduto et al., 2009).

AquaCrop (v4.0), used in this study, is structured so as to integrate the soil-plant-atmosphere continuum. It consists of five components: i) weather component, which requires five types of data input – daily maximum and minimum air temperatures (T), daily rainfall, daily reference evapotranspiration and the mean annual CO2 concentration in the atmosphere; ii) crop input parameters (planting dates, plant density, growth phenology and aerial canopy); iii) soil component, which is configured as an independent system of variable depth, with one or several horizons of varying texture compositions, and the hydraulic characteristics including hydraulic conductivity at saturation (Ks) and volumetric water content at saturation, FC and PWP; iv) field management conditions (fertilizer application and field-surface practices); and v) irrigation management (irrigation method, percentage of wetted surface, date and amount of water applied). AquaCrop provides default values, called conservative parameters, that are related to the crops being studied (Raes et al., 2009a). The non-conservative parameters depend on crop management and environmental conditions. In this study, these parameters were estimated using measured data from all three cropping seasons.

Default settings proposed by the model was used to estimate initial canopy cover (CCo) from sowing rate, seed weight, seed number and estimated germination rate. The canopy expansion rates were automatically estimated by the model after we provided phenological dates, i.e. (date of emergence, maximum canopy cover, senescence and maturity. The flowering date, length of flowering stage, HI reference and HI build-up period were specified in order to calculate grain yield production. AquaCrop was run on the basis of calendar days.

## Testing Aquacrop

In order to validate and calibrate the model for durum wheat (Triticum turgidum), we used the results of the experiments conducted from 2009-2010 to 2011-2012. Initially, the parameters were established using the whole dataset from 15 fields for the simulation of grain yield and biomass; for soil water content we used only the soil moisture measurements carried out in the E1 and E2 fields. Accuracy of the model was evaluated using the k-fold cross validation (k-fold CV) approach, given the small size of our dataset (< 30 observations). This approach uses k replicate samples of observation data, builds model with (k-1)/k of data and tests with the remaining 1/k. Where the number of observations was reduced, leave-one-out cross validation (LOOCV), which is a k-fold CV taken to its extreme, with K equal to the number of systematic repetitions, was used (Cawley & Talbot, 2003, 2004). In our study, it involved using a single observation from the 15 observations as the validation data, and the other observations as the training data. This was repeated such that each observation in the sample was used once for validation (Stone, 1974).

The conservative parameters of crop growth in winter soft wheat and durum wheat are presumed to be close as they display quite similar physiological behavior (El Hafid, El Mourid, Samir, & Bakoulou, 1996). On this basis, we initially used the conservative parameters of soft wheat following the AquaCrop manual annexes (Raes et al., 2009b) to derive the conservative parameters of durum wheat. This approach led us to adjust only parameters known to be very sensitive for wheat in AquaCrop, i.e. the parameters for canopy decline coefficient (CDC), canopy growth coefficient (CGC) and maximum root water extraction in the top and bottom quarters of the root zone in order to calibrate the model so that it could predict durum wheat yields under semi-arid irrigated conditions.

## Model Evaluation

For evaluating the performance of the AquaCrop, field measurements (observed data) were compared with outputs generated by the model (simulated data) in terms of crop yield, biomass and soil moisture in the root zone.

Different statistical indices were used to compare modeled with observed values. These indices were the coefficient of determination ($R^2$), the root mean square error (RMSE) and the normalized RMSE (nRMSE; expressed as a percentage; Loague and Green (1991).

$$RMSE = \left[ \sum_{i=1}^{n} \frac{\left(Si - Oi\right)^2}{n} \right]^{0.5} \tag{1}$$

$$nRMSE = \left[ \sum_{i=1}^{n} \frac{\left(Si - Oi\right)^2}{n} \right]^{0.5} \times \frac{100}{M} \tag{2}$$

Si and Oi refer to simulated and observed values of the studied variable, respectively; n is the number of observations; and M is the mean of the observed variable. The nRMSE indicates the accuracy of the model and the dispersion around the mean of the observed values. The simulation is considered to be excellent when the nRMSE value is lower than 10%, good if it is higher than 10% but lower than 20%, fair if it is higher than 20% but lower than 30%, and poor if it is higher than 30% (Jamieson, Porter, & Wilson, 1991).

## RESULTS AND DISCUSSIONS

### Grain Yield

Figure 6 presents the relationship between observed and modeled durum wheat grain yields. The simulated and measured yields showed a good correlation, with an $R^2$ value of 0.97. The calculated model evaluation criteria between the simulated and measured yields were: RMSE = 0.20 t.ha-1 and nRMSE = 4.06%.

Using the LOOCV method, we obtained $R^2$ values of 0.95. The difference between the values of the linear regression of the dataset and the values of the LOOCV method validation data were minimal, confirming the ability of the model to simulate yields. The differences between the observed and simulated outputs of LOOCV for durum wheat yield were: RMSE = 0.23 t.ha-1 and nRMSE = 4.67%.

Andarzian et al. (2011) simulated soft wheat grain yield using AquaCrop and obtained a $R^2$ of 0.95 and a nRMSE of 5%. The difference between the observed and modeled grain yield of soft wheat was 0.14 t.ha-1, indicating that the model overestimated the yield by 2.7%.

### Soil Water Content

Figures 7-E1 and 7-E2 present the comparison between observed and simulated soil water content of the whole profile for the two fields where this variable was measured. The soils in these plots have similar characteristics. However, when we apply the model at large-scale it advisable to integrate spatial variability of soil data, in order to improve simulation accuracy. An important point arising from these figures is that the variations in simulated soil moisture followed the variation of rainfall and the occurrence of irrigation events exactly, indicating that the model is sensitive enough to be used as an efficient tool to monitor irrigation in real time.

*Figure 6. Relationship between observed and simulated durum wheat grain yield (t.ha-1) using the whole datasets (Calibration)*

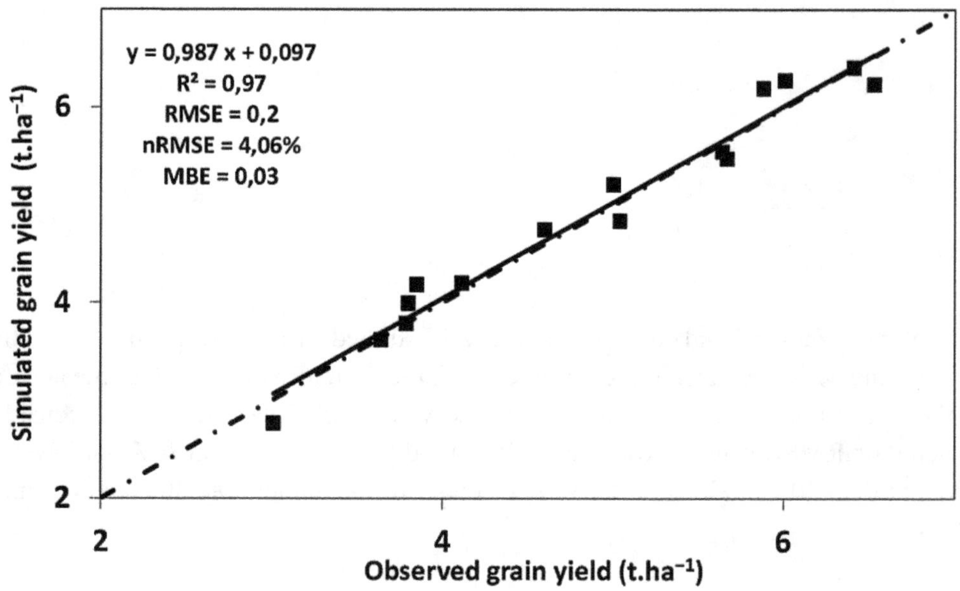

*Figure 7. Comparison between simulated and observed soil moisture measurements at 0-100 cm depth for plots E1 and E2 followed during the cropping seasons 2009/2010 and 2010/2011, respectively. Descendent arrows ↓ indicate irrigation water supply*

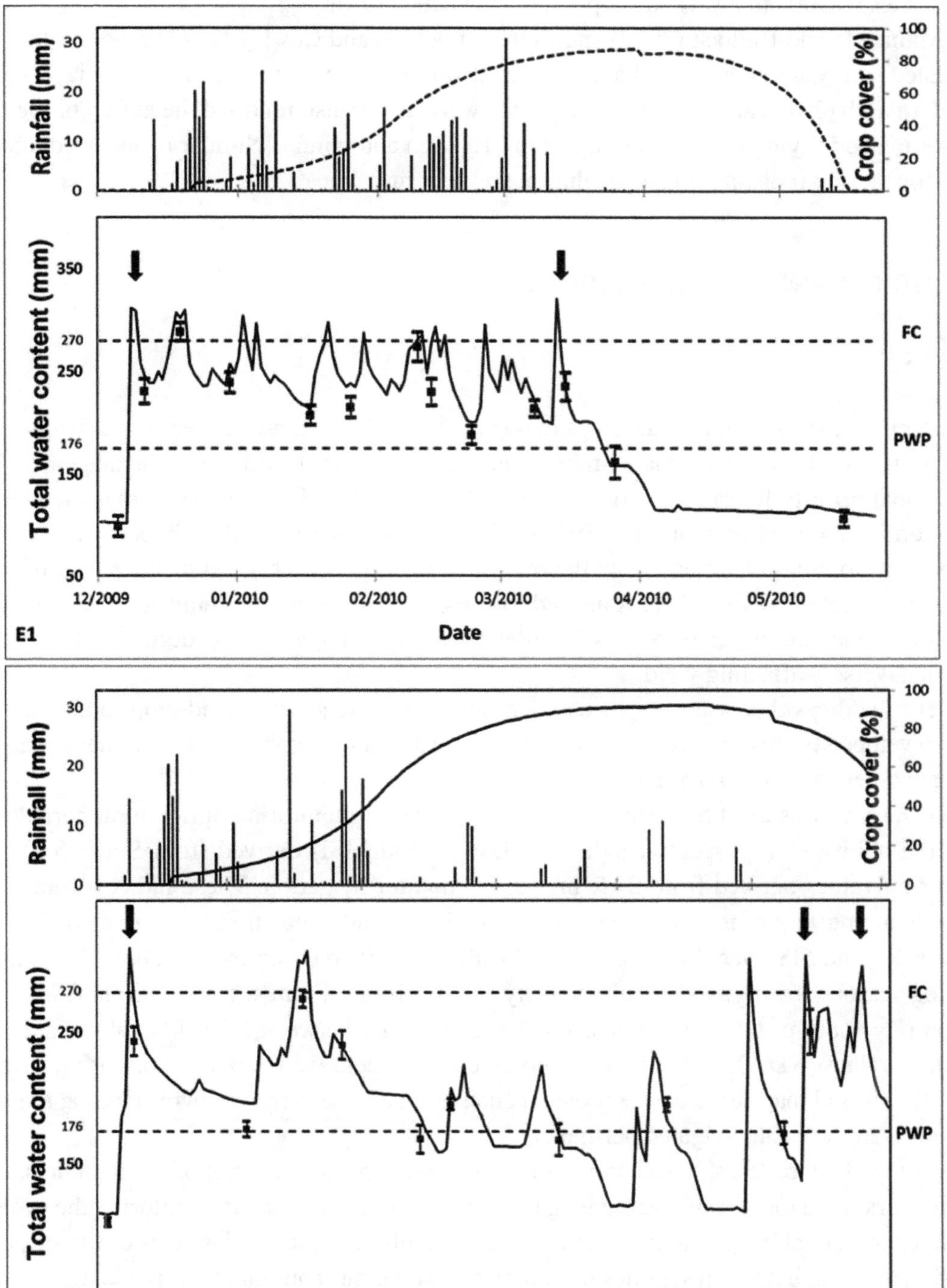

The statistical indices also showed the ability of the model to simulate total soil water content. The RMSE, nRMSE and R2 values were 17.68 mm, 8.55% and 0.95 for the E1 plot and 15.73 mm, 7.90% and 0.94 for the E2 plot, respectively. The model simulated well the variation in soil water content in the soil profiles (0-100 cm layer) and captured its tendency to vary.

Both Mkhabela and Bullock (2012) and Zeleke, Luckett, and Cowley (2011) reported that AquaCrop simulated soil water content well and the way it varied in the root zone (0-100 cm layer). Hussein, Janat, and Yakoub (2011) and Farahani, Izzi, and Oweis (2009) also reported the ability of the model to predict wetting and drying events resulting from irrigation and rainfall. Similar results were obtained in our study for the soil moisture profile evolution under durum wheat.

## CONCLUSION AND PERSPECTIVES

### Conclusion

Cereals are by far the most important crops in terms of added value, area covered and food security. In Morocco, water availability is the main limiting factor for cereal production and its management has long been a national priority for the agricultural sector. In terms of agricultural water management, flooding irrigation is practiced on more than 93% of the total area of irrigated perimeters in Morocco and mobilizes large volumes of water. Given the importance of this sector, good management of irrigation water at a large-scale is essential. This has led to work on developing optimum strategies for planning and managing available water resources in order to improve irrigation scheduling and prevent water stress from adversely affecting yield.

In order to address this issue, we focused on optical remote sensing; and crop modeling. The approaches developed in this research are intended primarily for decision-makers and managers of large-scale irrigated perimeters (40,000 ha).

Remote sensing was used because of its high potential in monitoring agricultural parameters. We analyzed the ability of two spectral indices ($NDWI_{Rog}$ and MSI) derived from SPOT-5 images and backscattering values derived from SAR images to monitor irrigation. These indices were compared with corresponding *in situ* measurements of soil moisture and vegetation water content in 30 wheat fields. $NDWI_{Rog}$ and MSI were highly correlated with the *in situ* measurements at both the beginning of the growing season (sowing) and at full maturity (grain filling). From sowing to grain filling, the best correlation ($R^2$=0.86; p<0.01) was found for the relationship between $NDWI_{Rog}$ values and observed soil moisture values. $NDWI_{Rog}$ can therefore be used as an operative index for monitoring irrigation in order to estimate and map surface water content changes at the main crop growth stages at the field and regional levels in the Tadla irrigated perimeter.

The developed approaches, based on remote sensing combined with crop growth models, could be used as an operational tool for managing irrigation and crops, as well as for monitoring the evolution of surface water content at the plot and irrigation scheme levels. The practical aspects of these approaches include: (i) triggering irrigation supplies in water stress situations and otherwise preventing excess supplies of irrigation water; and (ii) detecting illegal irrigation and pumping. This is relevant in irrigated areas where irrigation is not scheduled and uncontrolled water pumping is prohibited.

These approaches will help to improve irrigation monitoring and management in the Tadla irrigation. They will also directly contribute to the sustainability of agricultural production systems and preserve water resources (groundwater and surface).

The performance of the approaches presented need be checked in other contexts before judging their suitability for application in other areas.

The high resolution remote sensing approaches provide a means of synoptic monitoring for agricultural parameters at the time of image acquisition, but do not ensure their temporal monitoring. It is common for there to be more than 10 days between two satellites passes. In order to achieve maximum yield and improve the water-use efficiency in irrigation, it is necessary to control wheat production parameters in time and space. Resorting to crop models is justified mainly by the need to provide temporal monitoring and to assess the impact of the production parameters monitored on yields with a view to overcoming the limits of remote sensing.

The field crop model (AquaCrop v4.0) developed by FAO was chosen for this research. AquaCrop was adjusted and tested to simulate durum wheat yields and the temporal evolution of soil moisture status, as well as to assess the impact of irrigation water supplies (timing and quantity) on recorded yields. The required conservative parameters of the model were determined on the basis of predefined parameters for soft wheat. Only four of these parameters had to be adjusted. The use of the statistical leave-one-out cross-validation (LOOCV) method gave excellent results for the main model outputs. The comparison between observed and simulated grain yield and above-ground biomass gave a normalized root mean square error (RMSE) of 4.06% (200 $kg.ha^{-1}$) and 5.71% (820 $kg.ha^{-1}$), respectively. Similarly, the difference between observed and modeled soil water content was 16.7 mm on average.

The analysis of irrigation scenarios showed that the model can optimize the frequency and timing of irrigation water supplies. It could help to maintain good grain yields and increase water-use efficiency by avoiding periods of water stress and overwatering, thereby guaranteeing optimal growing conditions throughout the cropping season.

New prospects are opening up for improving crop modeling by integrating weather forecasts into the process of decision-making and by adopting spatial modeling that uses data layers on a grid format derived from satellite data. Decision-makers and managers need to be encouraged to adopt spatial and temporal monitoring techniques for better scheduling of irrigation supplies, improving water saving and avoiding supplying irrigation water when rainfall is sufficient. These approaches constitute an important step in establishing an effective irrigation management system across the studied irrigated perimeter.

Our research has provided methods to help stakeholders and policy-makers in their management and decision-making. Each method was applied independently.

## Perspectives

Our research sought to provide approaches to irrigation and surface water content monitoring in large-scale irrigated areas that would take account of the space-time dimension through the application of optical remote sensing and crop modeling tools.

With regard to remote sensing, we consider that these tools do have operational potential in irrigation management. The growing interest in satellite imagery is linked to the rise in acquisition frequency, the falling price of satellite images (optic and RADAR) and, very recently, the availability of free high-resolution satellite images and attractive revisit frequency.

*Figure 8. Schematic diagram illustrating the integration of remote sensing and crop modeling into crop yield forecasting models*

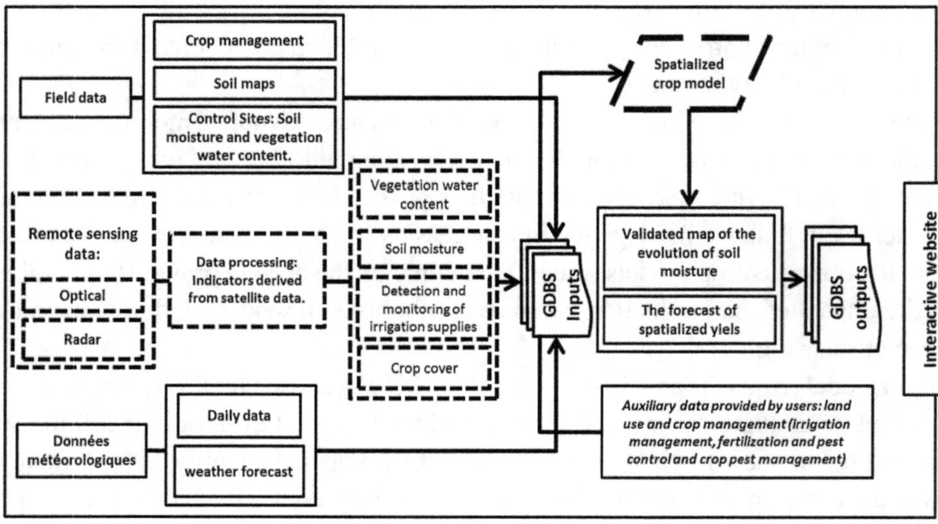


Remote sensing tools are one of the best ways to monitor large agricultural areas. Research should be done on improving the mastery and application of satellite imagery in agriculture. These tools would help in the control and monitoring of irrigation water supplies and in assessing surface water content changes.

Optical imagery could also be used for water stress warning by detecting plots with under-threshold water content that are probably suffering from water stress. This situation generally arises from the limited availability of water in the soil. In order to intervene in time, there needs to be a short gap between image acquisition and image processing and high acquisition frequency of satellite images.

This limitation of remote sensing data heightens the importance of modeling that will simulate the daily evolution of crop development and soil water content if daily input data are available. Crop growth models can also predict crop yield according to soil and weather conditions and irrigation schedule. At the plot level, AquaCrop could be a key factor in optimizing growing conditions throughout the growing season in order to maintain good grain yields and improve water-use efficiency, while avoiding excess water or water deficit situations.

The integration of remote sensing and crop modeling into crop yield forecasting models could lead to a significant improvement in wheat yield forecasts at the national, regional and local levels (Figure 8). In irrigated areas, where information on irrigation times and quantities supplied plot-by-plot is often lacking. Hence the interest of combining data on spatial (from satellite imagery) and temporal evolution (derived from crop models).

The possibilities for continuing to develop these tools and test their applicability to wheat and other important crops in other irrigated areas with different climatic and edaphic conditions are being studying. The outcome of this process could be an operational tool based on a system that integrates crop models and optical and radar remote sensing data. Such a system could contribute to improving the control of spatial variability and offer integrated management and monitoring in time and space, as well as providing valuable information to decision-makers, managers and stakeholders to ensure the proper monitoring of production parameters at the perimeter scale.

## ACKNOWLEDGMENT

This work was supported by the Belgian Technical Cooperation Agency and the National Institute for Agronomic Research of Morocco. The acquisition of SPOT-5 images was supported by the ISIS program of the National Centre for Space Studies (CNES). The authors are indebted to the staff of the Regional Centre for Agronomic Research of Tadla, for their collaboration, as well as to Mr Mohamed Saef, Mr Hamid Riani and Mr Rachid Amediaz from ORMVAT of Tadla for their valuable help and Mr Hassan Noury who participated to field measurements.

## REFERENCES

Allen, R. G., Pereira, L. S., Raes, D., & Smith, M. (1998). *Crop Evapotranspiration-Guidelines for Computing Crop Water Requirements* (Vol. 56). Rome: FAO.

Andarzian, B., Bannayan, M., Steduto, P., Mazraeh, H., Barati, M. E., Barati, M. A., & Rahnama, A. (2011). Validation and testing of the AquaCrop model under full and deficit irrigated wheat production in Iran. *Agricultural Water Management*, *100*(1), 1–8. doi:10.1016/j.agwat.2011.08.023

Ben-Gal, A., Kool, D., Agam, N., van Halsema, G. E., Yermiyahu, U., Yafe, A., ... Dag, A. (2010). Whole-tree water balance and indicators for short-term drought stress in non-bearing 'Barnea' olives. *Agricultural Water Management*, *98*(1), 124–133. doi:10.1016/j.agwat.2010.08.008

Benabdelouahab, T., Balaghi, R., Hadria, R., Lionboui, H., Djaby, B., & Tychon, B. (2016). Testing Aquacrop to Simulate Durum Wheat Yield and Schedule Irrigation in a Semi-Arid Irrigated Perimeter in Morocco. *Irrigation and Drainage*, *65*(5), 631–643. doi:10.1002/ird.1977

Benabdelouahab, T., Balaghi, R., Hadria, R., Lionboui, H., Minet, J., & Tychon, B. (2015). Monitoring surface water content using visible and short-wave infrared SPOT-5 data of wheat plots in irrigated semi-arid regions. *International Journal of Remote Sensing*, *36*(15), 4018–4036. doi:10.1080/01431161.2015.1072650

Bi, H., Li, X., Liu, X., Guo, M., & Li, J. (2009). A case study of spatial heterogeneity of soil moisture in the Loess Plateau, western China: A geostatistical approach. *International Journal of Sediment Research*, *24*(1), 63–73. doi:10.1016/S1001-6279(09)60016-0

Bouchaou, L., Michelot, J. L., Qurtobi, M., Zine, N., Gaye, C. B., Aggarwal, P. K., ... Vengosh, A. (2009). Origin and residence time of groundwater in the Tadla basin (Morocco) using multiple isotopic and geochemical tools. *Journal of Hydrology (Amsterdam)*, *379*(3–4), 323–338. doi:10.1016/j.jhydrol.2009.10.019

Cawley, G. C., & Talbot, N. L. C. (2003). Efficient leave-one-out cross-validation of kernel fisher discriminant classifiers. *Pattern Recognition*, *36*(11), 2585–2592. doi:10.1016/S0031-3203(03)00136-5

Cawley, G. C., & Talbot, N. L. C. (2004). Fast exact leave-one-out cross-validation of sparse least-squares support vector machines. *Neural Networks*, *17*(10), 1467–1475. doi:10.1016/j.neunet.2004.07.002 PMID:15541948

Ceccato, P., Flasse, S., & Grégoire, J. M. (2002). Designing a spectral index to estimate vegetation water content from remote sensing data: Part 2. Validation and applications. *Remote Sensing of Environment, 82*(2–3), 198–207. doi:10.1016/S0034-4257(02)00036-6

Ceccato, P., Flasse, S., Tarantola, S., Jacquemoud, S., & Grégoire, J. M. (2001). Detecting vegetation leaf water content using reflectance in the optical domain. *Remote Sensing of Environment*, *77*(1), 22–33. doi:10.1016/S0034-4257(01)00191-2

Ceccato, P., Gobron, N., Flasse, S., Pinty, B., & Tarantola, S. (2002). Designing a spectral index to estimate vegetation water content from remote sensing data: Part 1: Theoretical approach. *Remote Sensing of Environment*, *82*(2–3), 188–197. doi:10.1016/S0034-4257(02)00037-8

Cheng, T., Rivard, B., Sánchez-Azofeifa, A. G., Féret, J. B., Jacquemoud, S., & Ustin, S. L. (2012). Predicting leaf gravimetric water content from foliar reflectance across a range of plant species using continuous wavelet analysis. *Journal of Plant Physiology*, *169*(12), 1134–1142. doi:10.1016/j.jplph.2012.04.006 PMID:22608180

Doorenbos, J., & Kassam, A. H. (1979). Yield response to water. FAO Irrigation and Drainage, 33, 193.

Doorenbos, J., & Pruitt, W. O. (1977). *Guidelines for predicting crop water requirements* (Vol. 24). Rome: FAO.

Duchemin, B., Hadria, R., Erraki, S., Boulet, G., Maisongrande, P., Chehbouni, A., ... Simonneaux, V. (2006). Monitoring wheat phenology and irrigation in Central Morocco: On the use of relationships between evapotranspiration, crops coefficients, leaf area index and remotely-sensed vegetation indices. *Agricultural Water Management*, *79*(1), 1–27. doi:10.1016/j.agwat.2005.02.013

El Hafid, R., El Mourid, M., Samir, K., & Bakoulou, B. (1996). Caractérisation de certaines variétés de blé, orge et du triticale sous différentes situations hydriques en conditions de champ et simulées. *Al awamia, 93*, 7-26.

Farahani, H. J., Izzi, G., & Oweis, T. Y. (2009). Parameterization and Evaluation of the AquaCrop Model for Full and Deficit Irrigated Cotton All rights reserved. *Agronomy Journal*, *101*(3), 469–476. doi:10.2134/agronj2008.0182s

Feng, H., Chen, C., Dong, H., Wang, J., & Meng, Q. (2013). Modified Shortwave Infrared Perpendicular Water Stress Index: A Farmland Water Stress Monitoring Method. *Journal of Applied Meteorology and Climatology*, *52*(9), 2024–2032. doi:10.1175/JAMC-D-12-0164.1

Geerts, S., & Raes, D. (2009). Deficit irrigation as on-farm strategy to maximize crop water productivity in dry areas. *Agricultural Water Management*, *96*(9), 1275–1284. doi:10.1016/j.agwat.2009.04.009

Ghulam, A., Qin, Q., Teyip, T., & Li, Z. L. (2007). Modified perpendicular drought index (MPDI): A real-time drought monitoring method. *ISPRS Journal of Photogrammetry and Remote Sensing*, *62*(2), 150–164. doi:10.1016/j.isprsjprs.2007.03.002

Hadria, R., Duchemin, B., Jarlan, L., Dedieu, G., Baup, F., Khabba, S., ... Le Toan, T. (2010). Potentiality of optical and radar satellite data at high spatio-temporal resolutions for the monitoring of irrigated wheat crops in Morocco. *International Journal of Applied Earth Observation and Geoinformation, 12*(1), S32–S37. doi:10.1016/j.jag.2009.09.003

Hsiao, T. C., Heng, L., Steduto, P., Rojas-Lara, B., Raes, D., & Fereres, E. (2009). AquaCrop: The FAO crop model to simulate yield response to water. III. Parameterization and testing for maize. *Agronomy Journal, 101*(3), 448–459. doi:10.2134/agronj2008.0218s

Hunt, E. R. Jr, & Rock, B. N. (1989). Detection of changes in leaf water content using Near- and Middle-Infrared reflectances. *Remote Sensing of Environment, 30*(1), 43–54. doi:10.1016/0034-4257(89)90046-1

Hussein, F., Janat, M., & Yakoub, A. (2011). Simulating cotton yield response to deficit irrigation with the FAO AquaCrop model. *Spanish Journal of Agricultural Research, 9*(4), 1319–1330. doi:10.5424jar/20110904-358-10

Iqbal, M. A., Shen, Y., Stricevic, R., Pei, H., Sun, H., Amiri, E., ... del Rio, S. (2014). Evaluation of the FAO AquaCrop model for winter wheat on the North China Plain under deficit irrigation from field experiment to regional yield simulation. *Agricultural Water Management, 135*(0), 61–72. doi:10.1016/j.agwat.2013.12.012

Jamieson, P. D., Porter, J. R., & Wilson, D. R. (1991). A test of computer simulation model ARC-WHEAT1 on wheat crop grown in New Zealand. *Field Crops Research, 27*(4), 337–350. doi:10.1016/0378-4290(91)90040-3

Kogan, F. N. (2000). *Contribution of Remote Sensing to Drought Early Warning*. Paper presented at the Early Warning Systems for Drought Preparedness and Drought Management.

Lasaponara, R., & Masini, N. (2012). *Satellite Remote Sensing: A New Tool for Archaeology*. Springer. doi:10.1007/978-90-481-8801-7

Lionboui, H., Benabdelouahab, T., Elame, F., Hasib, A., & Boulli, A. (2016). Multi-year agro-economic modelling for predicting changes in irrigation water management indicators in the Tadla sub-basin. *International Journal of Agricultural Management, 5*(4), 96–105. doi:10.5836/ijam/2016-05-96

Lionboui, H., Benabdelouahab, T., Hasib, A., & Boulli, A. (2016). Analysis of Farms Performance Using Different Sources of Irrigation Water: A Case Study in a Semi-Arid Area. *International Journal of Agricultural Management and Development, 6*(2), 145–154.

Lionboui, H., Fadlaoui, A., Elame, F., & Benabdelouahab, T. (2014). Water pricing impact on the economic valuation of water resources. *International Journal of Education and Research, 2*(6), 147–166.

Loague, K., & Green, R. E. (1991). Statistical and graphical methods for evaluating solute transport models: Overview and application. *Journal of Contaminant Hydrology, 7*(1-2), 51–73. doi:10.1016/0169-7722(91)90038-3

Lobell, D. B., Asner, G. P., Ortiz-Monasterio, J. I., & Benning, T. L. (2003). Remote sensing of regional crop production in the Yaqui Valley, Mexico: Estimates and uncertainties. *Agriculture, Ecosystems & Environment, 94*(2), 205–220. doi:10.1016/S0167-8809(02)00021-X

Mkhabela, M. S., & Bullock, P. R. (2012). Performance of the FAO AquaCrop model for wheat grain yield and soil moisture simulation in Western Canada. *Agricultural Water Management, 110*(0), 16–24. doi:10.1016/j.agwat.2012.03.009

Moreno, A., Maselli, F., Chiesi, M., Genesio, L., Vaccari, F., Seufert, G., & Gilabert, M. A. (2014). Monitoring water stress in Mediterranean semi-natural vegetation with satellite and meteorological data. *International Journal of Applied Earth Observation and Geoinformation, 26*(0), 246–255. doi:10.1016/j.jag.2013.08.003

Muller, E., & Décamps, H. (2000). Modeling soil moisture-reflectance. *Remote Sensing of Environment, 76*(2), 173–180. doi:10.1016/S0034-4257(00)00198-X

Najine, A., Jaffal, M., Khammari, K. E., Aïfa, T., Khattach, D., Himi, M., ... Aqil, H. (2006). Contribution de la gravimétrie à l'étude de la structure du bassin de Tadla (Maroc): Implications hydrogéologiques. *Comptes Rendus Geoscience, 338*(10), 676–682. doi:10.1016/j.crte.2006.04.015

Ning, Z., Yang, H., Qiming, Q., & Lu, L. (2013). VSDI: A visible and shortwave infrared drought index for monitoring soil and vegetation moisture based on optical remote sensing. *International Journal of Remote Sensing, 34*(13), 4585–4609. doi:10.1080/01431161.2013.779046

ORMVAT. (2009). *Rapport annuel de l'Office de la mise en valeur agricole de Tadla.* ORMVAT.

Penuelas, J., Pinol, J., Ogaya, R., & Filella, I. (1997). Estimation of plant water concentration by the reflectance Water Index WI (R900/R970). *International Journal of Remote Sensing, 18*(13), 2869–2875. doi:10.1080/014311697217396

Raes, D., Steduto, P., Hsiao, T. C., & Fereres, E. (2009a). AquaCrop:The FAO Crop Model to Simulate Yield Response to Water: II. Main Algorithms and Software Description. *Agronomy Journal, 101*(3), 438–447. doi:10.2134/agronj2008.0140s

Raes, D., Steduto, P., Hsiao, T. C., & Fereres, E. (2009b). *AquaCrop:The FAO Crop Model to Simulate Yield Response to Water: Reference Manual Annexes.* Academic Press.

Rogers, A. S., & Kearney, M. S. (2004). Reducing signature variability in unmixing coastal marsh Thematic Mapper scenes using spectral indices. *International Journal of Remote Sensing, 25*(12), 2317–2335. doi:10.1080/01431160310001618103

Seckler, D., Barker, R., & Amarasinghe, U. (1999). Water scarcity in the twenty-first century. *Water: Research and Development, 15*, 29–42. doi:10.1080/07900629948916

Skidmore, E. L., Dickerson, J. D., & Shimmelpfennig, H. (1975). Evaluating surface-soil water content by measuring reflectance. *Soil Science Society of American Proceedings, 39*(2), 238–242. doi:10.2136 ssaj1975.03615995003900020009x

Song, X., Zhou, X., Li, X., & Li, X. (2009). *Spatial Heterogeneity Analysis of Soil Moisture Based on Geostatistics*. Paper presented at the Information Science and Engineering (ICISE), 2009 1st International Conference on.

Steduto, P., Hsiao, T. C., Raes, D., & Fereres, E. (2009). Aquacrop-the FAO crop model to simulate yield response to water: I. Concepts and underlying principles. *Agronomy Journal*, *101*(3), 426–437. doi:10.2134/agronj2008.0139s

Stone, M. (1974). Cross-Validatory Choice and Assessment of Statistical Predictions. *Journal of the Royal Statistical Society. Series B (Methodological), 36*(2), 111-147. doi: 10.2307/2984809

Tian, Q., Tong, Q., Pu, R., Guo, X., & Zhao, C. (2001). Spectroscopic determination of wheat water status using 1650-1850 nm spectral absorption features. *International Journal of Remote Sensing*, *22*(12), 2329–2338. doi:10.1080/01431160118199

Trombetti, M., Riaño, D., Rubio, M. A., Cheng, Y. B., & Ustin, S. L. (2008). Multi-temporal vegetation canopy water content retrieval and interpretation using artificial neural networks for the continental USA. *Remote Sensing of Environment*, *112*(1), 203–215. doi:10.1016/j.rse.2007.04.013

Wang, Y. G., Zhu, H., & Li, Y. (2013). Spatial heterogeneity of soil moisture, microbial biomass carbon and soil respiration at stand scale of an arid scrubland. *Environmental Earth Sciences*, *70*(7), 3217–3224. doi:10.100712665-013-2386-z

Zeleke, K. T., Luckett, D., & Cowley, R. (2011). Calibration and Testing of the FAO AquaCrop Model for Canola. *Agronomy Journal*, *103*(6), 1610–1618. doi:10.2134/agronj2011.0150

# Chapter 9

# Land Use Land Cover Dynamics in Ribb Watershed and Its Implication to the Sustainability of Ribb Dam: South Gondar, Ethiopia

**Yidnekachew Jember**
*Woldia University, Ethiopia*

## ABSTRACT

*Land use land cover dynamics is a widespread phenomenon in many parts of Ethiopia and in Ribb watershed. The main objective of the research was assessing land use land cover dynamics and its implication to the sustainability of Ribb Dam in 1973, 1986, 2001, and 2016 by using Landsat image and household questioner. During the last 44 years, cultivated and settlement land and forest cover showed an increment from 26.29% to 54.89% and 9.45% to 12.86%, respectively. The bush land, grazing land, water body, and wetland, however, showed a relative decrement from 29.48% to 17.09%, 21.45% to 12.70%, 4.64% to 2.39%, and 8.70% to 0.08%, respectively. Population pressure, poverty, weak policy and institutional enforcement, and tenure insecurity revealed as a major cause of the change in land use land cover. Soil erosion, lack of fuel wood, and impact on livelihood are major consequences of land use land cover change.*

## INTRODUCTION

Our use of land has a profound effect and results in an observable pattern in the land use land cover (LULC) over time (Opeyemi, 2006). As human kind alters the land surface, the landform, hydrologic budget, and stream processes are ultimately affected (DeBarry, 2004). The mix of landforms of Ethiopia, which ranges from high mountains to rugged and flatlands, and the rainfall regime coupled with various geological factors have contributed to the formation of various sizes of a watershed in the country.

DOI: 10.4018/978-1-5225-5939-9.ch009

Ethiopia has a total land area of about 1.1 million square kilometers and is divided into 12 river basins. Tana sub-basin comprises three main watersheds, namely Megech-Seraba (352,522.93ha), Gumara-rib (471,236.18ha), and Gilgel Abay (453,618.53 ha) (Tadesse and Solomon, 2014).

DeBarry (2004) said that a watershed is an area of land that captures water in any form, such as rain, snow, or dew, and drains it to a common water body. Watersheds face great challenges and they are subjected to LULC change. Watersheds face a range of degradation challenges associated with human activities. Also, the way in which they are managed can have a profound cascading effect on natural resources and communities in the wider basin (Bach *et al.*, 2011).

Analysis of LULC dynamics helps us to understand the characteristics and interdependencies of the components that constitute spatial systems. This research used Geographic information system (GIS) and Remote sensing (RS) techniques (in conjunction with a socio-economic and historical assessment) to analyze the dynamics of Ribb watershed LULC and its implications to the sustainability of Ribb dam.

The study attempted to gain a better understanding of LULC changes in Ribb watershed. The study is significant because the Ribb watershed is an area with many development activities, which includes Ribb dam construction. The irrigation scheme is planned to irrigate about 14,460 hectares of land and benefits about 11,500 households. There are also wetlands in the watershed, which made the research very important. The study recognizes that accurate information on LULC dynamicity, the cause (and consequence), the perception of households on LULC and the implication of LULC change on the sustainability of Ribb dam are essential to advance and support development interventions. These sorts of data are also important to decision makers in terms of developing strategies for natural resource management and monitoring of environmental changes in the study area and areas that have a similar geographic setting.

## STUDY AREA AND METHODOLOGY

### Study Area Description

The Ribb watershed, which is part of Lake Tana sub-basin, has an area of about 1972 km$^2$. Geographically Ribb Watershed is located in northwestern Ethiopia and stretches between 10°43'11.53'' N latitude and 37°35'50.22'' E. The area has a chain of the mountain, and at the low-lying area, it has wide flat floodplain of Fogera. Figure 1 shows the location of the Ribb watershed in Ethiopia, Amhara region, South Gondar Zone and five bounding *woredas* (districts).

Ribb irrigation project is intended to transform rain-fed subsistence agriculture into irrigated commercial agriculture. It is proposed to irrigate up to 14,460 ha of land found in the watershed and about 11,500 households are expected to be benefited from the irrigation scheme. The water stored behind the Ribb Dam (which is under construction) is planned to be released into the Ribb River and then distributed by gravity. This is expected to allow improved cropping in the dry season. Associated drainage and flood protection measures are also expected to improve conditions for wet season agriculture.

### Data Source and Method of Data Analysis

The major data that was used in this study was Landsat images, household data and ground control points collected using handheld GPS. To analyze the change in LULC in the watershed, Land sat MSS, TM,

*Figure 1. Location map of the study area*

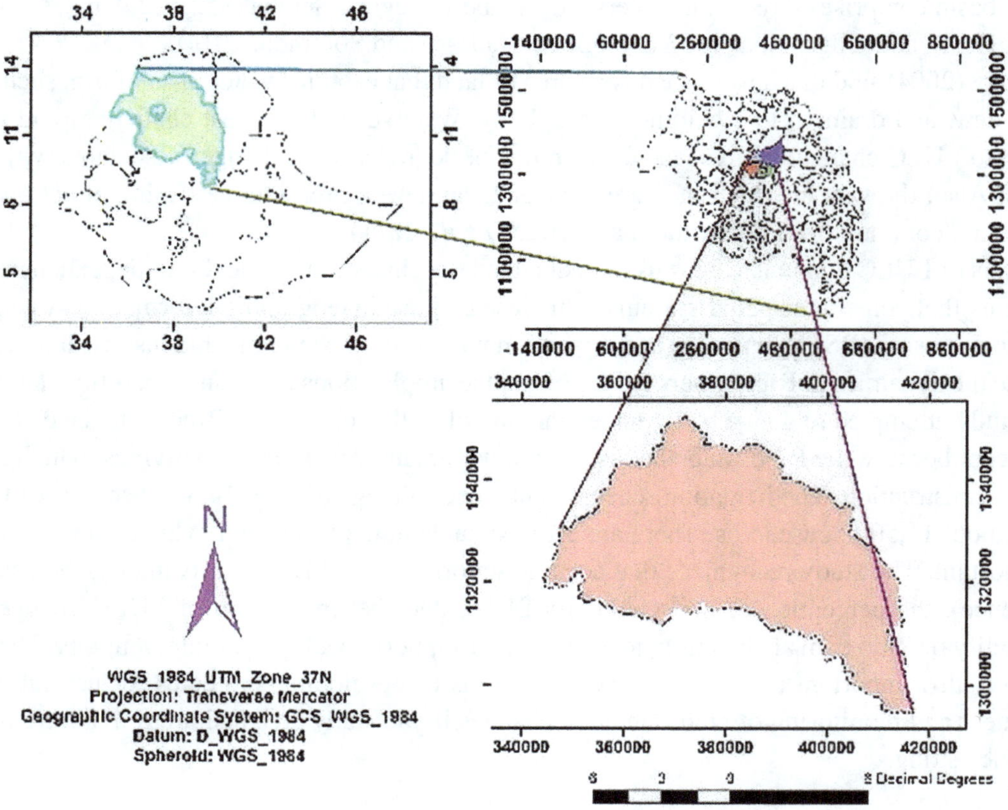


ETM+ and OLI satellite data of 1973, 1986, 2001 and 2016 (see Table 1) were downloaded freely from https://earthexplorer.usgs.gov/. The process of detecting LULC from the satellite images was prepared using these satellite data, together with field observation, general historical information and unsupervised classification. Each image has been layer stacked based on the band they have and since MSS and TM and OLI images have different resolutions, they were re-sampled to the same size using the nearest neighbor resampling technique. In order to obtain the real radiance or reflectance, General contrast adjustment with linear method was applied to each image. Based on the image characteristics the resolution merges image enhancement techniques were applied on 2001 and 2016 image. This technique was applied in the above selected year because they have panchromatic band 8 with the resolution of 15 m. Training sample for classification is 10 for each LULC class. The main reason to decide this number of samples

*Table 1. Description of Satellite Images*

| Landsat Types | Sensor | Resolution | Acquisition Date | Sources/Website |
|---|---|---|---|---|
| Landsat | MSS | 60m | 01/14/1973 | www.earthexplorer.usgs.gov |
| Landsat 5 | TM | 30m | 01/12/1986 | www.earthexplorer.usgs.gov |
| Landsat 7 | ETM+ | 30m, 15m pan | 01/01/2001 | www.earthexplorer.usgs.gov |
| Landsat 8 | OLI | 30m,15m pan | 02/16/2016 | www.earthexplorer.usgs.gov |

is the nature of the data. Parametric decision rule with maximum likelihood classifier was selected as it allows classification of all the pixels in scenes and avoids the probability of assigning a single pixel in more than one class. When finishing supervision classification, the manual-work interpretation was done in the image classified, and the precision of classification result was assessed.

## RESULT AND DISCUSSION

The LULC maps of the Ribb watershed for four reference years and description of the different LULC types were presented in (Figure 2 and Table 2). For the purpose of this study, coupled with field observation and general historical information FAO's land cover classification is used.

### Change in Bushland

Bushland in Ribb watershed occupied 58133.50ha in 1973 and this value decreased to 43389.60ha by 1986. Later in 2001 its decreeing also continued and reached 28800.7ha and showed increment by 33709.70ha at 2016. The net loss in bushland is 24423.80ha between 1973-2016 (Table 3). The improvement in coverage of bushland found during the third period of analysis (2001-2016) more likely related to the introduction and expansion of area closure introduced by the government. Other studies corroborate the finding of this research for Ribb watershed. Nurelegn and Amare (2014) showed that in Ribb watershed during the period between 1973-2011 bushland shown decreasing trend, however, increased in 2011. These changes were attributed to population growth which forced the farmers to till and expand their lands to a greater extent than before to cope up with the conditions and to sustain their life.

### Change in Cultivated and Settlement Land

Cultivated and settlement land occupied 5181.36ha in 1973 and increased to 70998.90ha in 1986 later the area coverage of cultivated and settlement land increased by 105645ha and 108234ha in 2001 and 2016 respectively. Cultivated and settlement land net gained 56390.40ha of land from all other land use land cover class (Table 3). The increasing change of cultivated and settlement land between the year

*Table 2. Land use land cover class and their description*

| Code | Class | General Description |
|---|---|---|
| 1 | BL | Shrub and sub- afro-alpine vegetation at the fringes of forest cover |
| 2 | C&S | Areas including intensively, moderately and sparsely cultivated land with standing crop, tree crops, and crop and include towns, villages, |
| 3 | FC | Areas of dense natural forest and plantation forest with tree cover |
| 4 | GL | Areas of natural/semi-natural grassland with shrub-grass |
| 5 | WB | Areas with surface water in the form of rivers, ponds, lakes, streams, and reservoirs. |
| 6 | WL | Areas covered by a marsh which is seasonal as well as permanent in nature |

Source: Di Gregorio, 2016.
BL= Bushland; C&S=Cultivated and settlement land; FC=Forest cover; GL=Grazing land; WB=Water body; WL=Wetland

*Figure 2. Land Use Land Cover Map from 1973- 2016*

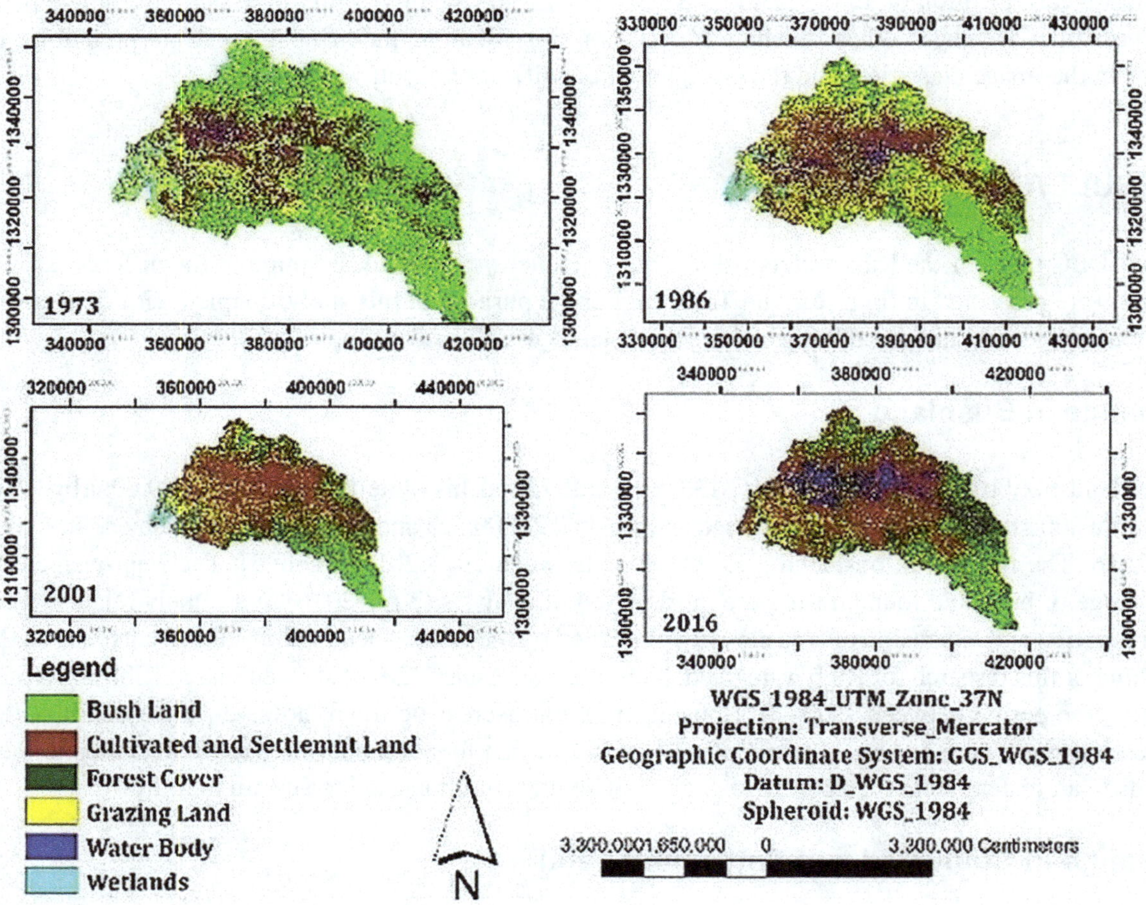


1973 and 1986 is more likely related to historical land reform proclamation of March 1975 by Derg's regime. The proclamation declares land as a collective property of people, redistributed land to farmers and abolishes the system of tenancy. Peasant Association was created and charged to distribute communal land to the farmers. The expansion of farmland is directly related to population growth, weak environmental conservation policy. As stated by Moges *et al.,* (2015) after studying the multi-temporal land use land cover change dynamics of Blue Nile basin; the trend of land use land cover has shown a tendency towards more land being brought under cultivated and built up between the 1973 and 2014.

Similarly, according to Nurelegn and Amare (2014) in the Ribb River watershed, there was a continued expansion of cultivated and settlement land over years which has brought a significant decrease in water bodies, forest, and bushland land use land cover classes.

## Change in Forest Cover

Between the year 1973 and 2016 forest cover showed great dynamics of increasing and decreasing. Forest cover showed a decrement in size between 1973 and 1986 study year (Table 3). The relative decrement in forest cover and bushland is more likely related to the restriction posed by the government on farmer not to cut their trees unless approved by peasant association committee. This restriction seems to have a positive effect on the protection of forest and bushlands. But this restriction and others discouraged private tree planting activities. After 1986 the area coverage of forest showed relative increment. The main reason for this related to private tree plantation for commercial purpose and forest conservation area closure practice by government and different NGOs in the watershed. This result also evident in another related study by Moges *et al.,* (2015) that forest land of Koga watershed happens to increase from 2001 classification map to 2014 and high amount of cultivation land converted to Eucalyptus tree plantation because of the framer's perception over the economic advantage of Eucalyptus.

## Change in Grazing Land

Grazing land showed a decrement between 1973 and 2016 study year. The net lost area by grazing land constituted 17259.94ha (Table 3). This result shows us the scarcity of farmland and illegal cultivation of grazing land related to population growth and lack of well-planned management in the area. The overall land use land cover change analysis of the Landsat image of the year 1986 and 2013 in Gumara-Ribb watershed by Tadesse and Solomon (2014) showed that grazing land has declined drastically by 30.52%. However, cultivated land has rapidly increased by 11.65%. The main driving forces for change in grazing land is illegal encroachment for agriculture, overgrazing leads to barren land, distributing grazing land to landless and construction of infrastructure on communal grazing lands.

## Change in Waterbody

Even if water body showed a considerable amount of increment in the year between 2001 and 2016 (Table 3), its area is decreasing in the study year between 1973 and 2016 from 9144.36ha to 4715.89ha. The net loss observed in this class is about 4428.47ha. There were farm ponds made by the farmers. Constructing an embankment across a water resource and excavation of a pit and conserving rainwater using geomembrane technology is the recent phenomenon practiced in the watershed. This is more likely the main reason for the increase of water body between 2001 and 2016.

*Table 3. Summary of land use land cover change between 1973 and 2016*

| Class | Area (ha) | | | | Net Gain/ Loss (ha) |
|---|---|---|---|---|---|
| | 1973 | 1986 | 2001 | 2016 | 1973-2016 |
| BL | 58133.50 | 43389.60 | 28800.7 | 33709.70 | -24423.80 |
| C&S | 51843.60 | 70998.90 | 105645 | 108234 | 56390.40 |
| FC | 18626.40 | 14695.60 | 22653.1 | 25352.80 | 6726.40 |
| GL | 42296.80 | 52929.57 | 32209.6 | 25036.86 | -17259.94 |
| WB | 9144.36 | 9091.98 | 1159.99 | 4715.89 | -4428.47 |
| WL | 17155.30 | 6094.35 | 6731.56 | 150.75 | -17004.55 |
| Total | 197200 | 197200 | 197200 | 197200 | |

## Change in Wetland

Wetland showed the dramatic change between the study year of 1973-2016. The net loss observed in this class is about 17004.55ha. The efforts of the government at Derg's regime to improve agricultural sector through the establishment of cooperatives and the cultivation of cash crops contribute to the increment of cultivated and settlement land in the watershed and decrease of wetlands found in Fogera plains.

Generally, in the course of the study year cultivated and settlement land, and forest cover showed increment. While bushland, grazing land, water body and wetland showed a relative decrement in the course of the study year from 1973-2016 (Figure 3).

## Accuracy Assessment

The most common and typical method used by researchers to assess classification accuracy is the use of an error matrix. Therefore, Table 4 shows details of the accuracy assessment done for the study areas. The results revealed that higher classification accuracies were obtained.

These accuracy levels are very close to those reported by Nurelegn and Amare (2014). The users' accuracy that was the percentage of correctly classified from total classified shown 100% for Bushland and 67% for the wetland. The overall classification accuracy was 87.18%. Conditional Kappa statistics(K^) for each Category were 1, 0.9378, 0.6777, 0.7341, 1, and 0.6486 bushlands, cultivated and settlement land, forest cover, grazing land, water body and wetland respectively and the overall kappa statistics was 0.8439 with the overall classification accuracy 87.18%. Since the results obtained from classification and the validation statistics for this study were higher than the minimum validation threshold, it was reasonable to employ the derived maps for further land use land cover dynamics assessment in the watershed.

*Figure 3. Land use land cover dynamics in Ribb watershed between 1973-2016*

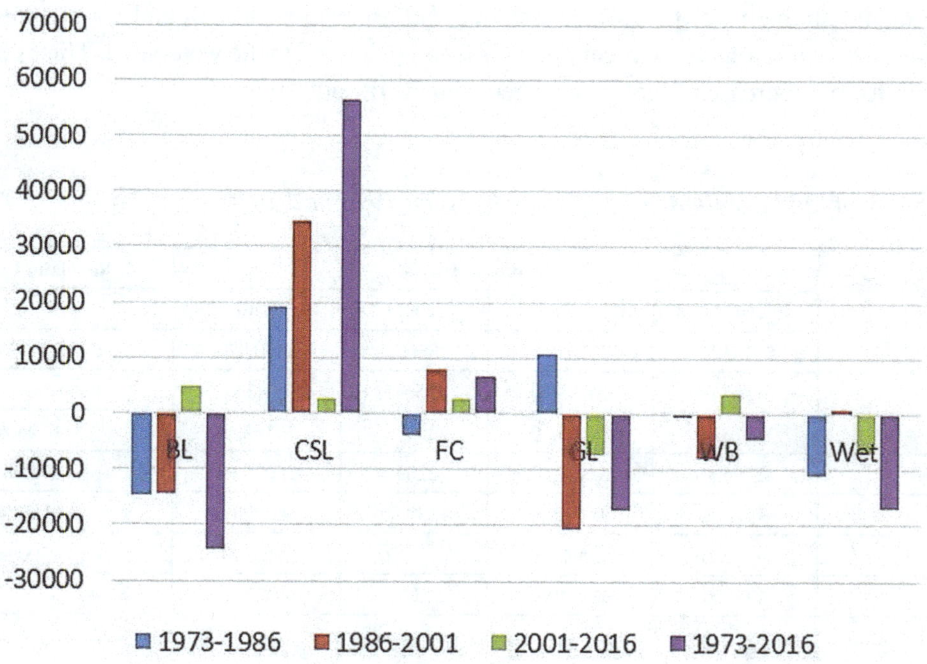

*Table 4. Confusion matrix and classification accuracy assessment*

| Classified Data | Reference Data | | | | | | | |
|---|---|---|---|---|---|---|---|---|
| | **BL** | **C&S** | **FC** | **GL** | **WB** | **WL** | **Total** | **User Accuracy** |
| BL | **20** | 0 | 0 | 0 | 0 | 0 | **20** | 100% |
| C&S | 1 | **19** | 0 | 0 | 0 | 0 | **20** | 95% |
| FC | 5 | 1 | **16** | 0 | 0 | 0 | **22** | 72.73% |
| GL | 1 | 2 | 2 | **17** | 0 | 0 | **22** | 77% |
| WB | 0 | 0 | 0 | 0 | **24** | 0 | **24** | 100% |
| WL | 1 | 1 | 0 | 0 | 1 | **6** | **9** | 67% |
| Total | **28** | **23** | **18** | **17** | **25** | **6** | **117** | |
| Producer Accuracy% | 71.43 | 82.61 | 89 | 100 | 96 | 100 | | |
| Overall accuracy | 87.18% | | | | | | | |

## RESPONSIBLE CAUSE FOR WATERSHED LAND USE LAND COVER DYNAMICS

### Population Pressure

The research by Moges *et al* (2015) in Koga watershed revealed that population pressure and limited income diversity are one of the major driving forces for the change of land use land cover.

It was observed that to get agricultural land the local community is encroaching into the grazing and bushland. Due to the high rate of population pressure and encouragement of government to expand agricultural production farmland expands. Every year a considerable amount of new farmland was created. Accordingly, the respondents witnessed that farmland expansion is the major factor of land use land cover change on the watershed due to population pressure. Based on the key informant interview and observation irrigation agriculture is increasing through time due to the introduction of water pumping technology and more farmers tend to irrigate using the streams found in the watershed.

This finding is also supported by other researchers. As confirmed by Malthus (1989), population growth is a function of agricultural productivity and is considered as a major cause of increasing demands for food, fuelwood, fodder, and other ecosystem services.

### Poverty

Poverty is linked to land use land cover change is that the poor overuse the natural resource to escape from poverty. Economic activity is the consequence of human efforts to improve the quality of life, the outputs of which are determined by natural resources (Nelson *et al.*, 2006). Land redistribution, which in recent years has been the only means of formally acquiring access to land to accommodate new households.

People often face additional challenges when population growth is recorded, including low economic growth and limited means of income generation outside the utilization of local ecosystem services (Pare, 2008). The amount of income generated by individual households ranged between the US $150 and the US $800/annum with a mean income of US $460 which is less than the US $2 a day. Additional income mostly generated by young farmers by working on commercial farms outsides the watershed during agricultural off-season. Overexploitation of natural resource without using proper management techniques cause depletion of natural resources. Therefore, poverty is one of the major reasons that caused land use land cover change in the watershed.

## Policy and Institutional Factor

Access to land, labor, capital, technology, and information is structured by local and national policies and institutions. Three different governments have been in power in Ethiopia since 1972, and the policies implemented by each have directly affected land use land cover. Before 1974, the relationship between land users and owners was based on a feudal system (Desta *et al.*, 2000). Under which the ownership of land was limited to a few individuals, and most inhabitants could access farmland only through sharecropping. The population density was relatively low, and the fallow land was common. Likely the classification result of 1973 land use land cover status of the watershed related to the system of the regime followed before the study year.

Much of the expansion of cultivated and settlement land and loss of forest cover and bushlands took place during the 1970s and 1980s. This is more likely related to the reform of 1975 which confiscated all rural lands and distributes to the rural tenants. At the time the arable communal land was abundant in most place of the watershed and the population pressure was low as a result of which poorly drained soils and hilly areas were not turned to cultivation. However, as result of frequent redistribution and the later land reform act of 1997 particularly enacted in Amhara region, more fragile land areas were converted to cultivation. The land redistribution of 1997 was mainly from communal grazing lands.

From the empirical evidence, farmer lack access to credit and most of the credit goes to buy agricultural input not to expand off-farm activities. Inaccessibility to market and road facilities coupled with the insecurity of tenure identified as the major institutional factor that causes land use land cover change.

Many land use land cover change is due to ill-defined policies and weak institutional enforcement, as explained by household respondents and focus group discussants in the Ribb watershed. Even if the current government enacted different policies and regulations to conserve the natural resource, their implementation is in the infant stage. Discussion with government officials revealed weak implementation of land administration and use policies. Among ill-defined policies in agricultural sector raised by the discussant is that the state policy which tends to encourage converting bush lands, grazing land and wetlands found in the watershed to attain food sufficiency. This policy implemented by concerned agricultural sector especially in the watershed by converting bushlands, grazing land and wetlands found in the watershed.

## Tenure Insecurity

Land tenure policy has a significant impact on the change of LULC for a specific area. The dynamics of land management has been aggravated by land tenure systems and the governmental agricultural policy that banned the sale and purchase of land. Farmers only have usufruct land rights. The present government allows access to the land only by lease, sharecropping, and borrowing. At present, the main way to gain access to land in the watershed is by inheritance and sharecropping. Because the land is owned by the state, farmers have feared to invest in long-term SWC measures, as the land might be taken by the government and redistributed to others. Even if all respondents confirmed having land holding certificate only 63% of them acknowledged holding of the certificate as an incentive for the management of their land.

## CONSEQUENCES OF LULC DYNAMICS ON THE SOCIOECONOMIC SETTING AND THE ENVIRONMENT

### Soil Erosion

The major problem faced in the watershed due to land use land cover dynamics especially by a decrement of bushland and forest cover are soil erosion and flood hazard problems. The work of Aklilu and de Graaff (2006), substantiate the above statement by stating the Ethiopian highlands have been experiencing declining soil fertility and severe soil erosion due to intensive farming on steep and fragile land.

The rate and severity of soil erosion and land degradation in the watershed depend on the pattern of land use. There is the removal of bushlands for the various purpose by the community.

In Ribb watershed, the majority of the cultivated land is used for the production of cereal crops. Unsustainable land use practices and improper management systems prevailed in the watershed create the dynamic change of land cover in the watershed and this, in turn, have played a significant role in causing high soil erosion rates, sediment transport and loss of agricultural nutrients.

### Lack of Fuelwood

In Ethiopia, the most important sources of energy are fuelwood, charcoal, cow dung and crop residues with fuelwood contributing about 85% of the energy source (EFAP, 1994). This shows that fuelwood and cow dung are the two principal sources of household energy. Obviously, agricultural production is negatively affected by the use of crop residue for these purposes instead of using it as a means of increasing organic fertilizer. Fuelwood collected from around the homestead, bushlands and forest lands were the main sources of fuelwood in the long past. Currently, these land cover types have greatly shrunk and are severely devastated based on the survey and GI and RS analysis result.

As population increases household energy consumption also increases. In the study area fuelwood is not only a source of energy but a means of income generation too.

Without adequate alternative sources of energy, population growth increases the demand for fuelwood, which in turn leads to the destruction of forests. It also contributes to the use of crop residues and dung for fuel rather than using them as sources of organic fertilizer to improve the already poor soils.

### Impact on Livelihood

Major environmental problem and livelihood change are the result of land use land cover change. The socio-economic activities of an area play an important role in the land use land cover type as well as on land degradation. In relation to this, economic activities of the study area are based on primary productions such as crop farming and animal husbandry, which indicate the agrarian background. The survey result revealed that the household livestock holding is low and more concentrated in sheep and goats. As the focus group, discussion confirmed there are feed scarcity and frequent drought that affect the livestock holding of the households. The existing livestock population is beyond the carrying capacity of the poorly managed communal resource pool. As a result, the productivity of livestock farming is low and in most cases, it is still in decline.

Agriculture which is the mainstay of the society is being largely practiced in garden and plotlands. Highly intensive farming is clearly observed in all part of the watershed. Crops such as teff, barley, wheat, maize, bean and different kinds of crops are grown in line with seasonal changes and depending on the normal condition of climate.

Reduction of cropland per capita and insecurity have led to a reduction in activities such as fallowing, planting trees and investing in conservation structures, while a reduction in cropland per capita has caused cropping and grazing activities to be shifted to hillsides and ecologically fragile areas.

### Implications of Watershed LULC Dynamics to the Sustainability of Ribb Dam

Watersheds are sensitive to LULC dynamics induced by human activities. This dynamic change will have strong implication on the sustainability of the dam by affecting river-flow and sediment-yield if the trend continues in its current way. This statement also supported by MoWR (2002) as the problem of land degradation is a threat and devastating challenge to the proposed Ribb dam reservoir and downstream areas due to generating high runoff discharges and imposing huge sediment yield, which may result in reducing water storage capacity of the dam reservoir, unless the upper watershed is treated with appropriate watershed management interventions and strategies.

Expansion of agriculture, urbanization, deforestation and the day to day activities of the community in the watershed resulted in a temporal and spatial change in land use land cover and land use and the cover condition can affect the hydrological balance of the watershed by changing magnitude and pattern of runoff, peak flow, and groundwater levels. It is well known that deforestation causes changes in soil properties and infiltration rates, which ultimately affects the soil erosion processes and hydrological cycle of the catchment. Average annual sediment deposition estimated by Tensay (2011) and Estifanos (2014) to the Ribb dam is 72.79 ton/Km and 39.8 tonnes/ha/year.

There was extreme sedimentation case in Ethiopia such as Borkena Dam in wollo, which cost $35 million US dollar in 1991 and Adrako Dam (Ebnat) South Gondar where the dead storage volume of the reservoir silted up before their construction ended (Nigussie *et al.*, 2006). The key informant and focus group discussant in Ebnat woreda also indicate the evidence of sedimentation and siltation on previously constructed irrigation dam in the woreda.

In the Ribb watershed, gully formation (with exposure of rock and stones on previously cultivated steep-upper slopes) are the most visible evidence that shows erosion problems. Unmitigated land use land cover change leads to degradation of forest or woodland and these have an impact on catchment processes and biochemical cycles and leads to soil erosion and water shortage in the watershed. It is urgent that this effect is stopped in the watershed, especially in areas already affected by water shortage, and areas extending to Lake Tana.

## CONCLUSION

The identification of Spatiotemporal watershed land use land cover change showed that Ribb watershed had experienced significant land use land cover dynamics over the past 44-years. Among the class of land use land covers, which show a large alteration, is cultivated and settlement land that has increased dramatically. This led to the conclusion that there is a high trend of changing bushlands, grazing lands, and wetlands into cultivated and settlement land (as a result of land developments by farmers and government).

The growing population in the watershed put high pressure on the existing land use land cover of the watershed. Reinforced by the underlying cause of poverty, weak policy and institutional enforcement and, tenure insecurity result in the problem of soil erosion, lack of fuelwood and deterioration of livelihood through time. Land use practices and improper management systems have played a significant role in causing high soil erosion rates, sediment transport and loss of agricultural nutrients. If not managed and monitored properly, the problem of land use land cover change in the watershed will have a great impact on the sustainability of Ribb dam.

Based on the findings of this study, it is recommended the land use land cover dataset to monitor the change should be developed. Policies related to population, agricultural productivity, and environment with the aim of sustainability should be devised to address the environmental, social and economic problem in the watershed.

## ACKNOWLEDGMENT

Thanks to God. His love, righteousness, mercy, and glory have kept me on the right track through so many struggles. I would like to express my genuine gratitude and thank Dr. Tadesse Amsalu for all immeasurable support, motivation, and guidance during the whole process of my studying. The support and guidance I received from Dr. Gebeyehu Belay, Dr. Berhanu Kefele and Birtukan Fikadie are highly appreciated.

## REFERENCE

Estifanos Abera. (2014). *Assessment of micro- watershed vulnerability for soil erosion in Ribb watershed using GIS and Remote Sensing*. Author.

Amsalu, A., & de Graaff, J. (2006). Farmers' views of soil erosion problems and their conservation knowledge at Beressa watershed, central highlands of Ethiopia. *Agriculture and Human Values*, *23*(1), 99–108. doi:10.100710460-005-5872-4

Amsalu, T., & Addisu, S. (2014). *Assessment of Grazing Land and Livestock Feed Balance in Gummara-Rib Watershed, Ethiopia*. Academic Press.

Atnafu, Dejen, & Vijverberg. (2011). Assessment of the Ecological Status and Threats of Welala and Shesher Wetlands, Lake Tana Sub-Basin (Ethiopia). *Journal of Water Resource and Protection*, *3*(7), 540–547.

Bach, H., Clausen, T. J., Trang, D. T., Emerton, L., Facon, T., Hofer, T., ... Whiting, L. (2011). *From local watershed management to integrated river basin management at National and Transboundary levels. Lao PDR*. Mekong River Commission.

DeBarry, P. A. (2004). *Watersheds: Processes, Assessment and Management*. John Wiley & Sons, Inc.

Desta, L., Kassie, M., Benin, S., & Pender, J. (2000). *Land degradation and strategies for sustainable development in the Ethiopian highlands: Amhara Region*. Socio-economic and Policy Research working paper 32. International Livestock Research Institute, Nairobi, Kenya.

Di Gregorio, A. (2016). *Land Cover Classification System: Classification concepts*. Software version 3. Available at: http://www.fao.org/3/a-i5232e.pdf

Donia, N. (2012). Application of remotely sensed imagery to watershed analysis a case study of Lake Karoun, Egypt. *Arabian Journal of Geosciences*, *6*(9), 3217–3228. doi:10.100712517-012-0621-7

Eldho, P. (2012). *Module 4 – (L12 -L18) Watershed modeling*. Available at http://nptel.ac.in/courses/105101010/downloads/Lecture12.pdf

Ethiopian Forestry Action Plan (EFAP). (1994). Final Report Volume II –The Challenges for Development and Volume III – Issues and Action. EFAP Secretariat.

Tensay Getnet. (2011). *Sedimentation Modeling for Ribb Dam*. Author.

Haregeweyn, N., Poesen, J., Nyssen, J., De Wit, J., Haile, M., Govers, G., & Deckers, S. (2006). Reservoirs in Tigray (Northern Ethiopia): Characteristics and sediment deposition problems. *Land Degradation & Development*, *17*(2), 211–230. doi:10.1002/ldr.698

Malthus, T. R. (1989). On Population (First Essay on Population, 1798, and Second Essay on Population, 1803). Modern Library, for Random House.

Moges Kidane, B., Amare Sewnet, M., & Alemu Bezie, D. (2015). Multi-temporal Land Use Land Cover Change and Dynamics of Blue Nile Basin by Using GIS and Remote Sensing Techniques, North-Western Ethiopia. *International Journal of Environmental Sciences*, *4*(2), 81–88.

MoWR (Ministry of Water Resources). (2002). *Water Sector Development Program*. Addis Ababa, Ethiopia: Author.

Nelson, G. C., Bennett, E., Berhe, A. A., Cassman, K., DeFries, R., Dietz, T., ... Zurek, M. (2006). Anthropogenic Drivers of Ecosystem Change: An Overview. *Ecology and Society*, *11*(2), 29p. doi:10.5751/ES-01826-110229

Nurelegn Mekuriaw, G., & Amare Sewnet, M. (2014). Land Use/Cover Dynamics in Ribb Watershed, North Western Ethiopia. *Journal of Natural Sciences Research, 4*.

Opeyemi, Z.A. (2006). *Change detection in land use land cover using Remote sensing data and GIS: A case study of Ilorin and its environs in Kwara State*. Academic Press.

Paré. (2008). *Land Use Dynamics, Tree Diversity, and Local Perception of Dry Forest Decline in Southern Burkina Faso, West Africa* (Doctoral Thesis). Department of Forest Genetics and Plant Physiology, Umeå, Sweden.

## ADDITIONAL READING

Bweuket, W. (2002). Land Cover Dynamics Since the 1950s in Chemoga Watershed, Blue Nile Basin, Ethiopia. *Mountain Research and Development*, *22*(3), 263–269. doi:10.1659/0276-4741(2002)022[0263:LCDSTI]2.0.CO;2

Rahmato, D. (1999). Revisiting the Land Issue: Options for Change. *Economic Focus*, *2*(4), 9–11.

## KEY TERMS AND DEFINITIONS

**Land:** Land is any definable area of the Earth's terrestrial surface involving all attributes of the biosphere immediately above and below this surface.

**Land Use:** The utilization of land cover type by human activities for the purpose of agriculture, forestry, settlement, and pasture.

**Watershed:** Area of land that drains water to a common point along a river, stream, pond, lake, estuary, ocean, or other water bodies.

# Chapter 10
# A Geo-Decisional Tool for the Management of the Agricultural Development Fund Under the Green Morocco Plan

**Fatima Ezzahra Adous**
*Institut Agronomique et Vétérinaire Hassan II, Morocco*

**Mohammed Abdelwafi El Ghouat**
*Institut Agronomique et Vétérinaire Hassan II, Morocco*

**Fatiha Ibannain**
*ANCFCC, Morocco*

## ABSTRACT

*Within the agricultural strategy called Green Morocco Plan launched by the Moroccan government, the funds are allocated by the agricultural development fund to encourage investments. In this context, to manage the grant requests, the agricultural department uses an information system, which processes the files from their submission until the payment of the farmers. However, this system does not take into account the spatial data. This research aims to develop a geospatial model to study the distribution and dynamics of the allocated fund. Through this approach, policymakers will have a strategic key to survey the behavior of the spatial component, to manage funding applications, and to better manage allowed funds. The methodological approach adopted consists of carrying out a mapping of the real world identifying the needs of the future system and modeling the process of the geo-spatial tool. The result of the model will be used to develop and implement a spatial tool for policymakers to improve land information management related to the funded agricultural project.*

DOI: 10.4018/978-1-5225-5939-9.ch010

## INTRODUCTION

According to its objectives, the Green Morocco Plan (PMV) aims to redesign the process of farmer support to focus on real actors. To allocate financial funds as grants to farmers, the PMV Department of Agriculture has adopted and adapted a distinction on the type of actors. These grants are assigned within the framework of the agricultural development fund (FDA) to promote agricultural projects in various sectors according to the addressed needs of each category of farmers.

The addressed dossiers requesting funds are managed with an information system, which deals exclusively with data attributes. It does not, however, consider spatial data needed to locate clients. The spatial data would carry out the thematic map analysis to establish a frame for decision making related to requested grants.

The projected solution aims to implement a geo-decisional tool to meet the needs of the decision makers. It will provide an overall visibility of the land subject to these grants, as well as enhance information management between stakeholders.

The approach for developing the tool requires new platforms to enable high performance. This includes the following geo-spatial technologies: (1) geographic information system (GIS) to collect, visualize, analyze, store, and interpret data, as well as understand relationships (ESRI, 2014); (2) global navigation satellite system (GNSS) to provide autonomous geo-spatial positioning; and (3) remote sensing. GIS technology deals with both informational and functional aspects to represent, manage, and manipulate spatial objects' links to attribute data. The purpose of the system is to provide users and decision makers with a strategy to deploy, share, and update data to facilitate decision-making. It will also offer up-to-date geographic information related to the FDA's investment over the territory.

The adopted approach focuses on the development of four areas. The first evaluates existing organizational and procedural structures, adopted grants, existing tools, and corresponding data. This defines the needs of the expected solution. The second is related to the analysis of the first area's results. The third concerns the design and modeling of the geospatial tool in the object-oriented approach. The final key develops the strategy to implement the geo-spatial decision support tool to facilitate the management of FDA grants within the PMV.

The FDA was established to enable incentive investments of agriculture projects and promote farmers' ambitions. It has faced several reforms to meet PMV requirements.

When managing and monitoring requested grants, the department of agriculture uses an information system to generate summary reports, planned investments, and grant dashboards. However, the system does not provide spatial-related data although the spatial information may afford tools to increase control of grant eligibility. With spatial information, managers could establish a spatiotemporal analysis to clearly understand grant allocation and distribution. This approach would require new skills and high-performance platforms based on geospatial technologies, including GNSS and Web mapping tools.

The expected result of this research is to highlight potential offerings of spatial data to obtain answers to the following challenges:

1. Lack of spatial information in running systems
2. Thematic analysis based on summary reports and statistics
3. Absence of the spatial component in existing funding-request systems (existing systems focus on attribute data through computerized information systems)

## METHODOLOGY AND MATERIALS

The PMV strategy involves several directions and partners with the farmer as the target. Two pillars are developed in the PMV framework (Ministry of Agriculture, Fisheries, Rural Development and Water and Forests, 2010):

1. **Modern Agriculture:** Based on new technology and modern strategies, the first pillar enhances productivity and improves competitiveness of the agricultural sector.
2. **Social Agriculture:** Based on the development of social agriculture through social projects, the second pillar improves existing traditional agriculture and reduces poverty, especially in hard areas.

To successfully implement these pillars, the Agricultural Department has defined support policies for sustainable growth. These sector policies are: (1) water; (2) land; (3) exchange policy; (4) public-private partnership; and (4) business that support famers through funding and subsidies.

### Agricultural Subsidies in Morocco

The FDA aims to improve the monitoring and management of accorded subsidies to agricultural projects. Editing has occurred to achieve PMV goals. Furthermore, the new agricultural subsidies system:

1. Introduces new grants in accordance with PMV goals
2. Reinforces grants allocated to agricultural sectors
3. Encourages farmers to invest in agriculture

Management of grant requests (through collection and assessment) is provided via a unique hub (UH) and collaboration with regional- and provincial-level technical services (TS). These local unit services offer efficient and improved services to the farmers (MAFRDWF-ISD, 2013).

When allocating grants, the following process has been adopted (MAFRDWF, 2011):

1. A farmer who wants to apply for a grant provides the UH with an application dossier.
2. The UH records and forwards the application to the TS.
3. The TS performs a pre-project inventory check.
4. If the request is eligible, the UH issues the farmer a prior authorization to carry out the project.
5. After completion of the project, the farmer returns supporting documents to the UH. These documents must justify an engaged investment.
6. The TS conducts an audit study through a field inspection of the project area.
7. After the UH validates the investment dossier, it performs a payment order to allow the farmer to get the grant.

## GOVERNANCE ANALYSIS OF THE GEO-DECISIONAL SYSTEM

### Strategic Guidelines

As mentioned, the strategic guidelines of the PMV are based on two pillars. The department of agriculture is responsible for developing and implementing this agricultural strategy by (Moroccan Government, 2013):

1. Establishing incentive measures to encourage investment in the agricultural sector
2. Collecting, analyzing, and distributing statistical and agricultural information

### Geo-Decisional Governance System of the Project Management

This study mainly used the control objectives for information and related technology (COBIT) related to the governance of information systems. This standardized method aims to conceive, develop, publish, and promote a globally-recognized and up-to-date framework for governance monitoring of information systems. It is based on major axes, including strategic integration, value creation, risk management, efficiency assessment, and resource management (www.afai.fr).

Governance of the developed geo-decisional system consists of defining and managing processes according to PMV guidelines.

1. **Strategic Integration:** The fundamental issue is to provide a system in line with the strategic objectives of the agricultural department. The strategic integration of the system is in accordance with the agricultural strategy. It develops a support system for farmers to finance their investments (based on the needs of each farmer segment) and encourage agricultural investment. The system fulfills this condition.
2. **Value Creation:** After implementing the system, grant application dossier management will be more practical and visible. Managers will be able to manipulate the grants' spatial visualization, which impacts the processing of the dossiers. It will save time due to the availability of data facilitated by the built maps. Financial tasks will be gained in a relevant and efficient manner because grants will be allocated according to regional needs. This creates value.
3. **Risk Management:** The use of passwords is a condition of a secure information system. Therefore, it is fundamental to implement a password policy to protect the user and the information system, as well as manage the risk of system vulnerability. Accordingly, the system will have a secured policy requiring user authentication prior to access.
4. **Efficiency Measurement:** To measure the efficiency of an information system, managers need indicators to identify the distance between results and objectives. Using generated spatial dashboards, the system will deduce concrete indicators to measure efficiency.
5. **Resource Management:** The system will offer Web services for each provincial directorate of agriculture and regional office within agricultural development. The responsibilities of each user will be clearly identified and applied.

## THE IMPORTANCE OF SPATIAL DATA

A strong link exists between agricultural subsidies granted to farmers and territory. Parcels, which are the subject of grants, take up a considerable part of useful agricultural areas. According to the chapter's authors, the spatial component is the base of the data type to consider in managing projects within the FDA framework. Spatial information is planned in the development of the current FDA project management system by introducing global positioning system (GPS) coordinates for the parcels.

To provide decision makers with new platforms, GNSS technology facilitates the following tasks using satellite images and Web-mapping based on techniques and principles of Web 2.0 (Lardon & Roche, 2008):

1. **Visualization of Farms:** Using aerial photographs or satellite images
2. **Spatial Distribution:** Analyzing territorial dimensions of financed applications by mapping the distribution of grants, projects at the identification phase, projects being formalized, and validated projects
3. **Inconsistencies:** Detecting discrepancies in case of a double request of the same parcel
4. **Spatial Location Queries:** Associating corresponding requests with spatial location via GNSS
5. **Decision Support:** Using processed data as indicators, statistical analysis, cartographic summaries, and geo-spatial information crosschecks through all levels of decision making (decision makers can have a global vision of perimeters subject to agricultural subsidies)

Visualizing objects of interest on a map is more explicit than the use of statistical data. It provides an additional dimension to monitor the evolution of project dynamics and measure spatial distribution at local, regional, and national levels. Spatial data allows users to evaluate the relevance of grants in at least two dimensions:

1. Comparing results and objectives in terms of production and amount of grants at provincial, regional, and national levels
2. Reassessing grants by performing a posteriori analysis of the projects to meet PMV objectives

## RESULTS AND DISCUSSION

This tool can be the geo-spatial baseline for adding data, including information to support smallholder farming. Linking data will provide the agricultural department with a comprehensive overview of grant distribution and allocation. This allows for continuous monitoring and evaluation to raise or decrease grants in a specific region of the country. Through the tool, the agricultural department can cooperate with other departments in the Moroccan government to discuss water, cadastre, and spatial data infrastructure. The tool will both improve the approach to grant distribution and help decisions makers determine if farms are already funded. Due to the lack of geographic views, the agreement to fund a farm is currently based on the declaration of owner.

## Phases Achieved Before Implementing the Geo-Spatial Tool

Before implementing the strategy, a modeling step has been carried out. Researchers developed several diagrams using the object-oriented model according to the unified modeling language (El-Ayachi, 2017). This model provides a comprehensive view of the functional aspect modeled by the use case diagram as a first aspect of the approach. It describes the users' point of view to display each system actor's behavior. The dynamics aspect is represented by the sequence diagram, which defines interactions between system objects. Finally, the static aspect, which is designed by the class diagram of a target system, identifies principal actors interacting with the system.

It is necessary to identify the functional and technical aspects when setting up the geo-decisional system. These aspects outline the real needs of the system. Moreover, the proposed solution will implement a geo-decision-making tool based on geo-spatial technology offering functionalities presented by spaces dedicated to the identified profiles: (1) decision-maker; (2) agricultural department; and (3) farmer or beneficiary of the agricultural subsidy. It will also meet the decision makers' requirements, enable better management of granting funds within the framework of the FDA, and improve the flow of information between stakeholders.

1. **Technical Aspect of the Geo-Decisional System:** The following elements should be considered for such an approach to work:
   a. Need for a cartographic repository for projection, data precision, and format
   b. Layers implemented by users
   c. Processing tools and general GIS software tools for raster or vector data processing (Web GIS provides customers with the tools to perform geographic data processing)

   The use of GPS by the TS is vital because it digitizes and integrates parcel limits.

2. **Functional Aspect of the Geo-Decisional System:** This aspect introduces the functionalities of the geo-decisional system. The following requirements specify the input and output behaviors of the system by allowing the user to:
   a. **Authenticating:** Users can exploit system resources according to access rights granted to the user's profile.
   b. **Managing System Users:** This involves the management of operations, including adding, deleting, modifying, and consulting the information of each user.
   c. **Managing Access Rights:** Administrator assign access rights and specific privileges to each profile.
   d. **Managing Beneficiaries:** Beneficiaries' personal information can be added to, deleted, modified, and consulted.
   e. **Managing Dossiers:** Request file types can be added to, deleted, modified, and consulted (using prior agreements or grant files).
   f. **Printing:** Dossier receipts can be obtained to trace administrators and applicants.
   g. **Extracting Reports:** Decision making is aided through report extractions on spatial analysis. Reports can be in the form of thematic maps, summaries, or dashboards.
   h. **Consulting Grant Distribution:** A thematic map consultation reviews active and paid projects according to agricultural sector.

i. **Reporting:** Monitoring and controlling applications for financial aid from the state is achieved through the generation of instantaneous situations (in the form of summaries and dashboards).
j. **Validating Inventory of Fixtures:** Project validation is achieved if conditions of the preliminary agreement are fulfilled. It draws the beneficiary's parcel in real time and adds it to the spatial database.
k. **Validating Accepted Projects:** The final step in applying for financial aid from the state consists of validating the project and granting the subsidy.

### Decision Maker Space

A mapping platform facilitates visibility and decisions on subsidized agricultural land. This platform carries out spatial analysis using the following functionalities:

1. Manipulation of geographical data with a geoportal
2. Improved knowledge and understanding of subsidized land, distribution, and dynamics at local, regional, and national levels
3. Extraction of reports and geo-analytical dashboards on project status
4. Innovative, powerful mapping platforms to facilitate spatial analysis and decision making (in particular, Web mapping and satellite images via GPS coordinates)

Implementation of this geo-decisional system clarifies the statistics generated by the existing tool. Managers will be able to easily display geographical views through a spatial analysis and data restoring tool. It will also provide a strategic environment for managers to develop individual reports and statistics according to their needs.

## The Geoportal

Figure 1 offers a multitude of spatial geo-processing widget tools for direct interaction with the map. These include a zoom tool, measurement tool, layer creation and modification tool, legend, and display of attribute tables.

The print tool in Figure 2 allows for the printing of cards (content visible in the geoportal) by selecting a zone, format, and the precision and scale of the card. Users can customize their printing by adding comments.

### The Geo-Analytic Dashboard Space

The geo-analytic dashboard in Figure 3 supports the decision process by including spatial dimensions while analyzing data (www.geotribu.net). It presents an explicit and structured graphic view to predict prospective behaviors and make quick decisions.

In this chapter, the authors opted for a prototype of three regions of Morocco using two agricultural grants. The subsidized land comparison of the regions allow the user to make a decision related to irrigation, improvement of land, and improvement of certified seeds and plants.

The thematic analysis aspect allows the users to:

*Figure 1. Geoportal of the geospatial tool*

*Figure 2. Printing maps using the geoportal*

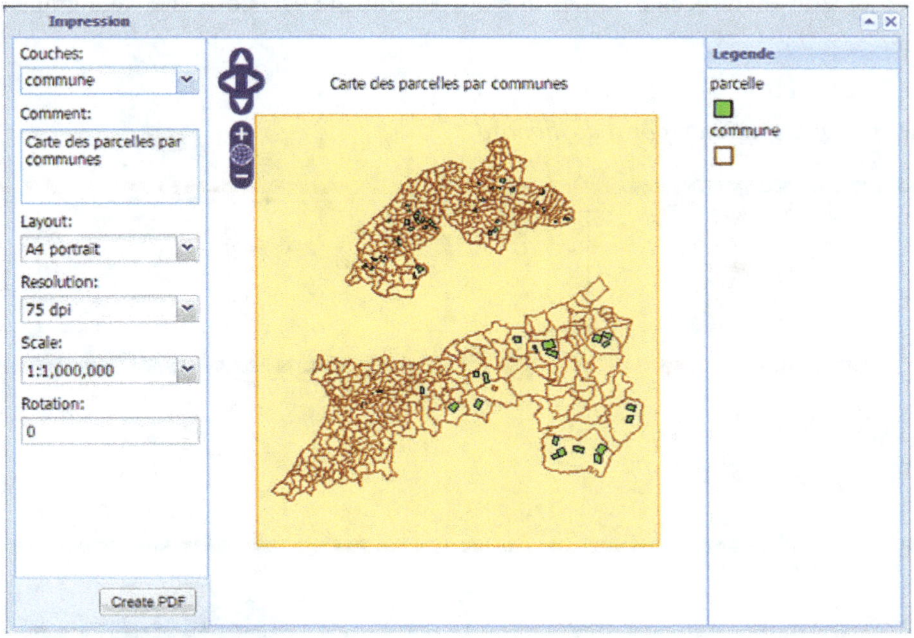


1. Monitor and anticipate progress of subsidies by year or project status (granted or paid) based on chosen regions
2. Compare rate growth of subsidies according to type of subsidized agricultural sectors
3. Make good decisions

Figure 3 is an example of dashboard graphs.

Dashboards provide spatiotemporal statistics, which allow the department of agriculture to:

1. Achieve the objectives of the FDA system (by region, year, agricultural sector, amounts of subsidies, etc.), support the PMV strategy, monitor and anticipate subsidy rates (regional increases or decreases), or direct subsidies to other sectors in their region
2. Provide managers with current and reliable databases to assess the evolution of dynamics, measure efficiency, and make decisions
3. Make the information accessible and exploitable to stakeholders to inform them of changes

## Agricultural Department Space

The department of agriculture can process spatial information to:

1. Identify agricultural farms based on spatiotemporal analysis of grant distribution
2. Visualize and consult space for farms benefiting from grants and use map drawing tools to add to, modify, search, and select a parcel without using database access (i.e., type of data, geographical coordinates, and project date)
3. Incorporate a spatial component into the dossier processing phase to increase control of request eligibility for subsidies
4. Improve thematic analysis capabilities to manage strategic decisions and state grants

*Figure 3. Example of a spatiotemporal dashboard*

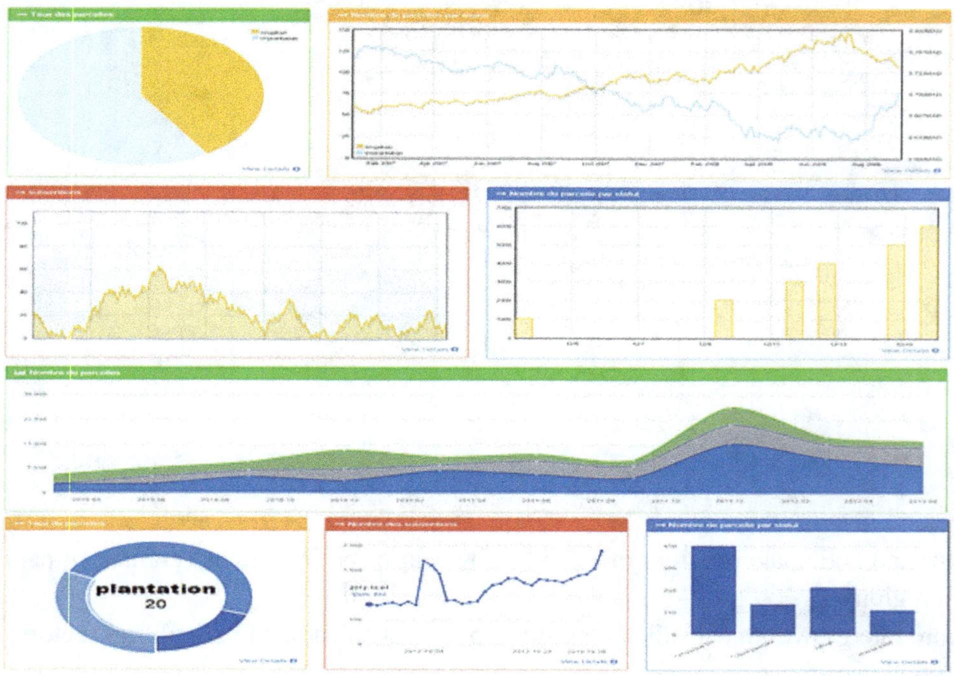

Space dedicated to the department in Figure 4 allows for:

1. Visualization of farms (object of the subsidy)
2. Handling of parcels to extract reports on project statutes
3. Comparisons of distribution of subsidies across regions (using thematic, statistical, and spatial analysis)
4. Tracing for the addition and modifications to parcels in the database

## Farmer Space

Once connected as a grant beneficiary profile, the homepage interface in Figure 5 changes the form and content. Through the menu, users can follow the progress of dossiers and submit complaints.

In Figure 6, farmers can participate in the process by reviewing their files and visualizing the distribution of their projects.

The farmer can also carry out project research through a filtered map (i.e., region, province, commune, etc.). Here, the map consultation allows the beneficiary to consult the geographical distribution of their projects. The geo-decisional system also allows the beneficiary to filter projects according to criteria (see Figure 7).

The second subsection of the progress state of dossiers space is the project consultation. It allows the beneficiary to consult the list of their projects in a dynamic table (see Figure 8).

The complaint space allows the beneficiary to submit a complaint to the system Webmaster (see Figure 9).

*Figure 4. Space reserved for the agricultural department*

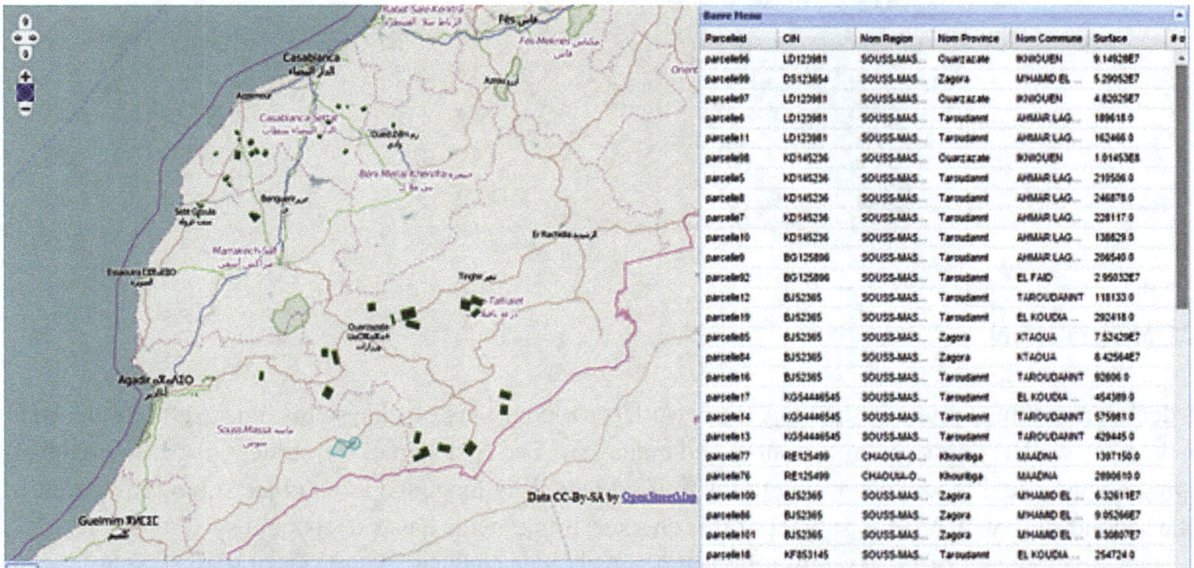

*Figure 5. Homepage of beneficiary profile*

*Figure 6. Example of project distribution*

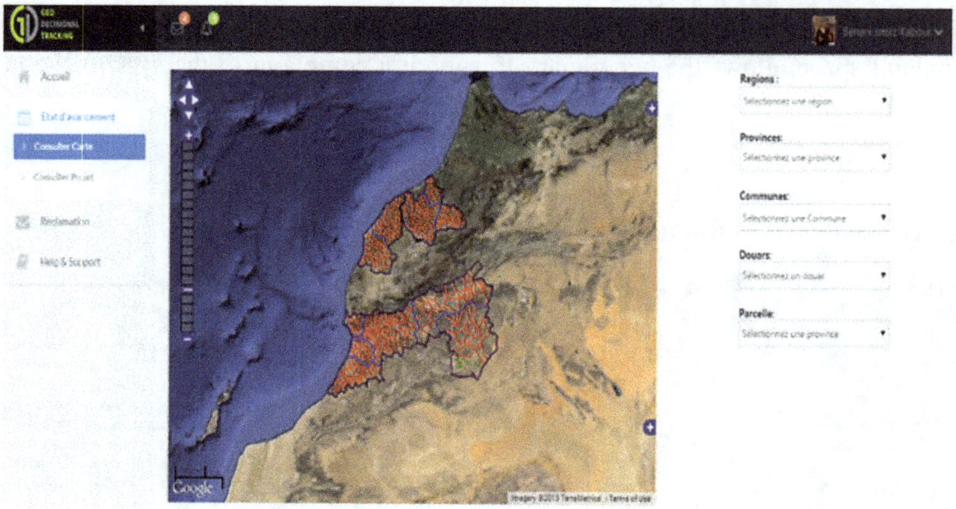


## CONCLUSION

A decision-making system in the FDA framework is a modern vision of systems. It is useful for decision makers, as well as dossier administrators and managers. The first step of the strategy is based on developing a conceptual model according to the unified modeling language as an object-oriented approach. The second step, which is the subject of this chapter, implements a geo-decisional system.

The development of this system followed a stakeholder analysis by considering all stakeholders at the FDA level. This included managers and decision makers from the department of agriculture, related offices, directorates, and beneficiaries of the agricultural subsidies.

*Figure 7. Example of parcel research*

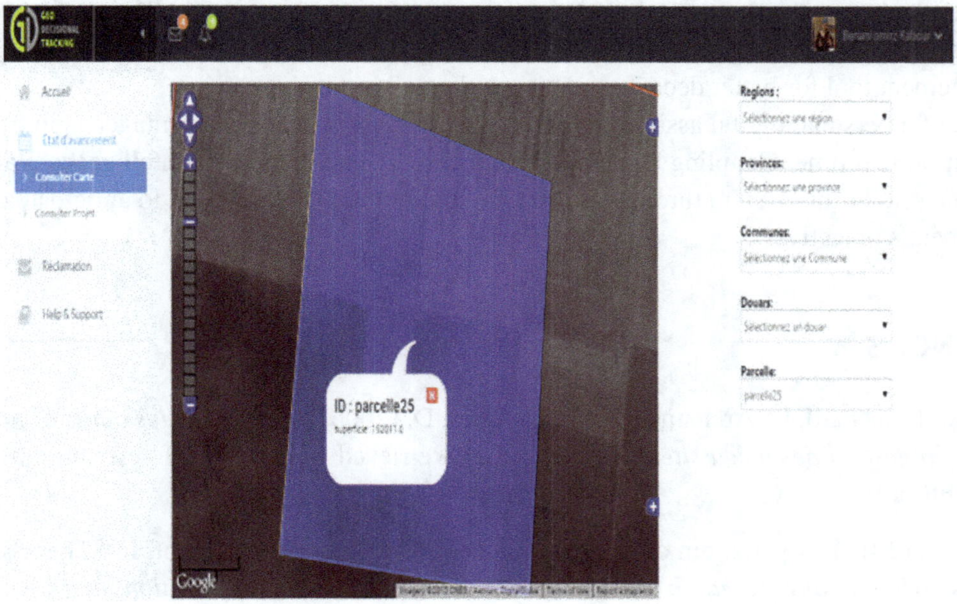


*Figure 8. Table showing checked projects*

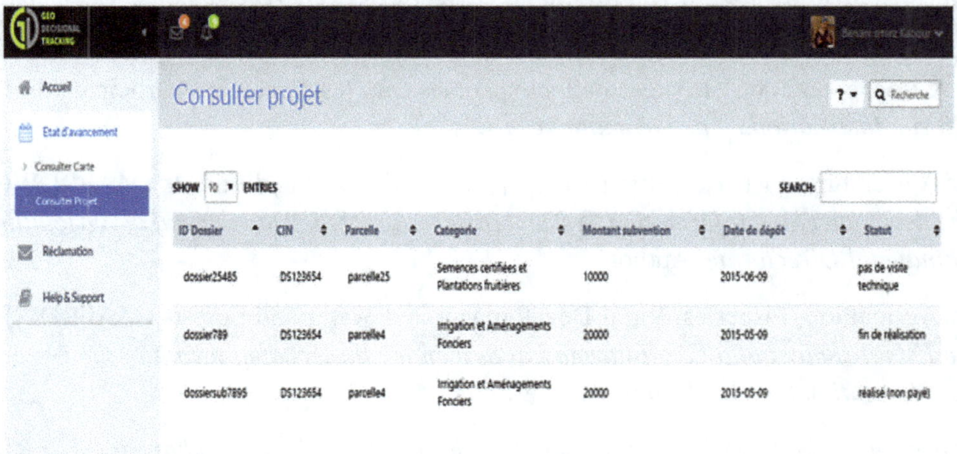


*Figure 9. Space reserved for farmers' complaints*

The use of geo-spatial technology for project management under the FDA is a timely decision. Due to the large number of applicants and tight timeframes, this technology will alleviate the work of the department's TS in scheduling field visits. In terms of decision-making, the geo-decisional system will be a management tool for better decision-making and precision agriculture.

The use of a personal digital assistant (PDA) for field visits would be advantageous in transmitting data on map in real time. Coupling a geo-decisional system with a business intelligence platform will complete the role of this system through advanced exploitation of information to anticipate changes to the department's objectives.

## REFERENCES

Béranger, S., Blanchard, F., Archambault, A., & Allier, D. (2006). *Utilisation des outils d'aide à la décision dans la gestion des méga sites* [Rapport final]. Retrieved from http://infoterre.brgm.fr/rapports/RP-55223-FR.pdf

El-Ayachi, M. (2017). A paradigm of improving land information management. In S. Faiz & K. Mahmoudi (Eds.), *Handbook of research on geographic information systems applications and advancements*. Hershey, PA: IGI Global. doi:10.4018/978-1-5225-0937-0.ch022

Joliveau, T. (2004). *Géomatique et gestion environnementale du territoire: Recherche sur un usage géographique des SIG. Mémoire d'Habilitation à Diriger des Recherches en Sciences Humaines*. Université de Rouen.

Lardon, S., & Roche, S. (2008). Représentations spatiales dans les démarches participatives. Production et usages. *Revue Internationale de Géomatique*, *18*(4).

Ministry of Agriculture, Fisheries, Rural Development and Water and Forests (MAFRDWF). (2010). *Study for the development of Information System Master Plan of the Department of Agriculture, carried out by the Financial Directorate*. Author.

Ministry of Agriculture, Fisheries, Rural Development and Water and Forests (MAFRDWF). (2011). *Study of the Establishment of a Geo-Information System for the Accompanying of the Green Morocco Plan by the Information Systems Department*. Author.

Ministry of Agriculture. (2013). *Fisheries, Rural Development and Water and Forests*. MAFRDWF-ISD.

Ministry of Agriculture. Fisheries, Rural Development and Water and Forest (MAFRDWF). (2017). *Le fond de développement Agricole*. Retrieved from http://www.agriculture.gov.ma/sites/default/files/fda2017/FDA_2017_VF.pdf

Moroccan Government. (2013, March 18). Official Bulletin n° 6135 (p. 2590).

Plumjeaud, C. (2011). *Modèles et méthodes pour l'information spatio-temporelle évolutive* (Unpublished doctoral dissertation). Université de Grenoble, France.

Schuler, M., Thévoz, L., Walser, O., Joost, S., Debarbieux, B., & Dao, H. (2011). Les visages et usages de l'information géographique dans le processus de décision territoriale. In F. Joerin (Ed.), *Les SIG au service du développement territorial*. Presses Polytechniques et Universitaires Romandes.

## KEY TERMS AND DEFINITIONS

**Agricultural Subsidies:** Funding farmers to encourage them investing in agriculture.

**Business Intelligence:** The information technology used by decision makers.

**Geo-Decisional Tool:** A spatial data processing system that help managers to take a decision.

**Geoportal:** The web platform that presents a global view of spatial data and allows users to access to offered services.

**Green Morocco Plan:** The adopted strategy of agriculture in Morocco since 2008.

**Technical Services:** The group of agronomists who check projects realization.

**Unique Hub:** The single point that farmers contact to get information about agricultural funds.

# Chapter 11
# Forest Inventory:
## Assessing Forest Resources for Sustaining Their Management – Contribution of Geospatial Technologies

**Said Lahssini**
*National School of Forestry Engineering, Morocco*

**Loubna El Mansouri**
*Institut Agronomique et Vétérinaire Hassan II, Morocco*

**Hicham Mharzi Alaoui**
*High Commission for Water and Forest and Combating Desertification, Morocco*

**Said Moukrim**
*Mohammed V University, Morocco*

## ABSTRACT

*Forest resources management requires a variety of information related to social systems and to land and its supported resources and their dynamics (land cover, forest stocking, and growth). Such information is, by nature, spatio-temporal and scale dependent and its quality relay on costs for obtaining it. Geosciences and forest geomatics offer valuable methods for ensuring a good compromise between the quality of required information and its costs. This chapter will review and discuss the contribution of geoscience to forest and land inventory. After presentation of information needed and their acquisition methods, through traditional forest inventory, the chapter will focus on technologies aiming at forest resources characterization and assessment such as aerial photogrammetry, satellite imagery, LiDAR data.*

DOI: 10.4018/978-1-5225-5939-9.ch011

## INTRODUCTION AND BACKGROUND

Forests contribute to provide ecosystem goods and services on which human communities depends, such as: biodiversity, carbon sequestration, regulation of water runoff (that ensure sustainable provision of water resources qualitatively and quantitatively), timber and non-timber products, as well as spiritual and cultural values (Johnston, Lindner, Parrotta, & Giessen, 2012). Forest ecosystems are then characterized by their multifunctionality which is expressed by the fact that forests may simultaneously contribute to the many requirements made by society to the main groups of functions organized and described in forest literature: production, option, regulation, information functions, or utility, realization, perception and protection (Cantiani, De Meo, Ferretti, & Paletto, 2010; Fernand, 1995; Paletto, Ferretti, Cantiani, & De Meo, 2012; Vos, 1996).

As the concept of multifunctional forestry become prerequisite, along with the changing perception of the relation men to nature, the raising environmental awareness (since the Rio de Janeiro Conference 1992) and the perspective of climate changes, forests and forest management around the world faces big issues (Farcy & Devillez, 2005; Johnston et al., 2012). Since then, forest management shifted from a model based mainly on wood and timber production to multipurpose planning with a lot of uncertainties (Cantiani et al., 2010; Hartikainen, Eyvindson, Miettinen, & Kangas, 2016).

Managing forest resources aims at maintaining the consumptive needs of society while caring for the integrity and function of ecological systems. Including, managing for wood production objectives, or promoting wildlife, fisheries, recreational dimension as well as other environmental and social services (Bettinger, Boston, Siry, & Grebner, 2009). According to Hartikainen et al. (2016), it could be simply described as the process that consists on selecting optimal harvest schedules including one or more treatment option(s) and their timing for forest stands (homogeneous sub-unit of the forest, with regards to tree species composition, and age); treatment options, within a stand, may include final felling (all trees harvesting) or thinning (harvesting a part of trees), planting new seedlings and tending them after a harvest has been carried out. With a multifucntionnality and sustainability perspectives, intensive forest management became necessary. Moreover, as human population expands, lands resources become scarce and the consumptive needs of society increase. Land planning become widely accepted as contributing to human well-being while maintaining ecosystems integrity and functions. Being a multidisciplinary process that involves multiple actors, land planning effectiveness depend on reliable, timely, and readily available information that is either quantitative or qualitative. It is a very data consumptive process needing accurate information on forest resources development (McCullagh & Nieuwenhuis, 2015). Within the scope of forest management, forest inventories remain the main tools used for obtaining information related to forest resources. Such information includes: biophysical context, natural resources (stands and its supported resources), and the predicted future forest conditions. Forest Inventory (FI) is then the foundation of forest planning and forest policy (Köhl, Magnussen, & Marchetti, 2006).

Within the scope of FI, gathering information on forest resources depends on the scale of interest, ranging from stand level to national or to global level, and on the forest evaluation objectives. Consequently, kind of information collected differ. Furthermore, FI have been reshaped in order to deal with the new trends of forest management such as multifunctionality, plurality and sustainability perspectives and moved forward to benefit from new advances related to data gathering tools, algorithms and statistical tools and other geosciences related techniques.

Further, beyond a long history of using aerial photography in forest inventory, the diversification of kinds and sources of remotely sensed data in addition to advances to image analysis, 'statistical' and learning algorithms and to the computers power, brought some new perspectives to traditional forest data acquisition methods. Such addition is critical for extensive environmental studies at multiple scales and is of high added value for monitoring forest resources, health, growth and sustainability. Then, this chapter will discuss the contribution of geoscience to forest inventory. It will be organized on information needed and their acquisition methods through traditional forest inventory for managing forest resources. Then, it will focus on reviewing and summarizing the contribution of geoscience to forest inventories either for describing the state of forest resources while focusing on the use of GPS and GIS, aerial photogrammetry, satellite imagery, LiDAR Data or for forecasting future forest conditions through the use of forest growth models.

## FOREST INVENTORY AS TOOL FOR ASSESSING FOREST RESOURCES

In order to manage forests, evaluation of stand resources, their desired and projected states are main pillars for forest management. It relays on measuring trees and stands characteristics.

As presented by the main forest measurement and forest inventory books (Kershaw, Ducey, Beers, & Husch, 2016; Köhl et al., 2006; West & West, 2009) main trees attributes are: age, diameter, height, form, volume or biomass and crown information. Age could be considered as elapsed time since the germination of the seed or the budding of the sprout. In relation of the other tree attributes, age is used to assess trees and stands yield and growth. Diameter, measured as the length of the straight line passing through the center of the tree, is measured while accepting that the cross section of the tree is circular. It may be measured for the main stem, for the branches or at multiple stem height levels and it is widely used for computing tree volume. Tree height is vertical distance from an interest point on tree (tree tip, crown level, merchantable wood level, …) to the surface of the earth. Tree form could have many acceptation (Burkhart & Tomé, 2012), within the scope of forest inventory, tree form refers to the taper that is expressed as the diameter diminution rate from the tree base to its tip. It could be expressed as single value (form quotient or other form expressions), as curves, or as formulas (taper equations). Its main purpose is to estimate tree cubic volume. Tree crown dimensions are used as surrogates for assessing foliage or resource use, for predicting suitability for wildlife or for linking tree crown attributes extracted from aerial photographs with other variable of interest such as volume (Kershaw et al., 2016). Furthermore, tree could be described using static relations (Weiskittel, 2011) or allometric relations at tree level linking tree height, volume or biomass to its diameter, or to its, both, diameter and height.

Considering trees within a given area lead to a stand that is a sub unit of forest having some common trees attributes such as origin, species composition, size or age. The state of the stand is then the result of site quality defined by its environmental conditions, the species growth and the management practices. The site is usually considered to be an area that is sufficiently homogenous with regards to its environmental factors (climate, soil fertility, and moisture) that it is meaningful to treat it as a single unit for management or research purposes, although it must always be expected that sites will show some internal heterogeneity. Site quality, widely accepted as measure of site fertility, is interpreted as potential productivity of a designated land area in terms of tree stem volume (Vanclay, 1994) . Site quality could be based on the site factors underlying its fertility such as climate, terrain, soils and is termed « geogenctric approach » or it could be based on the expression of such characteristics on plant communities and is called phytocentric approach (Skovsgaard & Vanclay, 2013).

Site quality assessment through dendrocentric approach is widely used for predicting, within homogeneous sites, the future conditions of the stand. Stand could be defined through the aggregation of trees characteristics and may be characterized using structure, age, species composition, diameter and basal area, height and crown closure, density and stocking, volume, weight, and site quality. Aggregating multiple trees characteristics could be achieved through central tendency (position and dispersion parameters) or through fitting probability distribution function to observed data. Traditionally based on the basis of tree age (Kershaw et al., 2016), stand structure could be described using species composition, diameter distribution, height distribution, and crown classes (McElhinny, Gibbons, Brack, & Bauhus, 2005; Oliver & Larson, 1996).

In addition, other stand characteristics are used such as volume or biomass, density and stocking, regeneration and understory which are important for ensuring the sustainability of the stand.

Forest inventory activities leading to forest resources assessment are based, traditionally, on aerial photo-interpretation associated to field sampling to collect tree and stand measurement (Mailly, 2017). During field measurement of a sample plot, diameter of all the trees are measured. For a sub-sample trees within the sampled plot and in addition to trees diameters, other attributes that are difficult to measure for all the trees (volume, biomass or height) may be collected in order to be used to develop predictive models that will be useful for inference purposes.

If, within a large area forest, measuring all trees seem to be very time consumptive and impractical (Thies & Spiecker, 2004), numerous inventory methods based on sample plots have been developed (De Vries, 2012; Johnson, 2000; Köhl et al., 2006; Schreuder, Gregoire, & Wood, 1993). Furthermore, with a fewer measurements, sampling may produce more reliable results than a complete tally (Kershaw et al., 2016).

Based on sampling theory (Cochran, 1977), many sampling references to forest inventory have been published (Freeze, 1962; Gregoire & Valentine, 2007; Johnson, 2000; Mandallaz, 2008; Schreuder et al., 1993). The idea behind sampling is to find the better compromise between the required precision and the used efforts (time, costs and human resources). Sampling design is defined by the kind and the number of sampling units, their form, disposition and the way of choosing them as the way for analyzing the gathered data (Kershaw et al., 2016). Moreover, inventories scales ranges from local forest to national, and to regional inventories (Rondeux, 1999). Modern National Forest Inventories based on statistical sampling dates back to the early years of the twentieth century (Barreiro et al., 2016) and the basic forest inventory designs falls within the two categories: i- probability sampling (simple random sampling, systematic sampling, Stratified sampling, Multistage sampling, Multiphase sampling, and Sampling with varying probabilities), and ii-nonrandom or selective sampling.

## CONTIBUTION OF GEOSPATAIAL TECHNOLOGIES TO FOREST INVENTORIES

Sustainable forest management, considered as multifaceted approach (White et al., 2016), is based on increasingly complex and wide ranging information, that must be accurate, spatially detailed, up to date, and must characterize forest composition, structure, and, ultimately, wood supply attributes(Groot, Cortini, & Wulder, 2015; Wulder et al., 2012). Within the scope of forest inventories, reducing costs and improving data quality or expanding it to new parameter categories rely on the use of new measurement technologies and instruments for forest inventory that have been continuously improved (Thies & Spiecker, 2004) and on methodologies for handling newly acquired data. Then, geosciences and forest geomatics and learning approaches offers valuable methods for ensuring a good compromise between the quality of required information and its underlying costs.

For forest inventory process, since the earlier stages many geosciences related tools could be involved for planning, acquiring, storing, handling and sharing forest inventories related data. As presented on Table 1, GPS, GIS, LiDAR, Digital Aerial photogrammetry, Satellite data in addition to processing and learning algorithms become widely used at many levels.

## GPS AND GIS

When executing a plot based forest inventory, locating sampling units is of high importance. Traditional methods based on compass and distance were widely used. Nowadays, GPS technology is being used for identifying borders, laying out trails or relocating surveys markers (Kershaw et al., 2016). As described by the same authors, GPS consists of: satellite constellation (network of satellite in orbit around earth), control segment (consisting on a control stations that receives data from the monitor stations in real time 24 hours a day in order to determine whether the satellites are experiencing clocks changes, and to detect equipment malfunctions), and a user segment consisting on user and its GPS receiver (such receivers are described by the number of receiver channels they have that determines the number of satellite signals they can lock on).

According to the spatial nature of sampling units within forest inventory, locating points, drawing lines, plot and measuring area are prerequisite for planning and executing forest inventory. GIS is widely used for such purposes. In addition, samples distribution that are systematic or random could be easily planed using the GIS abilities to generate random X-Y point or regular X-Y points that will be used as plot centers and which can be identified using GPS receiver. Furthermore, spatial analysis including proximity analysis or spatial overlays are useful for assessing site suitability's, for detecting changes or for planning activities. Moreover, the progress on open source software and open data resources enhanced the possibilities of their integration on forest inventories either as data provider, as data storage using spatial databases (such as PostgreSQL or SQLite), as tools for data handling using statistical methods, machine learning algorithms, or spatial analysis (using R language, QGIS and GRASS GIS, …etc.), or as business intelligence (BI) and as tool for data diffusion and sharing among involved actors (such as

*Table 1. Area where geospatial technologies are beneficial to forest inventories*

| Methodologies | Planning | Implementation | Data Quality and Management | Analysis | Reporting |
|---|---|---|---|---|---|
| Remote sensing | | X | | X | X |
| GPS | | | | X | |
| Measurement devices | | X | | | |
| Mobile Information technologies | | X | X | | |
| Algorithms | | X | X | X | X |
| Sampling techniques | X | | X | | |

Kleinn, 2003.

spatial data infrastructure and BI like GEONODE, SAIKU or PENTAHOO). The open foris initiative promoted by FAO (www.openforis.org) is an example of platform that integrates open source software in order to ensure environmental monitoring.

## AERIAL PHOTOGRAMERTY

Historically, forest inventory approaches are mainly based on manual interpretation of aerial photography (Kangas & Maltamo, 2006) in order to delineate stand according to some forest attributes such as species composition, age, height and density or cover. Based on subjective expert judgment, objects are delineated based on their attributes such as location, size, shape, shadow, pattern, tone/color, texture, height/depth, sites (Jensen, 2007).

Using optical imagery for stands composition stratification is mainly based on manual identification of species composition which remain a challenging task for manual photo-interpretation for which the accuracy depends on the scale of the photography, the species composition complexity, and the structural conditions at the location (Morgan & Gergel, 2013; White et al., 2016). In spite of its accuracy and due to its subjectivity, manual photo-interpretation can be difficult to replicate leading to inconsistent results (Morgan & Gergel, 2013; Wulder, White, Hay, & Castilla, 2008).

Furthermore, progress on object oriented image analysis algorithms made possible homogeneous object delineation through segmentation (Figure 1). Many algorithms have been used either within

*Figure 1. Aerial photogrametry for forest cover stratification (right: automatized process using mean shift algorithm, right: manual stand delineation)*

commercial software or within open source solutions (such as eCognition©, or mean shift algorithm). Involved approach are based on grouping neighboring pixels based on their similarity. Moreover, integrating contextual information and textural information could help to predict categorical or continuous variables through a modeling approach such as classification and regression trees or random forest (Laliberte, Fredrickson, & Rango, 2007; Puissant, Rougier, & Stumpf, 2014). Authors compared the quality of manual pho-interpretation to an automated process (Cheng, Bo, Zhu, & Ji, 2014; Morgan & Gergel, 2013; Radoux & Defourny, 2007) that seems to be scene dependent and affected by tree shadows effects.

Furthermore, advances in sensors and computing capacities and methods enabled recently to produce 3D spatial resolution from aerial images (Rahlf, Breidenbach, Solberg, Næsset, & Astrup, 2017). 3D point Cloud or DEM are extracted from overlapping images through image matching and then become an auxiliary information for forest inventory with similar accuracy then that extracted from Air Borne Laser Scanner (White et al., 2016). Used with digital terrain model, vegetation height elevation could be extracted; Further, other trees attributes such as species composition could be extracted from the spectral information of aerial images (Breidenbach & Astrup, 2012; Pulliti, Gobakken, Ørka, & Næsset, 2016; Stepper, Straub, & Pretzsch, 2015). 3D data extracted from digital aerial photograph are widely processed through an area based approach (Naesset, 2002) that is based on numerical metrics, describing the vertical distribution, of remote sensing-based vegetation height measurements within an area. Through predictive models linking such metrics with field measurements from the same area, it become easier to predict, for new areas having only remotely sensed data, forest stand attributes. Furthermore, repeated time elapsed stand images could help to monitor forest dynamics (Vastaranta et al., 2013).

According to White et al. (2016) review, digital aerial photogrammetry (DAP) has been used to predict forest attributes such as height, DBH, and Volume and the authors revealed that the main issue to DAP is a lack of standards or best practices surrounding appropriate image inputs for point-cloud generation (White et al., 2016; White, Stepper, Tompalski, Coops, & Wulder, 2015).

Moreover, the use of hemispherical photography (HP) is promising perspective in forest inventory and measurement. Jonckheere et al. (2004), Mailly (2017) et Weiss et al. (2004) have reviewed the use of HP in support of forest inventory. HP have been widely used for estimate crown dimensions using the fraction of sky that could be seen through the crown of an individual tree (Astrup & Larson, 2006) and Leaf Area Index (Jonckheere et al., 2004). Authors used standard HP or near infrared captured images to extract standing trees parameters (crown dimensions, and to differentiate between bole and branch elements) (Chapman, 2007; Ducrey & Bartoli, 1974).

Using single hemispheric image face limitations that are due to the required model assumptions (Nilson & Kuusk, 2004; Rodríguez-García, Montes, Ruiz, Cañellas, & Pita, 2014; Weiss et al., 2004). Furthermore, stereoscopic hemispherical photography was used to face such limitations and to measure identifiable feature (tree) attributes such as diameters along the stem, height and foliage elements and then to derive stand characteristics in addition to the fact that, as a permanent record of a sampling point, stereo hemispheric photos could be used for stand attributes change and growth monitoring over time (Rodríguez-García et al., 2014).

## SATELITE IMAGERY

Even if aerial imagery have been widely used for forest inventory purposes, the launch of high spatial resolution satellites brought a new possibilities and perspectives (White et al., 2016). Remotely sensed data for forest resource assessment become reliable alternative to traditional field sampling methodology (Hyyppä et al., 2000; Næsset, 1997, 2002). Indeed, the rapid increase in sensor resolution and computational power, and decreasing data acquisition costs, made automatic individual tree interpretation increasingly feasible.

Satellite Imagery uses for forest attribute extraction was reviewed (Xie, Sha, & Yu, 2008) and analyzed either for high satellite resolution (Wulder, Hall, Coops, & Franklin, 2004) and for very high spatial resolution (Falkowski, Wulder, White, & Gillis, 2009). Even if multispectral data may offer powerful possibilities for characterizing vegetation conditions (Xie et al., 2008), the higher spatial resolution possible from panchromatic imagery allows for both digital and manual interpretation approaches (Falkowski et al., 2009). White et al. (2016) reviewed the uses for satellite imagery for identifying dominant species, stand height, volume and biomass, basal area, crown closure, shadow fraction. Although the fact that using very high spatial resolution (VHSR) images could be associated with additional geometric and radiometric processing necessitating image co-registration and normalization (Falkowski et al., 2009), the significant gain in estimation accuracy at stand level could justify such processing (White et al., 2016). Furthermore, in addition to manual photo-interpretation, automated approaches to species identification using VHSR imagery have been demonstrated (Li, Hu, & Woods, 2015; Mora, Wulder, & White, 2010).

The processes, dealing with satellite imagery to extract vegetation information, could be qualified as supervised or unsupervised depending on the use of ground data as references. It consists on general steps: preprocessing, including all preparatory steps to enhance the raw images, and a classification system which use the preprocessed images in order to produce classified images (Xie et al., 2008). The output could be either classified land objects (vectors) or pixels. As example, figure 2 present land cover map of Ourika Watershed located in High Atlas-Morocco for 2014 produced through supervised classification (Rihane, 2015).

Monitoring changes between two dates could be achieved through comparison of multiple classified images. Change detection techniques have been reviewed (P. Coppin, Jonckheere, Nackaerts, Muys, & Lambin, 2004; P. R. Coppin & Bauer, 1996; Lu, Mausel, Brondizio, & Moran, 2004), they include: visual interpretation, pixel-based and map area to area analysis (Desclée, Bogaert, & Defourny, 2006; Franklin, 2001). Pixel-based methods use spectral information in order to classify images according to spectral similarities with predefined land cover classes (Yan, Mas, Maathuis, Xiangmin, & Dijk, 2006). Within the pixel-based methods (Coppin et al. 2004; Lu et al. 2010), change detection may be based on a differential analysis or a composite one (Sader, Hayes, Hepinstall, Coan, & Soza, 2001). Lu et al. (2010) classified change detection methods in two categories: i) binary methods detecting presence / absence of change as image difference, image ratio, principal component analysis and linear regression and ii) methods identifying the nature of the change as a post-classification comparison, hybrid change detection and vectors analysis. For area to area, algorithms may select homogeneous pixels (parcels) from both images and compare them and focus solely on area that not match according to classification (Franklin, 2001).

## LIDAR DATA

LiDAR Data used in forestry application are gathered through a scanner mounted on an aircraft (ALS) or mounted on a tripod (terrestrial laser scanner: TLS) (Dassot, Constant, & Fournier, 2011). The returns from laser scanner form a point cloud of [$x$, $y$, $z$] coordinates from which it is possible to extract the

*Figure 2. Land cover map of Ourika Watershed (High Atlas-Morocco) for 2014 produced using supervised classification of Landsat imagery*
*Rihane, 2015.*

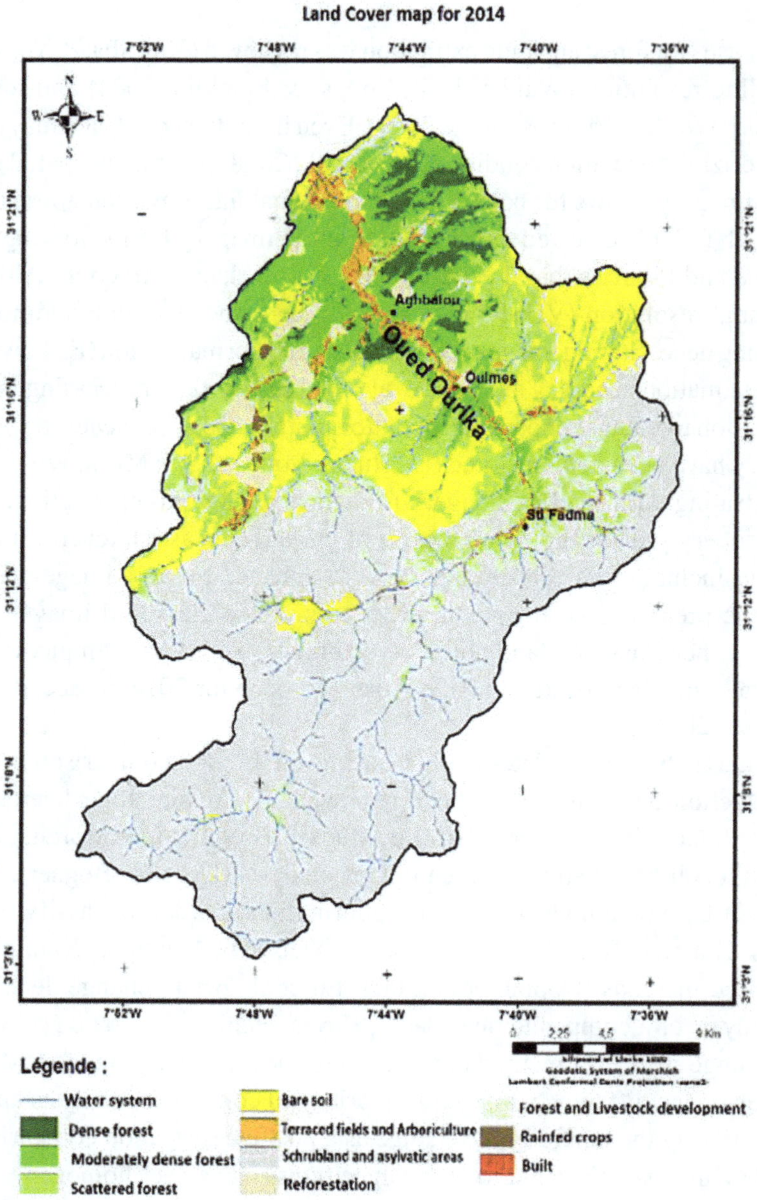


ground surface, the canopy surface and, if the density of points is high enough, recognizable attributes of individual trees (Kershaw et al., 2016).

Tree attributes are extracted from LiDAR Data according to process which consists on many steps: i-ground identification and filtering, ii-individual tree points identification according to a segmentation using many algorithms (Strimbu & Strimbu, 2015) and, iii-extraction of tree attributes from single tree points. Within such process two approach could be adopted for tree attributes extraction: area based (Næsset, 2002) or individual tree detection (ITD) approach (Koch, Heyder, & Weinacker, 2006).

As reviewed by White et al. (2016), area-based approach consists on developing statistical relationship between ground collected predictor variables and the vertical distribution of LiDAR returns (Næsset et al., 2011; Woods et al., 2011). ITD approach involves individual trees extraction from the canopy height model or the point cloud, and derivation of tree height, crown dimensions, and other attributes using ground-based measurements for calibration; information for individual trees can then be aggregated to provide estimates at plot or stand level to assess its basal area, density, area, biomass, …etc. (Breidenbach & Astrup, 2012; White et al., 2016).

LiDAR data filtering algorithms have been reviewed (Meng, Currit, & Zhao, 2010; Montealegre, Lamelas, & de la Riva, 2015; Sithole & Vosselman, 2003; Véga & Durrieu, 2011; Zhang & Whitman, 2005) and have been classified according to: i-number of point classified at time (neighborhood test), ii-measure of discontinuity, and iii-filter concept (which could be slope based, block minimum, surface based, cluster based), and iv-the involved steps (single vs. iterative steps). If the eight tested algorithms by Sithole and Vosselman (2003) did will for typically noncomplex data, surface based filtering algorithms outrank other algorithms. Moreover, vegetation density and terrain roughness must guide algorithm choice and the remaining challenges still: i- rough terrain or discontinuous slope, ii- dense forest canopies that reduces the penetration of laser beams and iii- regions with low vegetation that is ignored by many filters (Meng et al., 2010).

Strimbu and Strimbu (2015) reviewed tree detection and delineation algorithms that could be summarized as follow: Gougeon's valley following technique (starts at local minima and searches for adjacent pixels lying in between pixels of higher value, the resulting path is assumed to be the tree crown boundaries); Local maxima filtering (used for tree top detection and further for generating seed points for algorithms delineating entire tree crowns); Template matching segmentation methods (based on a generalized tree the choice of delineating a tree is based on many criteria such as illumination patterns, ellipses, ellipsoids, gaussian blobs, conic and parabolic surfaces); Region growing algorithms (start with seed pixels that are grown progressively to regions through iteratively inclusion of adjacent pixels until a stopping criteria are met); Watershed segmentation (by analogy of pouring water in the local minima, and considering the basins limits as the water level rises); Clustering algorithms (was used at pixel, voxel, and sub-segment level); Global optimization algorithms (segment trees simultaneously in an iterative process to segment forest as whole); Markov Random Field (was used to optimize spatial configuration of the trees) ; Gas of circles (was used to model tree crown distribution). Furthermore, Strimbu and Strimbu (2015) used hierarchized data structures and a weighted graph to construct an abstract representation of the forest that enables quantifying logical and spatial relations between tree crown components and used a system of weights for five different cohesion criteria in order to accommodate with different forest structures.

It is widely recognized that successful extraction of tree or stand attributes from LiDAR data depends on the density of point cloud and on the shape and form of the tree crown (Côté, Fournier, Frazer, & Olaf Niemann, 2012). Therefore, within dense shrublands, it has been found that LiDAR returns reflect the larger woody structures within the shrub layer, rather than the shrub canopy surface which lead to systemic height underestimation in shrublands (Hudak, Evans, & Stuart Smith, 2009).

*Figure 3. Point cloud LiDAR Data (left: raw data, right: filtered cloud data)*

## PROJECTION OF FUTURE FOREST STATES

In addition to accurate information on forest resources that become widely involving geospatial technologies, the projection of the future states and conditions of these resources are necessary to ensure sustainable forest management (McCullagh & Nieuwenhuis, 2015). Predictions are based on the uses of growth and yield models. Advances in informatics, statistical methods in addition to new silvicultural perspectives that become moving toward mixed species and not only focusing on even-aged monospecific stands, and beyond the yielded timber taking into account the causal relationship underplayed by the growth process (Porté & Bartelink, 2002) enhanced the possibilities and the models' perspectives become wider for forest managers to predict future forest conditions.

Growth and yield models are a projection systems that was used mainly to ensure sustainable provision of forest timber (Monserud, 2003). Growth and yield models could be classified in many categories according to classification criteria (Vanclay, 1994). They could be regrouped on two broad types empirical/statistical (based on observed relationship linking current climate conditions, tree attributes and time steps without any consideration of the underlying physiological processes) and process based models describing explicitly the involved biophysical and biochemical processes (Constable & Friend, 2000). Moreover, depending on the projection units, models could be classified into: individual tree level models, size class models, stand-level models, and distribution-based models (Burkhart & Tomé, 2012; Weiskittel, 2011).

Since the 18th century, yield tables were used as stand level growth models for predicting future forest conditions (Peng 2000). Since then, empirical growth models have been developed. Stand level growth and yield models share a common principle: yield is predicted over time for specific site (Monserud, 2003). Site is then assessed through phytocentric or geocentric approaches (Skovsgaard & Vanclay, 2008, 2013). Site Index, defined as dominant height at index age, is widely used for predicting stand productivity through height-age curves fitting (Rennolls, 1995). Within the scope of stand based models, stand attributes such as basal area, density and volume are projected. Size class models, may be based on transition probability matrices, predict future forest structure, i-e number of trees by size classes. Individual tree growth models project single tree attributes over time. Within individual tree growth models, competition between trees is either accounted through location of trees or through stand parameters. Such consideration defines two sub class of individual tree growth models: distance dependent and distance independent. Furthermore, shifting from stand to single tree level could be achieved through the use of probability distribution function that links stand attribute to trees distribution (Burkhart & Tomé, 2012), probability distribution function with a wide range of forms such as Gamma, Beta, Weibull and Johnsons' SB where widely used in forestry (Gorgoso, Rojo, Cámara-Obregón, & Diéguez-Aranda, 2012; Merganič & Sterba, 2006; Palahí, Pukkala, Blasco, & Trasobares, 2007).

Process based Models involve physiological process underlying tree growth including photosynthetic equations, stomatal model in addition to allocation and turnover processes (Constable & Friend, 2000). Moreover, integrating process-based model with empirical-based growth and yield model become of great interest to simulate climate change impacts on forest resources (Baldwin, Burkhart, Westfall, & Peterson, 2001; Reyer et al., 2014; Schelhaas et al., 2015) or to assess new silvicultural treatments (Söderbergh & Ledermann, 2003) .

If they are widely useful for forest management, growth and yield models face challenges related to spatio-temporal database that needs re-measurement of tree populations (Barreiro et al., 2016), which is expensive and usually not available (Rennolls et al., 2007). Moreover, process based models are difficult to calibrate and then to use at operational scale (Mäkelä et al., 2000). Availability and easiness of acquiring geospatial data expanded possibilities for forest resources assessment or for predicting the resources dynamics.

## CONCLUSION

Evaluating forest resources, trough forest inventories, is perquisite for ensuring their sustainable management. If, within the scope of forest inventories, innovations seem to be crucial since the development of angle count sampling by Bitterlich (1948). Nowadays and if forest inventory approaches still widely based on manual photo-interpretation of aerial photographs for delineating stands and then, through field measurement of limited ground sampling plots, for identifying stand attribute, the availability and affordability of geospatial data from LiDAR, Digital photogrammetry and Satellite Images brought new perspectives to ground based forest inventories.

If the contribution of these technologies has been proven, their effectiveness depends on the scale and the kind of information of interest, the data mobilized and the treatment approaches. Indeed, surface-based approaches are widely used to estimate forest stand attribute (such as height, basal area and volume) using LiDAR or satellite imagery. Nonetheless, automatic tree species identification remain a common drawback to LiDAR, high resolution satellite and airborne optical sensors (Leckie et al., 2005).

Moreover, at tree level, if the TLS have been showed opportunities, the use of stereo-fisheye images expanded the opportunities and made possible collecting tree attributes without measuring it. Moreover, according to white et al., 2016, integration of LiDAR and digital photogrammetry offer synergies, through the integration of structural and functional canopy attributes, *"that will likely to have the greatest impact on forest inventories during the next decade"*.

Sustainable management of forest resources depends on availability of accurate information that is related to forest resources and their future conditions. Advances in remote sensing brought new data sources and processing approach that are valuable and highly useful to assess such resources. During such assessment, appropriate data sources must be chosen according to their effectiveness, to data needs and to the processing capacities.

## REFERENCES

Astrup, R., & Larson, B. C. (2006). Regional variability of species-specific crown openness for aspen and spruce in western boreal Canada. *Forest Ecology and Management*, *228*(1), 241–250. doi:10.1016/j.foreco.2006.02.048

Baldwin, V. C., Burkhart, H. E., Westfall, J. A., & Peterson, K. D. (2001). Linking Growth and Yield and Process Models to Estimate Impact of Environmental Changes on Growth of Loblolly Pine. *Forest Science*, *47*(1), 77–82.

Barreiro, S., Schelhaas, M.-J., Kändler, G., Antón-Fernández, C., Colin, A., Bontemps, J.-D., ... Ferezliev, A. (2016). Overview of methods and tools for evaluating future woody biomass availability in European countries. *Annals of Forest Science*, *73*(4), 823–837. doi:10.100713595-016-0564-3

Bettinger, P., Boston, K., Siry, J., & Grebner, D. (2009). *Forest management and planning*. Amsterdam: Academic.

Breidenbach, J., & Astrup, R. (2012). Small area estimation of forest attri- butes in the Norwegian National Forest Inventory. *European Journal of Forest Research*, *131*(4), 1–13. doi:10.100710342-012-0596-7

Burkhart, H. E., & Tomé, M. (2012). *Modeling forest trees and stands*. Springer Science & Business Media. doi:10.1007/978-90-481-3170-9

Cantiani, P., De Meo, I., Ferretti, F., & Paletto, A. (2010). Forest functions evaluation to support forest landscape management planning. *Forestry Ideas*, *16*(1), 39.

Chapman, L. (2007). Potential applications of near infrared hemispherical imagery in forest environments. *Agricultural and Forest Meteorology*, *143*(1), 151–156. doi:10.1016/j.agrformet.2006.12.006

Cheng, J., Bo, Y., Zhu, Y., & Ji, X. (2014). A novel method for assessing the segmentation quality of high-spatial resolution remote-sensing images. *International Journal of Remote Sensing*, *35*(10), 3816–3839. doi:10.1080/01431161.2014.919678

Cochran, W. G. (1977). *Sampling techniques* (3rd ed.). Academic Press.

Constable, J. V. H., & Friend, A. L. (2000). Suitability of process-based tree growth models for addressing tree response to climate change. *Environmental Pollution, 110*(1), 47–59. doi:10.1016/S0269-7491(99)00289-4 PMID:15092855

Coppin, P., Jonckheere, I., Nackaerts, K., Muys, B., & Lambin, E. (2004). Review ArticleDigital change detection methods in ecosystem monitoring: A review. *International Journal of Remote Sensing, 25*(9), 1565–1596. doi:10.1080/0143116031000101675

Coppin, P. R., & Bauer, M. E. (1996). Digital change detection in forest ecosystems with remote sensing imagery. *Remote Sensing Reviews, 13*(3–4), 207–234. doi:10.1080/02757259609532305

Côté, J.-F., Fournier, R. A., Frazer, G. W., & Olaf Niemann, K. (2012). A fine-scale architectural model of trees to enhance LiDAR-derived measurements of forest canopy structure. *Agricultural and Forest Meteorology, 166–167*, 72–85. doi:10.1016/j.agrformet.2012.06.007

Dassot, M., Constant, T., & Fournier, M. (2011). The use of terrestrial LiDAR technology in forest science: Application fields, benefits and challenges. *Annals of Forest Science, 68*(5), 959–974. doi:10.100713595-011-0102-2

De Vries, P. G. (2012). *Sampling theory for forest inventory: a teach-yourself course*. Springer Science & Business Media.

Desclée, B., Bogaert, P., & Defourny, P. (2006). Forest change detection by statistical object-based method. *Remote Sensing of Environment, 102*(1), 1–11. doi:10.1016/j.rse.2006.01.013

Ducrey, M., & Bartoli, M. (1974). Utilisation des photographies hémisphériques pour le cubage des arbres sur pied. *Annales des Sciences Forestieres, 31*(4), 229–248. doi:10.1051/forest/19740403

Falkowski, M. J., Wulder, M. A., White, J. C., & Gillis, M. D. (2009). Supporting large-area, sample-based forest inventories with very high spatial resolution satellite imagery. *Progress in Physical Geography, 33*(3), 403–423. doi:10.1177/0309133309342643

Farcy, C., & Devillez, F. (2005). New orientations of forest management planning from an historical perspective of the relations between man and nature2. *Forest Policy and Economics, 7*(1), 85–95. doi:10.1016/S1389-9341(03)00013-3

Fernand, J. (1995). Multiple-use forestry-a concept of communication. *Multiple-Use Forestry in the Nordic Countries*, 67–80.

Franklin, S. E. (2001). *Remote sensing for sustainable forest management*. CRC Press. doi:10.1201/9781420032857

Freeze, F. (1962). *Elementary Forest Sampling*. USDA.

Gorgoso, J. J., Rojo, A., Cámara-Obregón, A., & Diéguez-Aranda, U. (2012). A comparison of estimation methods for fitting Weibull, Johnson's SB and beta functions to Pinus pinaster, Pinus radiate and Pinus sylvestris stands in northwest Spain. *Forest Systems, 21*(3), 446. doi:10.5424/fs/2012213-02736

Gregoire, T. G., & Valentine, H. T. (2007). *Sampling strategies for natural resources and the environment*. CRC Press.

Groot, A., Cortini, F., & Wulder, M. A. (2015). Crown-fibre attribute relationships for enhanced forest inventory: Progress and prospects. *Forestry Chronicle*, *91*(3), 266–279. doi:10.5558/tfc2015-048

Hartikainen, M., Eyvindson, K., Miettinen, K., & Kangas, A. (2016). Data-Based Forest Management with Uncertainties and Multiple Objectives. In *International Workshop on Machine Learning, Optimization and Big Data* (pp. 16–29). Springer. 10.1007/978-3-319-51469-7_2

Hudak, A. T., Evans, J. S., & Stuart Smith, A. M. (2009). LiDAR Utility for Natural Resource Managers. *Remote Sensing*, *1*(4), 934–951. doi:10.3390/rs1040934

Jensen, J. R. (2007). *Remote Sensing of the Environment: An Earth resource perspective*. Prentice Hall.

Johnson, E. W. (2000). *Forest sampling desk reference*. CRC Press. doi:10.1201/9781420042498

Johnston, M., Lindner, M., Parrotta, J., & Giessen, L. (2012). Adaptation and mitigation options for forests and forest management in a changing climate. *Forest Policy and Economics*, *24*(Supplement C), 1–2. doi:10.1016/j.forpol.2012.09.007

Jonckheere, I., Fleck, S., Nackaerts, K., Muys, B., Coppin, P., Weiss, M., & Baret, F. (2004). Review of methods for in situ leaf area index determination: Part I. Theories, sensors and hemispherical photography. *Agricultural and Forest Meteorology*, *121*(1), 19–35. doi:10.1016/j.agrformet.2003.08.027

Kangas, A., & Maltamo, M. (2006). *Forest inventory methodology and applications*. Dordrecht: Springer. doi:10.1007/1-4020-4381-3

Kershaw, J. A., Ducey, M. J., Beers, T. W., & Husch, B. (2016). *Forest mensuration* (5th ed.). Chichester, UK: John Wiley & Sons. doi:10.1002/9781118902028

Kleinn, C. (2003). New technologies and methodologies for national forest inventories. *UNASYLVA-FAO*, 10–15.

Koch, B., Heyder, U., & Weinacker, H. (2006). Detection of Individual Tree Crowns in Airborne Lidar Data. *Photogrammetric Engineering and Remote Sensing*, *72*(4), 357–363. doi:10.14358/PERS.72.4.357

Köhl, M., Magnussen, S., & Marchetti, M. (2006). *Sampling methods, remote sensing and GIS multiresource forest inventory*. Berlin: Springer. doi:10.1007/978-3-540-32572-7

Laliberte, A. S., Fredrickson, E. L., & Rango, A. (2007). Combining decision trees with hierarchical object-oriented image analysis for mapping arid rangelands. *Photogrammetric Engineering and Remote Sensing*, *73*(2), 197–207. doi:10.14358/PERS.73.2.197

Leckie, D. G., Tinis, S., Nelson, T., Burnett, C., Gougeon, F. A., Cloney, E., & Paradine, D. (2005). Issues in species classification of trees in old growth conifer stands. *Canadian Journal of Remote Sensing*, *31*(2), 175–190. doi:10.5589/m05-004

Li, J., Hu, B., & Woods, M. (2015). A two-level approach for species identification of coniferous trees in central Ontario forests based on multispectral images. *IEEE Journal of Selected Topics in Applied Earth Observations and Remote Sensing*, *8*(4), 1487–1497. doi:10.1109/JSTARS.2015.2423272

Lu, D., Mausel, P., Brondizio, E., & Moran, E. (2004). Change detection techniques. *International Journal of Remote Sensing*, *25*(12), 2365–2401. doi:10.1080/0143116031000139863

Mailly, D. (2017). Hemispherical Photography in Support of Forest Inventory and Silviculture. In *Hemispherical Photography in Forest Science: Theory, Methods, Applications* (pp. 227–252). Dordrecht: Springer. doi:10.1007/978-94-024-1098-3_8

Mäkelä, A., Landsberg, J., Ek, A. R., Burk, T. E., Ter-Mikaelian, M., Ågren, G. I., ... Puttonen, P. (2000). Process-based models for forest ecosystem management: Current state of the art and challenges for practical implementation. *Tree Physiology*, *20*(5–6), 289–298. doi:10.1093/treephys/20.5-6.289 PMID:12651445

Mandallaz, D. (2008). *Sampling Techniques for Forest*. Academic Press.

McCullagh, A., & Nieuwenhuis, M. (2015). *Review of tools for growth forecasting and productivity assessment in forestry in Ireland*. IrIsh Forestry.

McElhinny, C., Gibbons, P., Brack, C., & Bauhus, J. (2005). Forest and woodland stand structural complexity: Its definition and measurement. *Forest Ecology and Management*, *218*(1), 1–24. doi:10.1016/j.foreco.2005.08.034

Meng, X., Currit, N., & Zhao, K. (2010). Ground filtering algorithms for airborne LiDAR data: A review of critical issues. *Remote Sensing*, *2*(3), 833–860. doi:10.3390/rs2030833

Merganič, J., & Sterba, H. (2006). Characterisation of diameter distribution using the Weibull function: Method of moments. *European Journal of Forest Research*, *125*(4), 427–439. doi:10.100710342-006-0138-2

Monserud, R. A. (2003). Evaluating forest models in a sustainable forest management context. *Forest Biometry. Modelling and Information Sciences*, *1*(1), 35–47.

Montealegre, A. L., Lamelas, M. T., & de la Riva, J. (2015). A comparison of open-source LiDAR filtering algorithms in a mediterranean forest environment. *IEEE Journal of Selected Topics in Applied Earth Observations and Remote Sensing*, *8*(8), 4072–4085. doi:10.1109/JSTARS.2015.2436974

Mora, B., Wulder, M. A., & White, J. C. (2010). Identifying leading species using tree crown metrics derived from very high spatial resolution imagery in a boreal forest environment. *Canadian Journal of Remote Sensing*, *36*(4), 332–344. doi:10.5589/m10-052

Morgan, J. L., & Gergel, S. E. (2013). Automated analysis of aerial photographs and potential for historic forest mapping. *Canadian Journal of Forest Research*, *43*(8), 699–710. doi:10.1139/cjfr-2012-0492

Naesset, E. (2002). Determination of mean tree height of forest stands by digital photogrammetry. *Scandinavian Journal of Forest Research*, *17*(5), 446–459. doi:10.1080/028275802320435469

Næsset, E. (2002). Predicting forest stand characteristics with airborne scanning laser using a practical two-stage procedure and field data. *Remote Sensing of Environment, 80*(1), 88–99. doi:10.1016/S0034-4257(01)00290-5

Næsset, E., Gobakken, T., Solberg, S., Gregoire, T. G., Nelson, R., Ståhl, G., & Weydahl, D. (2011). Model-assisted regional forest biomass estimation using LiDAR and InSAR as auxiliary data: A case study from a boreal forest area. *Remote Sensing of Environment, 115*(12), 3599–3614. doi:10.1016/j.rse.2011.08.021

Nilson, T., & Kuusk, A. (2004). Improved algorithm for estimating canopy indices from gap fraction data in forest canopies. *Agricultural and Forest Meteorology, 124*(3), 157–169. doi:10.1016/j.agrformet.2004.01.008

Oliver, C. D., & Larson, B. C. (1996). *Forest stand dynamics: updated edition*. John Wiley and sons.

Palahí, M., Pukkala, T., Blasco, E., & Trasobares, A. (2007). Comparison of beta, Johnson's SB, Weibull and truncated Weibull functions for modeling the diameter distribution of forest stands in Catalonia (north-east of Spain). *European Journal of Forest Research, 126*(4), 563–571. doi:10.100710342-007-0177-3

Paletto, A., Ferretti, F., Cantiani, P., & De Meo, I. (2012). Multi-functional approach in forest landscape management planning: An application in Southern Italy. *Forest Systems, 21*(1), 68–80. doi:10.5424/fs/2112211-11066

Porté, A., & Bartelink, H. H. (2002). Modelling mixed forest growth: A review of models for forest management. *Ecological Modelling, 150*(1), 141–188. doi:10.1016/S0304-3800(01)00476-8

Puissant, A., Rougier, S., & Stumpf, A. (2014). Object-oriented mapping of urban trees using Random Forest classifiers. *International Journal of Applied Earth Observation and Geoinformation, 26*, 235–245. doi:10.1016/j.jag.2013.07.002

Pulliti, S., Gobakken, T., Ørka, H. O., & Næsset, E. (2016). Assessing 3D point clouds from aerial photographs for species-specific forest inventories. *Scandinavian Journal of Forest Research, 32*(1), 68–79. doi:10.1080/02827581.2016.1186727

Radoux, J., & Defourny, P. (2007). A quantitative assessment of boundaries in automated forest stand delineation using very high resolution imagery. *Remote Sensing of Environment, 110*(4), 468–475. doi:10.1016/j.rse.2007.02.031

Rahlf, J., Breidenbach, J., Solberg, S., Næsset, E., & Astrup, R. (2017). Digital aerial photogrammetry can efficiently support large-area forest inventories in Norway. *Forestry, 00*, 1–19.

Rennolls, K. (1995). Forest height growth modelling. *Forest Ecology and Management, 71*(3), 217–225. doi:10.1016/0378-1127(94)06102-O

Rennolls, K., Tomé, M., McRoberts, R. E., Vanclay, J. K., LeMay, V., Guan, B. T., & Gertner, G. Z. (2007). Potential contributions of statistics and modelling to sustainable forest management: review and synthesis. *Sustainable Forestry: From Monitoring and Modelling to Knowledge Management and Policy Science*, 314–341.

Reyer, C., Lasch-Born, P., Suckow, F., Gutsch, M., Murawski, A., & Pilz, T. (2014). Projections of regional changes in forest net primary productivity for different tree species in Europe driven by climate change and carbon dioxide. *Annals of Forest Science, 71*(2), 211–225. doi:10.100713595-013-0306-8

Rihane, R. (2015). *Contribution à l'évaluation des services hydrologiques de la forêt et des aménagements antiérosifs dans le bassin versant de l'Ourika (Mémoire de 3ème cycle)*. Ecole Nationale Forestière d'Ingénieurs, Salé.

Rodríguez-García, C., Montes, F., Ruiz, F., Cañellas, I., & Pita, P. (2014). Stem mapping and estimating standing volume from stereoscopic hemispherical images. *European Journal of Forest Research, 133*(5), 895–904. doi:10.100710342-014-0806-6

Rondeux, J. (1999). *La mesure des arbres et des peuplements forestiers*. Les presses agronomiques de Gembloux.

Sader, S. A., Hayes, D. J., Hepinstall, J. A., Coan, M., & Soza, C. (2001). Forest change monitoring of a remote biosphere reserve. *International Journal of Remote Sensing, 22*(10), 1937–1950. doi:10.1080/01431160117141

Schelhaas, M.-J., Nabuurs, G.-J., Hengeveld, G., Reyer, C., Hanewinkel, M., Zimmermann, N. E., & Cullmann, D. (2015). Alternative forest management strategies to account for climate change-induced productivity and species suitability changes in Europe. *Regional Environmental Change, 15*(8), 1581–1594. doi:10.100710113-015-0788-z

Schreuder, H. T., Gregoire, T. G., & Wood, G. B. (1993). *Sampling methods for multiresource forest inventory*. John Wiley & Sons.

Sithole, G., & Vosselman, G. (2003). Comparison of filtering algorithms. In *Proceedings of the ISPRS working group III/3 workshop* (pp. 71–78). Academic Press.

Skovsgaard, J. P., & Vanclay, J. K. (2008). Forest site productivity: A review of the evolution of dendrometric concepts for even-aged stands. *Forestry, 81*(1), 13–31. doi:10.1093/forestry/cpm041

Skovsgaard, J. P., & Vanclay, J. K. (2013). Forest site productivity: A review of spatial and temporal variability in natural site conditions. *Forestry, 86*(3), 305–315. doi:10.1093/forestry/cpt010

Söderbergh, I., & Ledermann, T. (2003). Algorithms for simulating thinning and harvesting in five European individual-tree growth simulators: A review. *Computers and Electronics in Agriculture, 39*(2), 115–140. doi:10.1016/S0168-1699(03)00022-X

Stepper, C., Straub, C., & Pretzsch, H. (2015). Assessing height changes in a highly structured forest using regularly acquired aerial image data. *Forestry, 88*(3), 304–316. doi:10.1093/forestry/cpu050

Strimbu, V., & Strimbu, B. (2015). *A graph-based segmentation algorithm for tree crown extraction using airborne LiDAR data* (Vol. 104). Academic Press. 10.1016/j.isprsjprs.2015.01.018

Thies, M., & Spiecker, H. (2004). Evaluation and future prospects of terrestrial laser scanning for standardized forest inventories. *Forest, 2*(2), 1.

Vanclay, J. K. (1994). *Modelling forest growth and yield: applications to mixed tropical forests*. Wallingford, UK: CAB International.

Vastaranta, M., Wulder, M. A., White, J. C., Pekkarinen, A., Tuominen, S., Ginzler, C., ... Hyyppä, H. (2013). Airborne laser scanning and digital stereo imagery measures of forest structure: Comparative results and implications to forest mapping and inventory update. *Canadian Journal of Remote Sensing*, *39*(5), 382–395. doi:10.5589/m13-046

Véga, C., & Durrieu, S. (2011). Multi-level filtering segmentation to measure individual tree parameters based on Lidar data: Application to a mountainous forest with heterogeneous stands. *International Journal of Applied Earth Observation and Geoinformation*, *13*(4), 646–656. doi:10.1016/j.jag.2011.04.002

Vos, W. (1996). *Changing forest functions in NW Europe: From alienation to involvement*. Academic Press.

Weiskittel, A. R. (2011). *Forest growth and yield modeling*. Hoboken, NJ: Wiley. doi:10.1002/9781119998518

Weiss, M., Baret, F., Smith, G. J., Jonckheere, I., & Coppin, P. (2004). Review of methods for in situ leaf area index (LAI) determination: Part II. Estimation of LAI, errors and sampling. *Agricultural and Forest Meteorology*, *121*(1), 37–53. doi:10.1016/j.agrformet.2003.08.001

West, P. W., & West, P. W. (2009). *Tree and forest measurement*. Springer. Retrieved from http://link.springer.com/content/pdf/10.1007/978-3-540-95966-3.pdf

White, J. C., Coops, N. C., Wulder, M. A., Vastaranta, M., Hilker, T., & Tompalski, P. (2016). Remote Sensing Technologies for Enhancing Forest Inventories: A Review. *Canadian Journal of Remote Sensing*, *42*(5), 619–641. doi:10.1080/07038992.2016.1207484

White, J. C., Stepper, C., Tompalski, P., Coops, N. C., & Wulder, M. A. (2015). Comparing ALS and image-based point cloud metrics and modelled forest inventory attributes in a complex coastal forest environment. *Forests*, *6*(10), 3704–3732. doi:10.3390/f6103704

Woods, M., Pitt, D., Penner, M., Lim, K., Nesbitt, D., Etheridge, D., & Treitz, P. (2011). Operational implementation of a LiDAR inventory in Boreal Ontario. *Forestry Chronicle*, *87*(04), 512–528. doi:10.5558/tfc2011-050

Wulder, M. A., Hall, R. J., Coops, N. C., & Franklin, S. E. (2004). High spatial resolution remotely sensed data for ecosystem characterization. *A.I.B.S. Bulletin*, *54*(6), 511–521.

Wulder, M. A., White, J. C., Hay, G. J., & Castilla, G. (2008). Towards automated segmentation of forest inventory polygons on high spatial resolution satellite imagery. *Forestry Chronicle*, *84*(2), 221–230. doi:10.5558/tfc84221-2

Wulder, M. A., White, J. C., Nelson, R. F., Næsset, E., Ørka, H. O., Coops, N. C., ... Gobakken, T. (2012). Lidar sampling for large-area forest characterization: A review. *Remote Sensing of Environment*, *121*(Supplement C), 196–209. doi:10.1016/j.rse.2012.02.001

Xie, Y., Sha, Z., & Yu, M. (2008). Remote sensing imagery in vegetation mapping: A review. *Journal of Plant Ecology*, *1*(1), 9–23. doi:10.1093/jpe/rtm005

Yan, G., Mas, J.-F., Maathuis, B. H. P., Xiangmin, Z., & Dijk, P. M. V. (2006). Comparison of pixel-based and object-oriented image classification approaches—a case study in a coal fire area, Wuda, Inner Mongolia, China. *International Journal of Remote Sensing*, *27*(18), 4039–4055. doi:10.1080/01431160600702632

Zhang, K., & Whitman, D. (2005). Comparison of three algorithms for filtering airborne lidar data. *Photogrammetric Engineering and Remote Sensing*, *71*(3), 313–324. doi:10.14358/PERS.71.3.313

# Chapter 12
# Modeling Hydrological Functioning of a Drainage Basin With Relation to Land Use Change in the Context of Climate Change:
## Ourika Watershed Case Study

**Reda Rihane**
*High Commission for Water and Forest and Combating Desertification, Morocco*

**Abdellatif Khattabi**
*National School of Forestry Engineering, Morocco*

**Nabil Rifai**
*High Commission for Water and Forest and Combating Desertification, Morocco*

**Said Lahssini**
*National School of Forestry Engineering, Morocco*

## ABSTRACT

*Ourika basin in Morocco has very steep slopes with impermeable ground favoring water flows and flooding. History has shown deadly flood events. Floods are becoming recurrent and exacerbated not only by human activities that degrade soil and vegetation cover, accelerating erosion and quick water flows, but also by climate change. In fact, the basin has experienced a very strong dynamic of its vegetation cover, during the last 30 years, and has been subject to climate change impacts. This study is devoted to evaluating the impact of land cover change, mainly vegetation cover, on hydrological functioning of the basin. The HEC-HMS model was used to simulate basin hydrological response, according to two scenarios of land cover change. The first scenario simulates deforestation and urbanization impacts on peak flows, showing an increase of the peak flow by 31.68%. The second evaluates the impact of both reforestation actions and proscription of forest harvesting in the region. The simulated results showed a decline of 17.25% in peak flows, except for heavy precipitation events.*

DOI: 10.4018/978-1-5225-5939-9.ch012

## INTRODUCTION

In semi-arid areas, in general, precipitation is concentrated in time and is often torrential leading to river floods (Sibari et al., 2001). Climate change is likely to produce more frequent and more intense extreme events (Allan and Soden 2008; Driouech, 2010). Indeed, possible amplification of extreme events in Morocco associated to a decrease in total precipitation, could lead to an increased probability of occurrence of events inducing floods and droughts (World Bank, 2013).

Indeed, Morocco is confronted with recurrent rainfall events that can generate flooding likely to cause significant damage. In the past, the country has already experienced recurrent floods (e.g., Ourika in 1984, 1987, 1995, 1999 and 2014; Oued El Maleh in 1996, 2001 and 2002) which have affected both human and animal lives and infrastructures. On the other hand, an increasing aridity, which is manifested by a negative water balance and a reduction in plant activity, is considered as an amplifier factor of forest degradation in fragile environments, which will lead in the future to a migration of some species and their replacement by the desert (World Bank, 2013) exacerbating flood risks.

In this climate change perspective, Morocco, as a Mediterranean country, is expected to be more exposed to flooding; thereby rational management of water levels is a key factor for the safety and security of inhabitants and human wellbeing (Aldosari, 2006). Such management requires a holistic approach to address the various components of the water system (Letcher and Gioppini, 2004).

Unusual floods remain and become stressful with respect to a climate change context in which adverse climate conditions are recurrent and will reflect a structural component to be taken into account for natural resources management. Extensive care should be given in the operation and management of river basins and dam reservoirs used for water supply, flood control or both, so that they are able to overcome the water related problems (Yener et al. 2007). These activities include physical intervention and other soft practices in order to maintain and enhance a forest cover. Even if this later choice is largely adopted and accepted by local managers and the scientific community (Andréassian et al., 2001; Andréassian, 2002; Andréassian, 2004; He, 2003; Choi and Deal, 2008; Gerrits, 2010; Ohana-Levi et al., 2015), the question of how a forest cover contributes to moderate floods remain stressful for natural resources manager.

In order to address that question in a Moroccan environament, the Ourika basin is a good case study. It is known for alarming storms characterized by heavy precipitation localized in space and time (El Alaoui El Fels and Saidi, 2014) with a frequent and brutal flows of the main river and tributaries. e,g, in 1995, such phenomena lead to more than 200 lives lost (Said et al. 2003). Furthermore, vulnerability of this basin to climate change has been shown by many studies (Babqiqi and Messouli, 2013; Salama and Tahiri, 2010; Rochdane et al., 2014). Projected trends show an increasing frequency of extreme events and flooding.

On the other hand, the forest cover has been reduced during the period 1984 to 2000, then, due to the past events and prohibition of forest harvesting starting on the year 2000, the forest cover has been fairly restored (Rihane, 2015). In addition when using climate data from the MPI model (the fifth generation of the European Community Hamburg general circulation model developed at MPI, Germany (Roeckner et al. 2003), it is estimated that the basin biome will shift to bare land by 2070 to 2099. Nonetheless, impacts of forest cover on flood moderation in Ourika basin has not previously been assessed.

This work aims to assess forest cover change on flood moderation through modeling the hydrological behavior of the Ourika basin. Thus, floods have been simulated according to forest cover change scenarios using HEC-HMS solution.

## Study Area and Methodology

### Study Area

Ourika basin is located in the Central High Atlas (Figure 1), and is a sub-basin of the great Tensift catchment. This basin covers an area of 576 km$^2$. It is bounded to the south by the Souss basin, to the north by the Haouz plain, to the east by the Zat basin and to the west by the Rhéraya basin. On the bioclimatic situation, the Ourika basin is subject to semi-arid conditions with a cool winter and a maximum precipitation during winter and spring (Saidi et al., 2010).

Data on land cover in 1984, 2000 and 2014 were analyzed and processed using remote sensing and Geographic Information System (GIS) (Rihane, 2015). The main features of this evolution are summarized in Table 1. The dynamics of land cover systems is characterized by an increase in cultivated

*Figure 1. Location of the Ourika basin*

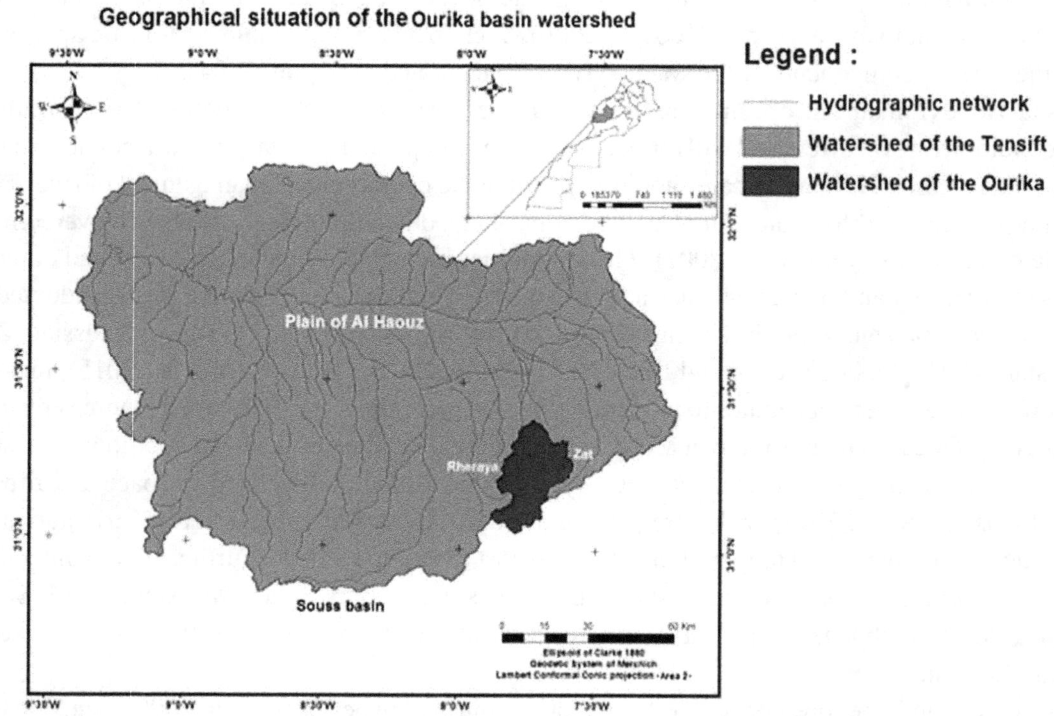


*Table 1. Land cover in the Ourika basin between for the years 1984, 2000 and 2014*

| | 1984 | 2000 | 2014 |
|---|---|---|---|
| Forest | 32.44% | 25.89% | 31.41% |
| Crops | 1.51% | 1.89% | 3.91% |
| Bare soils | 66.05% | 72.22% | 64.68% |

Rihane, 2015.

lands and a decrease of forest area during the period 1984-2000, and a recovery of forested land during the period 2000 – 2014, respectfully. In comparison with the situation of 1984, the land cover map of the year 2000 shows a decline in forest area equivalent to 20.16%. Land cover in 2014 has known an increase in forest cover of approximately 21.28% compared to 2000.

Subsequently, these data were implemented to simulate the hydrological response of the catchment basin according to two scenarios of land cover change, and to evaluate, for a further work, the extent of the centennial flood of the Ourika River through hydraulic modeling.

## Methodology

In order to assess impact of land cover change on the hydrological behavior of the Ourika basin, hydrological responses assessing stream flows were computed using the Hydrologic Engineering Center's Hydrologic Modeling System (HEC-HMS 3.4) model and the data prepared (the basin characteristics and meteorological information). Such data consist of the catchment area, land-use patterns, daily rainfall data, daily river flow data, monthly evaporation data, base flow, peaking coefficient, imperviousness, standard lag, initial deficit, constant rate, time of concentration, storage coefficient and curve number (Halwatura and Najim, 2013). Precipitation-runoff processes of watershed systems were modeled and simulated using HEC-HMS. Then, within HEC-HMS, the basin model comprises three vital processes: the loss, the transform and the base flow.

*Figure 2. Land cover map in Ourika basin for 2014*
*Rihane, 2015.*

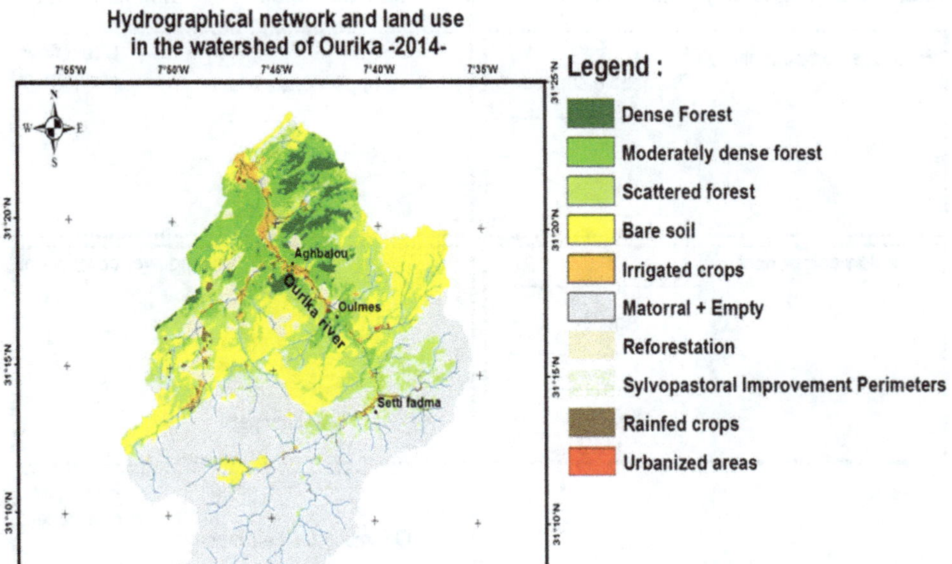

## Data

The spatial database used includes physiographic data (topography and soil maps), land cover maps rainfall and River Flow. Land cover maps were established for 1984, 2000 and 2014 based on Landsat satellite images (Rihane, 2015). Hydrometeorological data were collected from the Hydraulic Basin Agency of Tensift for the period 1984-2013 (30 years) for the Aghbalou station located at the river outlet. They were used to assess spatial and temporal variability of climate.

## Modeling With HEC-HMS

Due to the lack of continuous precipitation data, event modeling was preferred over continuous modeling (Taha and Mualla, 2003). Figure 3 summarizes the operating steps of the model in an event-driven structure.

The HEC-HMS model considers the basin as a uniform system with multiple components that ensures each part of the hydrological processes produce flow hydrographs. It works on the principle of separate tasks.

*Figure 3. HEC-HMS model functioning in event-driven structure*

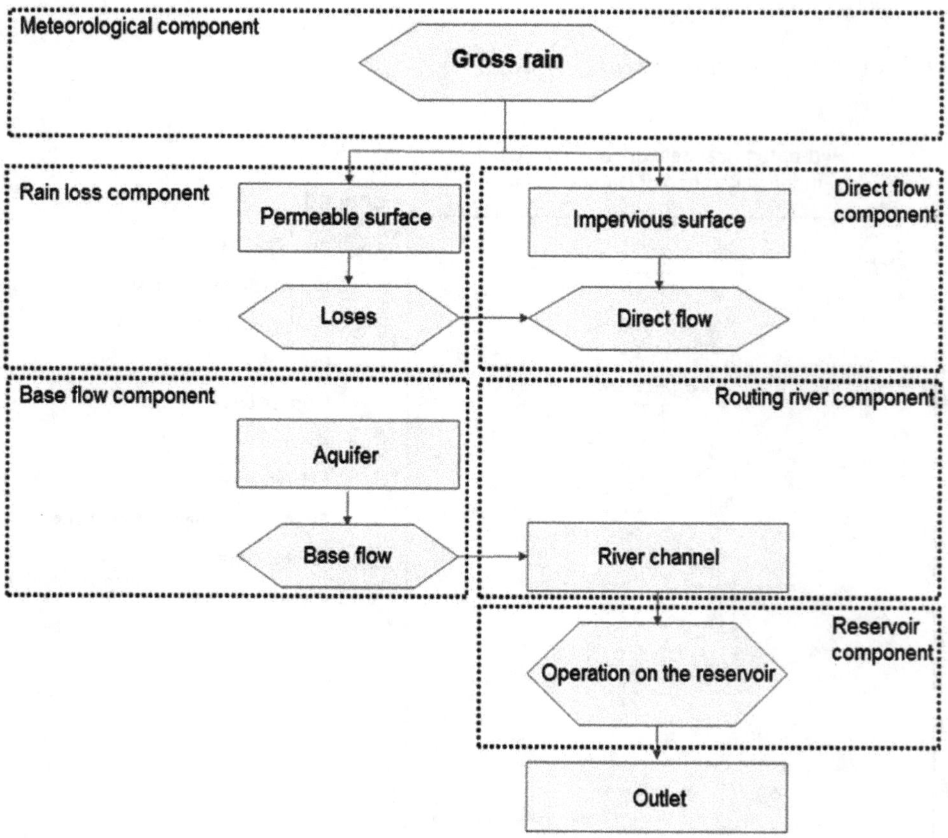

Meteorological module NRCS for flooding (Motevalli et al., 2012) was used to model the precipitation temporal distributions based on daily rainfall quantity. Depending on the geographical location (dominant climate), four floods types can be defined and distinguished by intensity, peak time and form. In terms of maximum intensity, the descending order starting with the type II, then type III, then type I and finally type AI. These four types were used; the most suitable was identified comparing observation with the simulatd results.

Production function chosen, NRCS CN connects the participating net rainfall-runoff to three main factors: land cover, soil type and antecedent moisture.

Further, numerous infiltration loss parameterizations could be chosen (USACE, 2000). Spatially distributed infiltration could be quantified using a parameter derived by the Soil Conservation Service (SCS) called a curve number (CN). CN takes into account the effects of land use/cover, soil types, and hydrologic conditions on surface runoff, and relates direct surface runoff to rainfall. The SCS-CN method has been widely used for estimating rainfall-generated surface runoff in watershed hydrologic modeling (Oleyiblo and Li, 2010). CN can be determined from NRCS tables for each combination of land cover, hydrological conditions and soil type (Musy and Higy, 1998).

Land cover classes were recoded to comply with those defined by the chosen production function (Jeníček, 2007). Soil map was used due to the lack of soil texture map. Then, soil types have been linked to NRCS classes: type A (sandy soils and brown soils on schists), type B (shallow clay sandy soils), type C (altitude skeletal soils) and finally impermeable soils corresponding to "bare soil". Consequently, a NRCS code and a CN parameter value were assigned to each land cover class based on soil type and hydrological condition (Table 2).

Initial abstraction is a variable parameter that takes into account losses prior to the start of runoff, such as, interception and depression storage. Evapotranspiration losses are considered negligible for the preliminary model due to several factors: the intensity of the storm being modeled, the continuous saturation of the air, and the resulting assumption that evapotranspiration volume is negligible compared to runoff volume.

Transformation of excess precipitation to runoff is accomplished using the NRCS unit hydrograph and Clark unit hydrograph (Clark, 1945). This method convolves precipitation increments with the unit hydrograph ordinates to determine the basin hydrograph.

*Table 2. NRCS Table for land cover classes identified on the Ourika basin*

| Code | Land Cover Description | | CN | | | |
|---|---|---|---|---|---|---|
| | Land Cover Type | Hydrological Condition | A | B | C | D |
| 11 | Urbanized areas / Bare soil | | 77 | 86 | 91 | 94 |
| 225 | Crops | Bad | 72 | 81 | 88 | 91 |
| 323 | Matorral | Bad | 48 | 67 | 77 | 83 |
| 343 | Sparse forest | Bad | 45 | 66 | 77 | 83 |
| 342 | Moderately dense forest | Moderate | 36 | 60 | 73 | 79 |
| 341 | Dense forest | Good | 30 | 55 | 70 | 77 |

Rihane, 2015.

Base flow is an important parameter in flood studies because it defines a minimum river depth over which additional runoff accumulates. Models that neglect base flow may underestimate water levels and therefore fail to identify inundated reaches. Base flow is modeled using an exponential decrease function. The model requires also two parameters recession constant (Rc) and threshold (Td). However, given unavailability of information to assign an initial value and their value can be calibrated, a value from literature has been used until calibration of these parameters (Tramblay et al., 2012).

## Modeling Process

After preparing geodatabase and characterizing spatial flow structures, simulation files were created. As illustrated in Figure 4, for each considered event, simulation files have been prepared taking into account flood type and the transfer function formalism. Then, each file has been used in order to run the simulation. The models were calibrated and validated.

### Model Calibration

Calibration consists of fitting a model that could reproduce the observed flow data. It was carried out using univariate gradient method, *i.e.,* each of the model parameters has been fitted as the sole variable while maintaining the others as constants. During this process, fitting quality was appreciated using the objective function "root square of the weighted mean error." It allows considering both magnitude and time synchronization of flooding.

### Model Validation

Validation aims to expose a calibrated model to a real phenomenon different from that used for calibration, in order to assess its response and its ability to reproduce the peak flow properly.

*Figure 4. Simulation files prepared for each event*

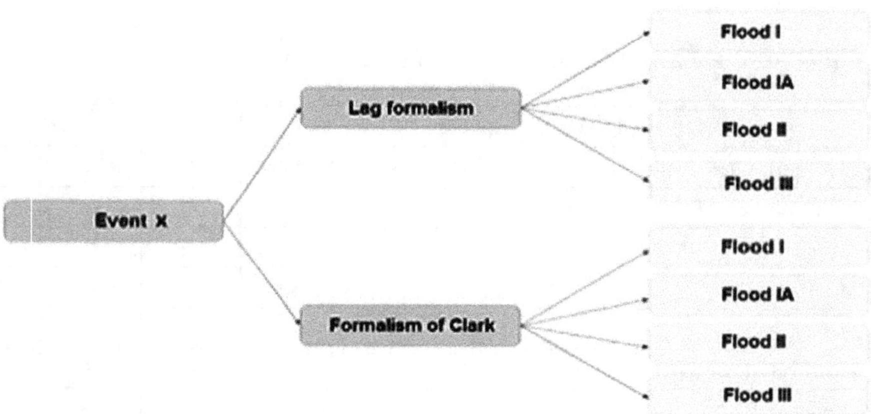

### Impact of Land Cover on Hydrological Functioning of the Ourika Basin

Fitting a probability distribution to observed flooding determines the floods quantiles. Probability distribution has been fitted using maximum likelihood method. Thus, corresponding floods at different return periods can be simulated.

Once the hydrological model has been validated, the impact of land cover on hydrological functioning of the Ourika basin was assessed through simulation of the hydrological response of the catchment basin after a given flood.

## RESULTS AND DISCUSSION

### Rainfall Characterization

#### Evaluation of Rainfall and the Resulting River Flow

Annual rainfall during the period (1976-2011) is shown in Figure 5. It ranges from 525.9 mm at the Aghbalou station to 321.2 mm at the Agouns station. Before the year 2000, the area has experienced two major dry periods, 1982-1988 and 1992-1996. However, from the 2000s, rainfall has become increasingly low and droughts are more frequent.

During the period 1984-2013, peak flows decreased from year 2000. However, after 2007 these peaks become more important (Figure 6).

Furthermore, peak flow analysis has identified thirty flood events, of which nine have the greatest peak flows. These events are retained for the hydrological modeling to calibrate or validate the model (Table 3).

*Figure 5. Annual rainfall in the Ourika basin (1976-2011)*

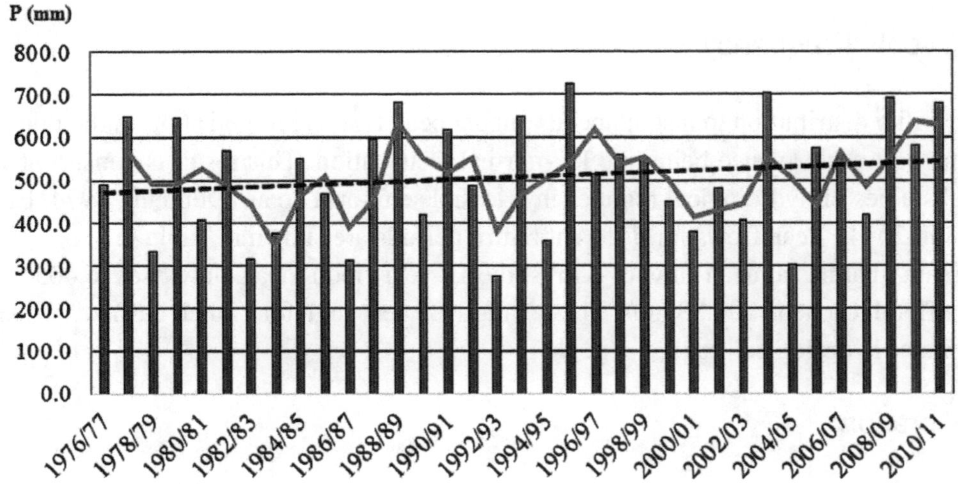

*Figure 6. Evolution of the peak flows between 1984 and 2013 (from Aghbalou station)*

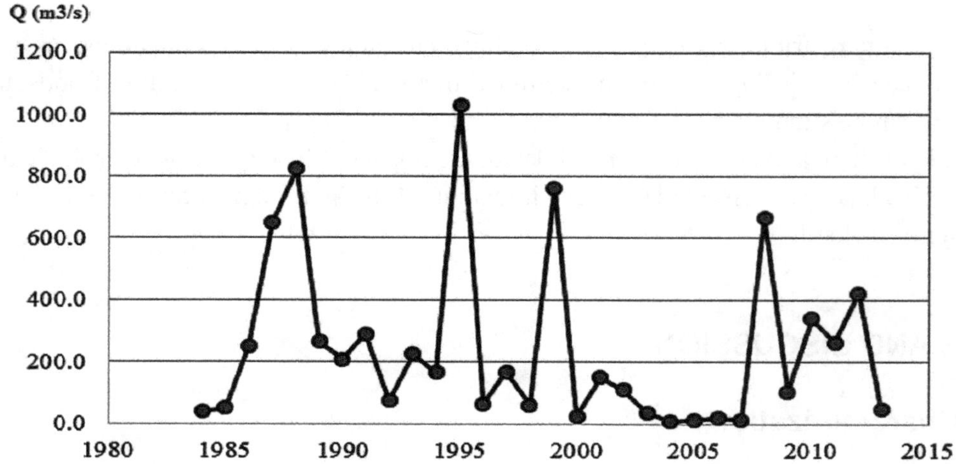


*Table 3. Nine (9) events selected for modeling the basin hydrological functioning*

| Event | Start Date | End Date | Function | Flow (m3/s) |
|---|---|---|---|---|
| 3 | 09/02/1986 | 12/02/1986 | Calibration | 249 |
| 4 | 01/11/1987 | 03/11/1987 | Calibration | 650 |
| 5 | 05/05/1984 | 10/05/1984 | Calibration/Validation | 823 |
| 6 | 10/03/1989 | 12/03/1989 | Calibration | 265 |
| 7 | 13/09/1990 | 15/09/1990 | Calibration | 207 |
| 8 | 01/08/1992 | 03/08/1992 | Calibration | 290 |
| 10 | 06/03/1993 | 09/03/1993 | Calibration | 226 |
| 12 | 15/08/1995 | 20/08/1995 | Calibration/Validation | 1030 |
| 16 | 26/10/1999 | 30/10/1999 | Calibration | 762 |

## Extreme Events Frequency

Fitting probability distribution to instantaneous annual peak flows recorded at the Aghbalou station has identified the best performance being the log-normal distribution. This result is consistent with those from several studies of hydrological regimes in arid and semi-arid areas (Dubreuil, 1974, El Alaoui El Fels and Saidi, 2014). Peak flows for different return periods are shown in Table 4.

For a stream, that has a mean flow of 5 $m^3/s$, the 10-year flood mean flow rises to 405.21 $m^3/s$ and the 100-year flood reached 1290.33 $m^3/s$. These values are very high for a small basin of 576 $km^2$, which indicates a major hydrological risk.

## Model Calibration

The values achieved by the base flow volume and the peak flow volume objective functions- together with the fitted models parameters, vary depending on the event, flood type and transfer function. It ap-

*Table 4. Peak flow simulated values for floods quantiles*

| Return Periods (Year) | Qp ($m^3/s$) |
|---|---|
| T=2 | 93.38 |
| T=5 | 253.04 |
| T=10 | 405.21 |
| T=20 | 611.82 |
| T=50 | 945.60 |
| T=100 | 1 290.33 |

pears from the first results that the objective function is never null (=0) in the case of the Lag formalism and for the four flooding types (except for events 4 and 6). Furthermore, flood peaks seem to be well predicted in the case of flood types I and IA.

The simulated volumes are always overestimated whatever the event and the flood type; overestimation is less in the case of flood type I and IA. Base flows appear to be poorly simulated. They are either overestimated (case of events 3, 6, 7 and 10) or underestimated (case of 16 events, 12 and 5).

Concerning optimized values of the parameters, "Td" is unchanged regardless of the event, while the other parameters change depending on the event. Indeed, the parameters have values that do not allow minimization of the objective function and reduction of the overestimation of base flow and flood volumes.

In the case of the Clark formalism and with the exception of the two events 7 and 10 who were eliminated because of their poor results, hydrograph restitution has been significantly improved for all events. Indeed, the objective function is null regardless of the event and flood type. Floods volumes overestimation still persists and could be explained by the fact that the transfer function has no effect on the flood volume. The same could be concluded for base flow that follows the volumes change.

In our context, the Clark hydrograph formalism seems to be more suitable than the Lag formalism. Furthermore, flood types I and IA can better reproduce the hydrograph characteristics and have a good performance. However, it does not reproduce exactly the volumes and base flows. This may be due to the misjudgment of the input variables of the production function or its parameters, or even the selection of the objective function. Consequently, and to reach an optimum set of parameters, various methods have been tried on the two flood types I and IA to detect one that improves the model performance. These methods are:

**Method 1:** Average of the most frequent parameter values;
**Method 2:** Maximum of the most frequent parameter values;
**Method 3:** Maximum of 7 events parameter values;
**Method 4:** Average annual parameter values;
**Method 5:** Most probably annual parameter values.

For the flood type I as illustrated in Figure 7, the difference between the peak flow observed and simulated varies from 0.1% (method 1, event 6) to 39.2% (method 3, event 16). The best-reproduced events are 8 and 12, while the weak simulated events are 4 and 16. Therefore, regardless of the method used, the model efficiency on peak flow of these two events is weak.

*Figure 7. Comparison of simulated peak flows and observed peak flows after validation by the five used methods*

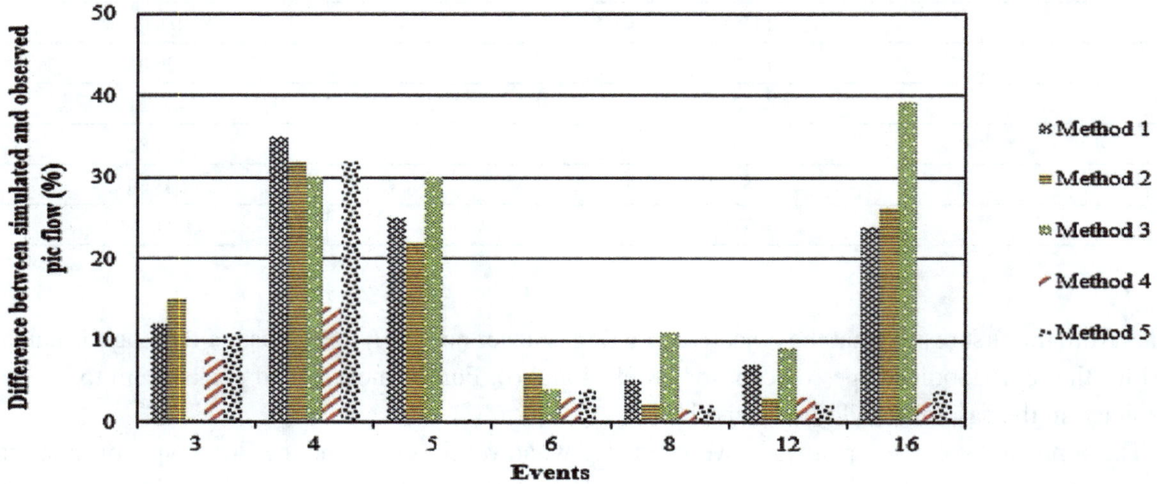


Concerning the accuracy of the simulated volume, Figure 8 shows that the behavior is different depending on the method and the event. It turns out that the problem of overestimated volume still persists.

In addition, the base flow is almost overestimated by all methods and for all events. This problem seems to be related to the unchanged value of the "Td" parameter (Figure 9).

Validation results for the flood type IA show a good performance compared to those obtained by the flood type I.

This validation test state that method 4 (Average annual parameter values) was one that gave the best results. Indeed, for both flood types, this method did not indicate large differences between simulated peak flows and those observed (between 0.2% and 16%). The same finding is applied to volumes and

*Figure 8. Comparison of simulated volumes and observed volumes after validation by the five used methods*

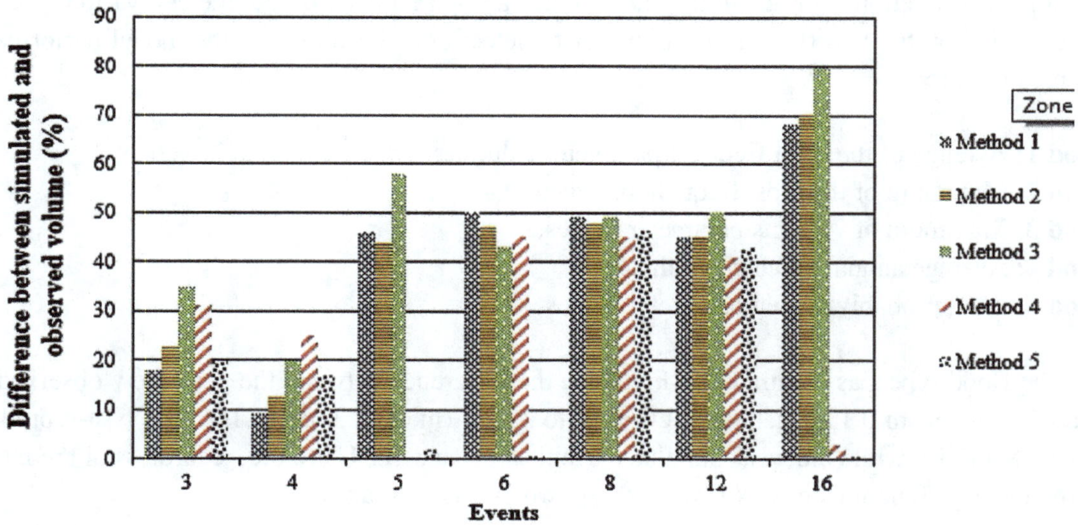

*Figure 9. Comparison between simulated and observed base flows after validation by the five used methods*

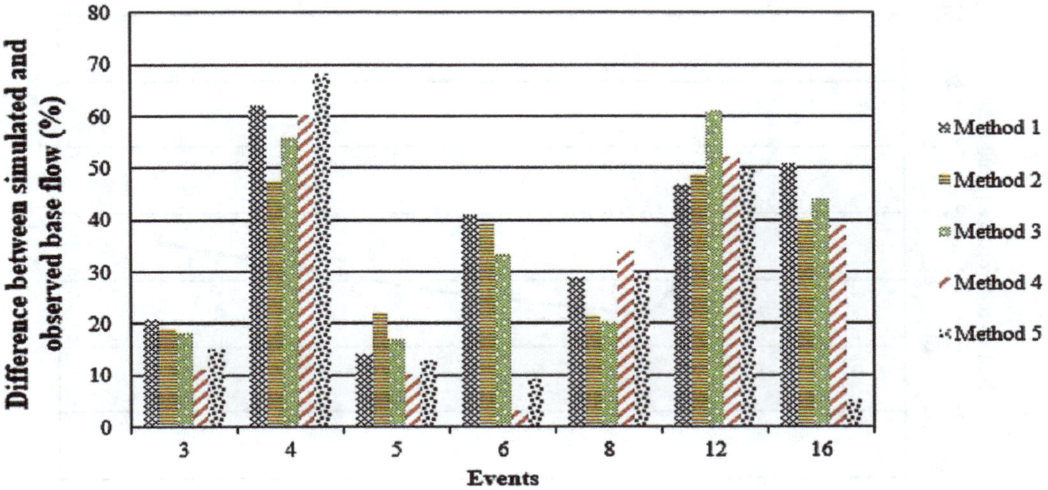


base flows. This can be justified by the fact that this model is closely linked to basin surface condition, *i.e.* if land use changes over short periods; optimal parameters set will follow such variations in a continuous manner.

Nevertheless, based on this first test, one cannot speak of a model validation, firstly because there is not enough events each year, and secondly, the weak performance obtained during validation cannot be linked only to land use change, but requires verification of the assumptions related to the objective function choice, estimated impermeability rate and daily precipitation distribution.

## Model Sensitivity to the Objective Function

To explain parameter values instability, and assuming that the objective function influences the peak flow in favor for volume and base flow, the objective volume function was used. However, nothing changed in flood volume simulations. Hence, the need to test the production function part was a weakness.

## Model Sensitivity to the Production Function

After eliminating the objective function as the probable causes of the model performance, its sensitivity to the production function to estimate the runoff volume induced by a given rainfall amount was assessed. In this implementation, the impermeability rate of the basin, derived from data of the Tensift Hydraulic Basin Agency (ABHT), was estimated at 55%. To justify this value, the event 16 was selected because it is the only event that had after calibration, a parameters set close to initial values.

Figure 10 shows the simulated volume according to basin impermeability. It indicates that the simulated volume and impermeability are proportional until to impermeability rate values found are less than 40%.

This can be explained by the fact that in addition to basin impermeability, other influencing factors occur; when impermeability decreases, its influence on flood volume decreases, this also increases this influence

*Figure 10. Change in the simulated volume for event 16 according to impermeability*

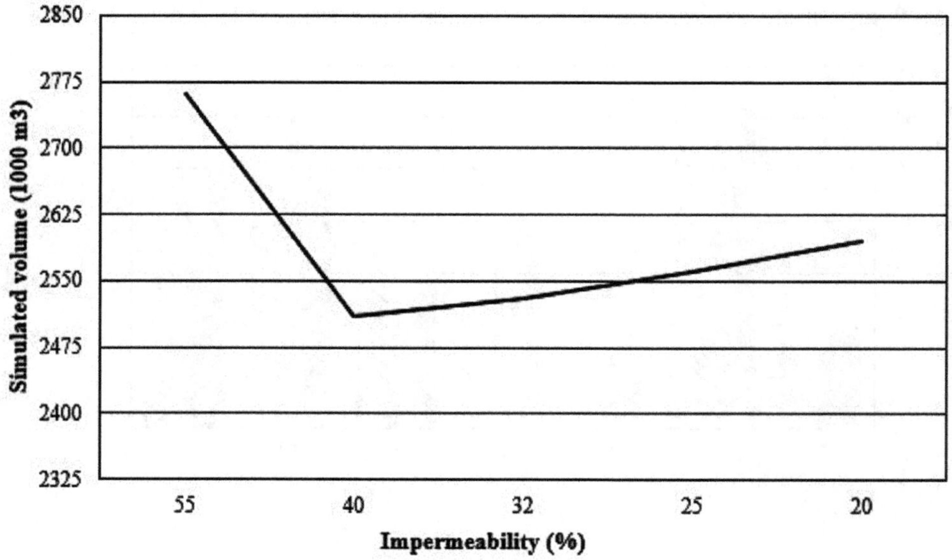


For an impermeability rate less than 40%, differences between the simulated volume and the real volume are observed induced by the production function effect on runoff and requires changing of the parameters function. Also, after eliminating the impermeability rate effect, the daily precipitation distribution seems to be the only factor that explains the weakness of the model performance which is due to the lack of data (measured with rain gauges) that could substitute the NRCS flood types responsible for model imperfection.

The last calibration performed on event 16 was chosen to validate the model. In this process, the impermeability rate was set at 40%, for flood type IA (the optimal parameters set is shown on Table 5).

## Model Validation

The simulation results, using the optimal parameters set, are shown in Table 6. The model was able to acceptably reproduce the peak flow to all events, despite the underestimation that varied between 13.19% and 32.82%. Overestimation of the simulated volumes ranged from 1.49% for event 16 and 58.94% for event 8.

For base flow, event 12 has a better simulation comparing to other events whose error rate is high.

*Table 5. Optimal parameter set used for validation*

| Parameter | Production Function | | Transfer Function | | Base Flow | | Impermeability | Flood Type |
|---|---|---|---|---|---|---|---|---|
| | Ia | CN | $T_c$ | $S_t$ | $R_c$ | $T_d$ | | |
| Optimal value | 8.39 | 70.01 | 5.05 | 4.89 | 0.79 | 0.1 | 40 | IA |

*Table 6. Results of validating the optimal parameters set*

| Event | Qp Observed ($m^3/s$) | Qp Simulated ($m^3/s$) | Diff. (%) | Observed Volume ($10^3$ $m^3$) | Simulated Volume ($10^3 m^3$) | Diff. (%) | Qb Observed ($m^3/s$) | Qb Simulated ($m^3/s$) | Diff. (%) |
|---|---|---|---|---|---|---|---|---|---|
| 3 | 249.0 | 201.12 | -19.23 | 10 104 | 10 769.11 | +6.58 | 19.4 | 10.4 | -46.39 |
| 4 | 650.0 | 436.67 | -32.82 | 34 365 | 35 161.52 | +2.32 | 12.5 | 23.58 | +88.64 |
| 5 | 823.0 | 700.41 | -14.90 | 34 417 | 34 998.01 | +1.69 | 26 | 13.9 | -46.54 |
| 6 | 265.0 | 187.92 | -29.09 | 7 708 | 8 100.97 | +5.10 | 1.2 | 2.07 | +72.50 |
| 8 | 290.0 | 221.59 | -23.59 | 4 412 | 7 012.3 | +58.94 | 2.4 | 3.93 | +63.75 |
| 12 | 1 030.0 | 894.16 | -13.19 | 3 651 | 3 911.15 | +7.12 | 50 | 61.1 | +22.20 |
| 16 | 762.0 | 592.19 | -22.28 | 25 987 | 26 375.05 | +1.49 | 12.72 | 20.03 | +57.47 |

These results can be explained by the ability of the exponential recession method to simulate the total flow after the hydrograph peak. It can be due either to the fixed Td parameter value or the Rc value. Indeed, an Rc value of 0.1 suggests that the underground aquifer resume its contribution to the hydrograph flow shape once the peak regressed by 10%. However, results show that over the observed base flow post peak moves away from 10% of peak flow, the error rate on the base flow increases. This is attributed to either weak quality of the measured flow data or the variability of aquifer-river relationship in the Ourika basin.

Despite anomalies, the model has been validated on events 5 and 12 for which the simulated flows *vs.* observed flows are shown in Figures 11 and 12. These events were used to simulate the basin response in different scenarios of land cover change.

*Figure 11. Model validation by event 5*

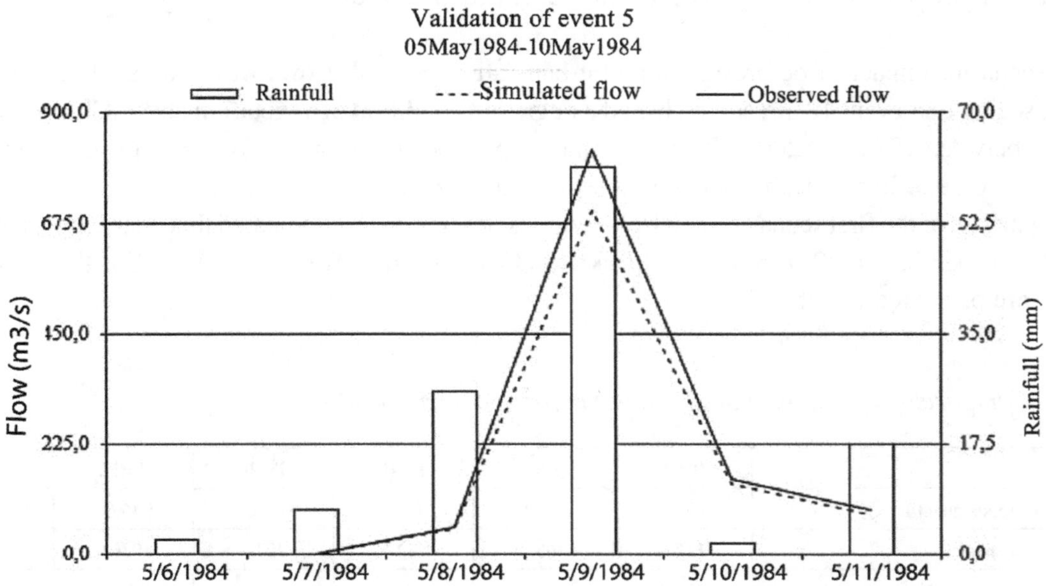

*Figure 12. Model validation by event 12*

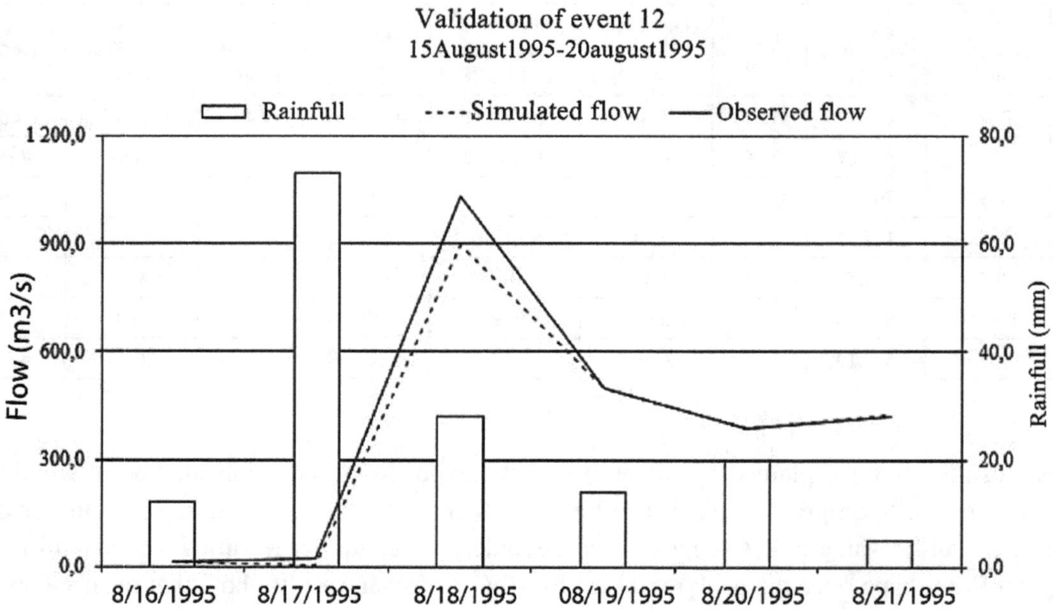


## Impact of Land Cover on Hydrological Behavior of the Ourika Basin

The impact of land cover on hydrological behavior of the Ourika basin was assessed through events 5 and 12 for which the model was able to reproduce the peak flow. The observed peak flow for event 5 was linked to the land cover condition in 1984. As for the event 12, which corresponded to the famous flood of 1995, this was linked to the land cover condition in 2000 in the absence of land cover map in 1995.

### Scenario 1: Impact of a Negative Change in Land Cover

In this scenario, impact of deforestation and urbanization on peak flows was simulated according to event 5, seeing that event 12 already reflected a degraded condition of vegetation cover. Changes in forest cover between 1984 and 2000 affected the model parameters, because CN value for the basin spent from 70.01 to 79.5. Impermeability rate was estimated at 48%.

Comparison of the first scenario peak flows with its observed value indicated that peak flow increased by 31.68% (1083.73 vs. 823). Simulated peak flows associated to floods quantiles (for the six return periods) are presented in Table 7.

*Table 7. Comparison of floods quantiles and the first scenario results*

| | T=2 | T=5 | T=10 | T=20 | T=50 | T=100 |
|---|---|---|---|---|---|---|
| Qp (m3/s): floods quantiles | 93.38 | 253.04 | 405.21 | 611.82 | 945.60 | 1 290.33 |
| Qp (m3/s): scenario 1 | 117.1 | 303.5 | 484.5 | 702.7 | 1056.9 | 1 390.59 |
| Difference (%) | 25.40 | 19.94 | 19.57 | 14.85 | 11.77 | 7.77 |

Comparison of these results with those of the first scenario, indicate that when the return period is greater, influence of land cover on peak flow decreases. In other words, the dampening effect of forest cover decreases when the event becomes extreme. This corroborates with the results found by Jeníček (2007).

## Scenario 2: Impact of a Positive Change in Land Cover

The idea of this scenario is to assess the impact of forest activities in the Ourika basin, particularly reforestation and to stop logging and socio-spatial dynamics in the region during the last fourteen (14) years. The new simulation parameters recalculated using the land cover map of 2014 is CN value of 72.09 and impermeability rate nearly to 43%. Simulating the impact of a positive dynamics of forest cover on the hydrological behavior of the basin was based on event 12, referring to a degraded forest cover condition. Comparison of the second scenario peak flows with its observed value indicates a decrease of approximately 17.25% in peak flow (852.34 vs. 1030m3/s).

However, for extreme events, there is a decrease in influence of land cover on flows, as shown in Table 8.

Precipitation data analysis and stream flows indicate a decreasing trend in rainfall volume associated with increasing river flows, which concord with findings of climate studies in the basin (Babqiqi and Messouli, 2013). In this climate change context characterized by water scarcity and recurrence of extreme events, adoption of adaptation measures by land managers are needed to limit disastrous impact on property and human lives.

Reinforcing the forest cover is one management option. However, questions are remaining about the role of forest cover on the hydrological behavior of the basin. Validation of hydrological model for different conditions of forest cover and for different events helped calculate the rivers peak flows that are as accurate as possible to those really measured.

Several issues relating to the quality of the model results emerge. The particular context of the basin characterized with a greater diversity of watershed characteristics, including topography, soils, and land use, seem, in comparison with homogeneous basins, to produce poorer results (Knebl et al., (2005). Furthermore, due to lack of information, averaging basin properties over larger areas may decrease model accuracy.

On the other hand, the used SCC-CN is controversial and its limitations and application conditions have been widely discussed (Knebl et al., 2005; Ponce and Hawkins 1996; Garen and Moore 2005). It was selected in order to fully utilize the available spatially distributed GIS data for the Ourika watershed. Furthermore, the infiltration coefficient (CN) is an empirical parameter and has a limitation in the present model.

*Table 8. Comparison of floods quantiles and the second scenario results*

| | T=2 | T=5 | T=10 | T=20 | T=50 | T=100 |
|---|---|---|---|---|---|---|
| Qp (m3/s): floods quantiles | 93.38 | 253.04 | 405.21 | 611.82 | 945.60 | 1 290.33 |
| Qp (m3/s): scenario 2 | 78.36 | 220.92 | 357.12 | 560.18 | 881.68 | 1 205.69 |
| Difference (%) | 16.08 | 12.69 | 11.87 | 8.44 | 6.76 | 6.55 |

In addition, due to the highly heterogeneous geological nature of the study area, groundwater recharge is difficult to quantify and a general parameterization of infiltration based on soil data, such as, that originally used by the SCS, may be inadequate to portray this heterogeneity. Indeed, the complex hydrological processes in the model may attenuate the error; alternatively, the model may exacerbate the input error (Knebl et al.,2005),

Thus, model results should be considered taking into account these limitations in order to better understand the contribution of forest cover / land use in hydrological functioning of the basin, in order to inform decision-makers. In any climate change perspective, such information is very valuable to mitigate adverse weathers conditions.

## CONCLUSION

The present study assessed the impact of forest cover on peak flows. Simulation results indicated that the forest cover degradation and increased bare soils has the consequence of increasing flood peak flow. Inversely, increasing forest cover results in a moderation of flood peak flow. However, this effect regresses in extreme events. This becomes not sufficiently clear as the return period increases. These results will help to make planning choices taking into account the role of forest cover and to promote better land use planning considering flooding risk, mainly in the context of climate change influencing both extremes events and land cover density and extent.

## REFERENCES

Aldosari, A. (2006). Middle East, Western Asia, and North Africa. In C. Carpenter (Ed.), *World and its Peoples* (p. 1258). New York: Marshall Cavendish Corporation.

Allan, R. P., & Soden, B. J. (2008). Atmospheric warming and the amplification of precipitation extremes. *Science*, *321*(5895), 1481–1483. doi:10.1126cience.1160787 PMID:18687921

Andréassian, V. (2002). *Impact de l'évolution du couvert forestier sur le comportement hydrologique des bassins versants. Tome 1.* Thèse de Doctorat de l'Université Paris 6. Ecole doctorale: Géosciences et Ressources Naturelles Spécialité: Hydrologie.

Andréassian, V, (2004). *Couvert forestier et comportement hydrologique des bassins versants.* La houille blanche/N° 2-2004-31.

Andréassian, V., Tangara, M., & Muraz J. (2001). Evaluer l'impact de l'évolution du couvert forestier sur le comportement hydrologique des bassins versants: méthodologie et premiers résultats fondés sur les données de l'IFN. *Rev. for. fr. liii -3-4-2001*, 475-480.

Babqiqi, A., & Messouli, M. (2013). Simulation of climate and its implication on agriculture in Morocco using Statistical DownScaling. *International Journal of Latest Research in Science and Technology,* 2(5), 83-96.

Choi, W., & Deal, B. (2008). Assessing hydrological impact of potential land use change through hydrological and land use change modeling for the Kishwaukee River basin (USA). *Journal of Environmental Management, 88*(4), 1119–1130. doi:10.1016/j.jenvman.2007.06.001 PMID:17689854

Clark, C. O. (1945). Storage and the unit hydrograph. *Transactions of the American Society of Civil Engineers, 110*, 1419–1488.

Driouech, F. (2010). *Distribution des précipitations hivernales sur le Maroc dans le cadre d'un changement climatique: descente d'échelle et incertitudes*. Thèse de Doctorat. Université de Toulouse.

Dubreuil, P. (1974). *Introduction à l'analyse hydrologique*. Paris: Ed. Masson-ORSTOM.

El Alaoui El Fels, A., & Saidi, M. E. (2014). Simulation et spatialisation du risque d'inondation dans une vallée anthropisée, le cas de la vallée de l'Ourika (Haut Atlas, Maroc). European Scientific Journal, 10(17).

Garen, D. C., & Moore, D. S. (2005). Curve number hydrology in water quality modeling: Uses, abuses, and future directions. *Journal of the American Water Resources Association, 41*(2), 377–388. doi:10.1111/j.1752-1688.2005.tb03742.x

Gerrits, A. M. J. (2010). The role of interception in the hydrological cycle. Academic Press.

Halwatura, D., & Najim, M. M. M. (2013). Application of the HEC-HMS model for runoff simulation in a tropical catchment. *Environmental Modelling & Software, 46*, 155–162. doi:10.1016/j.envsoft.2013.03.006

He, C. (2003). Integration of geographic information systems and simulation model for watershed management. *Environmental Modelling & Software, 18*(8-9), 809–813. doi:10.1016/S1364-8152(03)00080-X

Jeníček, M. (2007). Effects of land cover on runoff process using SCS CN method in the upper Chomutovka catchment. In *Proceedings of the first Scientific Conference on Integrated catchment management for hazard mitigation*. Remote Sensing Department, University of Trier.

Knebl, M. R., Yanga, Z. L., Hutchisonb, K., & Maidment, D. R. (2005). Regional scale flood modeling using NEXRAD rainfall, GIS, and HEC-HMS/RAS: A case study for the San Antonio River Basin Summer 2002 storm event. *Journal of Environmental Management, 75*(4), 325–336. doi:10.1016/j.jenvman.2004.11.024 PMID:15854726

Legrouri, A., Kettani, D., & Kalpakian, J. (2012). *Using Demand Side Management to Adapt to Water Scarcity and Climate Change in the Saïss Basin, Morocco*. Ifrane, Morocco: Al Akhawayn University.

Letcher, R. A., & Gioppini, C. (2005). Policies and tools for sustainable water management in the European Union. *Environmental Modelling & Software, 20*(2), 93–98. doi:10.1016/j.envsoft.2004.01.003

Motevalli, S., Hosseinzadeh, M., Esmaili, R., Derafshi, K., & Gharehchahi, S. (2012). Assessing the Effects of Land use Change on Hydrologic Balance of Kan Watershed using SCS and HEC-HMS Hydrological Models – Tehran, IRAN. *Australian Journal of Basic and Applied Sciences, 6*(8), 510-519.

Musy, A., & Higy, C. (1998). Hydrologie appliquée. Edition *H*G*A*.

Ohana-Levi, N., Karnieli, A., Egozi, R., Givati, A., & Peeters, A. (2015). Modeling the Effects of Land-Cover Change on Rainfall-Runoff Relationships in a Semiarid, Eastern Mediterranean Watershed. Hindawi Publishing Corporation. *Advances in Meteorology.*

Oleyiblo, J. O., & Li, Z. J. (2010). Application of HEC-HMS for flood forecasting in Misai and Wan'an catchments in China. *Water Science and Engineering, 3*(1), 14-22. 10.3882/j.issn.1674-2370.2010.01.002

Ponce, V. M., & Hawkins, R. H. (1996). Runoff curve number: Has it reached maturity? *Journal of Hydrologic Engineering, 11*(1), 11–19. doi:10.1061/(ASCE)1084-0699(1996)1:1(11)

Rihane, R. (2015). *Contribution à l'évaluation des services hydrologiques de la forêt et des aménagements antiérosifs dans le bassin versant de l'Ourika. Mémoire de 3ème cycle.* Maroc: Ecole Nationale Forestière d'Ingénieurs.

Rochdane, S., Bounoua, L., Zhang, P., Imhoff, M. L., Messouli, M., & Yacoubi-Khebiza, M. (2014). Combining Satellite Data and Models to Assess Vulnerability to Climate Change and Its Impact on Food Security in Morocco. *Sustainability, 6*(4), 1729–1746. doi:10.3390u6041729

Roeckner, E., Bäuml, G., Bonaventura, L., Brokopf, R., Esch, M., Giorgetta, M., . . . Tompkins, A. (2003). *The atmospheric general circulation model ECHAM5. Part I: Model description.* Max Planck Institute for Meteorology Report n° 349.

Roy, T., Dinar, A., Tsur, Y., & Diao, X, (2005). Feedback links between economy-wide and farm-level policies: With application to irrigation water management in Morocco. *Journal of Policy Modeling, 27,* 905–928.

Saidi, M. E., Daoudi, L., Aresmouk, M. E., & Blali, A. (2003). *Rôle du milieu physique dans l'amplification des crues en milieu montagnard,exemple de la crue du 17 août 1995 dans la vallée de l'Ourika* (Vol. 14-2). Haut Atlas, Maroc: Sécheresse.

Saidi, M. E., Daoudi, L., Aresmouk, M. E., Fniguire, F., & Boukrim, S. (2010). Les crues de l'oued Ourika (Haut Atlas,Maroc): Événements extrêmes en contexte montagnard semi-aride. Comunicações Geológicas, t. 97. *Lisbonne, 2010,* 113–128.

Salama, H., & Tahiri, M. (2010). La gestion des ressources en eau face aux changements climatiques. Cas du bassin Tensift (Maroc). *Larhyss Journal, 8,* 127-138.

Secrétariat d'Etat Chargé de l'Etat de l'Environnement (SEEE). (2010). *Seconde communication nationale à la convention cadre des nations unies sur les changements climatiques.* Retrieved from http://www.abhatoo.net.ma/content/download/20617/375478/version/1/file/seconde_communication_nationale_convention_cadre_NU_chang_clim.pdf

Sibari, T., Haidi, S., & Ait Foura, A. (2001). Typologie des crues et érosion mécanique dans un bassin versant de zone semi-aride: Bassin versant de I'Inaouene, Maroc. Science et changements planétaires. *Sécheresse, 12*(3), 187–193.

Taha, A., & Mualla, W. (2003). *La prédiction du comportement hydrologique du bassin Awage en utilisant le modèle WMS (Watershed Modelling System).* Publi. N° 278, 2003.

Tramblay, Y., Badi, W., Driouech, F., Al Adlouni, S., Neppel, L., & Servat, E. (2012). Modélisation des crues dans le bassin du barrage Makhazine. Rapport interne de l'ABH de Loukous.

U.S. Army Corps of Engineers (USACE). (2000). *Hydrologic Modeling System: Technical Reference Manual.* Davis, CA: Hydrologic Engineering Center.

World Bank. (2013). *Climate-Induced Spatio-Temporal Shifts in Natural and Agro-ecosystems in the Middle East and North Africa Region.* A synthesis report. World Bank.

Yener, M. K., Şorman, A. Ü., Şorman, A. A., Şensoy, A., & Gezgin, T. (2007). Modeling studies with HEC-HMS and runoff scenarios in Yuvacik basin, *Turkey. International Congress on River Basin Management.*

# Section 3

# Land Policy Assessment

# Chapter 13
# The Cadastral and Land Information Systems for an Effective Land Governance

**Abdeslam Moulay Adad**
*ANCFCC, Morocco*

## ABSTRACT

*The cadastral system as a civilian register of a property has a fundamental role in every society because it ensures the relationship between the land and the humankind. This relationship is evident in the form of property rights and established in many different manners from the full control, through communal forms of tenure, to the individual property rights. This chapter has the aim to deal with the nature and categories of the existing cadastral systems. It will highlight the new approaches and visions to establish cadastral systems that will meet the requirements of re-engineering the framework of land development projects. An overview of the state of the art of the modern concepts of cadastral systems will be provided such as Cadastre 2014, 3D Cadastre, and Marine Cadastre. The goal behind this approach is to enable decision makers to have an idea on the existing paradigms of land administration and to master the relationships between humankind and land, infrastructures and business systems, and between the human being and technological issues.*

## INTRODUCTION

The cadastral system is the basis used for the protection of the property by means of registration procedures and cadastral plans made from geodetic operations and surveying. Each parcel or property and its owners are registered and all the spatial structures consisting of location, boundaries and contents are graphically described in a cadastral map or digitally stored in a database. It may be a systematic or a sporadic and official description of properties affording information on real estate. The cadastral data need to be collected, managed, analyzed, and rationally stored to facilitate their retrieval. After being

DOI: 10.4018/978-1-5225-5939-9.ch013

processed, these data become land information that allow users to make analysis and plan future projects. Recently, there is pressure to develop a modern cadastral system based on the information technology to create a unique reference for land properties as geometrical features and land tenure as attribute information. For many reasons, the cadastral system is a fundamental key providing accurate information about the lands and their attributes. It is a basis for establishing national geo-information infrastructure. The data are the basic resources of the information that is the fundamental key for all decision-making. The collected data in a manual or digital form are stored and processed in order to produce information, which may be analytical or graphical result. The aim of the information is to facilitate analysis, retrieval, and updating processes. The quality of the information depends largely upon both the quality of data and the quality of processing systems. The technological tools used to manage and process data include the so-called Geographic Information System (GIS) that facilitates collecting, modeling, and managing, analyzing and storing spatial and non-spatial data. Spatial data are lines, surfaces, or volumes and they have their positions located by a coordinate system. Whereas non-spatial data are specific attributes characterizing spatial data. The GIS tools enable spatial searching, overlay operations, spatial analysis and modeling. Several GIS applications have been developed dealing with many different data. A kind of applications or computerized systems based on this technology is Land Information System (LIS). The land information systems have the benefits to serve society on a long-term basis and on a long-term return on investment by taking in account both the static and dynamic components of land. The static aspects are the norms and values adopted in society to define rights and to record existing land tenure, value, and use ascertained by adjudication process and mapping. While dynamic aspects are the changes that affect the relationship between man and land according to the changing nature of land rights, land value; and land use.

## CADASTRE AND LAND INFORMATION SYSTEMS

### Meanings of the Cadastral Systems

Generally, the cadastre means the civilian state of a property. Its role becomes fundamental in every society because it ensures the relationship between the land and the humankind. This relationship is evident in the form of property rights and established in many different manners of life; from the full state control, through communal forms of tenure, to the individual property rights (Dale and McLaughlin, 1988).

According to Hensen [1995], the cadastre is described as a methodically arranged public inventory of data concerning properties within a jurisdiction, based on a survey of their boundaries. In certain countries, for example Portugal, the word cadastre means a land register system. Silva L. (2001) stated that this system is an inventory system consisting of data that characterizes and identifies existing land parcels. Its main objective is the unique identification of every land parcel by giving it a unique code as well as a unique characterization through an administrative description, geographic location, shape and area.

Dowson and Sheppard [1952] remark that it is impossible to give a definition of a cadastre that is both terse and comprehensive. Its distinctive character is readily recognized and may be expressed as the marriage of (a) technical record of the parcellation of the land through any given territory, usually represented on plans of suitable scale, with (b) authoritative documentary record, whether of a fiscal or proprietary nature or for two combined. The International Federation of Surveyors (FIG, 1995) definition is that: "a cadastre is normally a parcel-based and up-to-date land information system containing a record of interests in the land (e.g., rights, restriction, and responsibilities). It usually includes a geometric description of land parcels linked to other records describing the nature of the interests, and often the value of the parcel and its improvements".

Larsson (1991, p.16) added: "cadastre thus has a distinct meaning as a specific type of land record supported by maps [...]. Essentially, a cadastre is thus a systematic description of the land units within an area. The description is made by maps that identify the location and boundaries of every unit, and by records. In the records, the most essential information is the identification number and the area of the unit [...]. Additional information may sometimes be found in the cadastral records or in adjacent records". Sometimes the term cadastre has the same meaning as the cadastral system. Then the concept of cadastre or cadastral system is the combination between a surveying and mapping system with its spatial focus and a land registration system with its juridical focus.

To summarize, the cadastre or the cadastral system, is the basis used for the protection of the property by means of registration procedures and cadastral plans made from geodetic operations and surveying. Each parcel or property and its owners are registered and all the spatial structures consisting of location, boundaries and contents are graphically described in a cadastral map or digitally stored in a database. It may be a systematic or a sporadic and official description of properties. It is seen as land information system affording information on real estate of a property (El Ayachi and Semlali, 2001).

## The Cadastral Information Systems

Cadastral systems furnish data focusing on the following components:

- Geometrical features, concerning the geometrical form of land parcels and properties and the positions of such data.
- Cartographical features, showing the generalized and symbolized data and the spatial relationships between data (topology).
- Attribute or textual features, regarding land tenure, value and use.

These data need to be collected, managed, analyzed, and rationally stored to facilitate their retrieval. After these processes, data become information in cadastral system. Their characteristic is that they allow users to make analysis and plan future projects, if they are gathered in a efficient and rational manner. While the cadastral system refers to the analogue or analytical system, the Cadastral Information System (CIS) refers to the computerized cadastral system. By definition, this means that all its components must be computerized.

One of the fundamental units in cadastral system is the cadastral map, which is the geometric description of land properties. In its conventional form, it cannot reflect the daily changes in the cadastral framework arising from land development. In addition, legal registers of rights and interests in land properties are manually described in separate analogous files.

There is pressure to develop a modern cadastral system based on information technology so as to create a unique reference for land properties as geometrical features and land tenure as attribute information. The modern cadastral system, which is the Cadastral Information System, offers the new following possibilities:

- The cadastral information system, cadastre in its digital format, enhances the process of maintaining digital cadastral maps, ensures the integrity of cadastral data, and enables the ability to integrate the spatial data with other land related spatial and non spatial data sets (Effenberg, 2001).
- The fundamental importance of the modern cadastral system is the possibility of linkage with other subsystems, leading to their integration in the frame of LIS and GIS aspects. All local and regional activities are coordinated to create a countrywide cadastral system (Cichocinski, 1999).
- The new system allows efficient information services with attribute data as well as spatial data on a map (Tuomaala and Uimonen, 1998).
- The modernized cadastral system is considered to be an important tool both from the information management and survey system perspective (Campbell, 1994).
- The modern cadastres are not ends in themselves, instead they support land markets, increased agricultural productivity, sustainable economic development, political stability and justice (Williamson, 1997).
- The new system serves not only the purpose of land registration but also any other land information systems.

For these reasons, the Cadastral Information System will be a fundamental key to provide accurate information about the lands and their attributes. It will be an essential basis for establishing national geo-information infrastructure. It needs to be designed in a good manner, rationally managed, and safely updated to serve security and sustainability.

## The Land Information Systems (LIS)

The data are the basic resources of the information that is the fundamental key for all decision-making. The collected data in a manual or digital form are stored and processed in order to produce information, which may be analytical or graphical result. The aim of the information is to facilitate analysis, retrieval, and updating processes. The quality of the information depends largely upon both the quality of data and the quality of processing systems.

The system "technology" managing and processing data includes the so-called Geographic Information System (hereafter noted GIS). In terms of computer technology, the GIS is a tool that facilitates collecting, modeling, managing, analyzing and storing spatial and non-spatial data. Spatial data are lines, surfaces, or volumes and they have their positions located by a coordinate system. Whereas non-spatial data are specific attributes characterizing spatial data. The GIS enables us spatial searching, overlay operations, spatial analysis and modeling. Several GIS applications have been developed dealing with many different data. A kind of applications or computerized systems based on this technology is Land Information System (LIS). In addition we can say that every system producing information on land and property related and spatially referenced data constitutes a Land Information System.

Dale and McLaughlin [1988, p.8] added that the Land Information System is a combination of human and technical resources, with a set of organizing procedures, to produce information in support of management activities [...]. In particular, LIS are concerned with detailed information recorded at the local level so they may be mapped at large scales. The land information systems serve the decision-maker to develop and to up-grade the national trends for land management. Their ultimate objective is to become a support to land administration by providing land and referenced information.

The land information systems have the benefits to serve society on a long-term basis and on a long-term return on investment. In the same way, we should take in account both the static and dynamic components. The static components are the norms and values adopted in society to define rights and to record existing land tenure, value, and use ascertained by adjudication process and mapping.

While dynamic components are the faces of effective changes that affect the relationship between man and land because there will be evolving forms of land rights, value and use, indigenous rights (natives titles and common properties), and public interests (Van der Molen, 2002). The dynamic aspects of the land information systems can be listed as below:

- Transferring land right, value, and use through the land market,
- Changing land tenure, value, and use through planning and development,
- Changing land tenure, value, and use through formalization processes,
- Changing land tenure, value, and use by integrating the indigenous right in a formal system,
- Changing land tenure, value, and use through land reform processes.

In order to meet the needs of the society, the Land Information Systems should have in their downstream reliable and accurate maps, which come from cadastre and survey mapping, and in their upstream good management (Figure 1).

Despite the LIS, as a computerized tool, enables to execute, manage, and analyze land tasks more expeditiously and efficiently; it could not be a magical key to overcome bureaucratic or administrative obstacles. According to Dale and McLaughlin [1999, p.94], the LIS are the subset of general information systems (Figure 1) and are designed to support strategic planning, management monitoring, and operational control.

The authors distinguish at least four types of LIS:

*Figure 1. Taxonomy of Land Information System*
*Adapted from Dale and McLaughlin, 1999.*

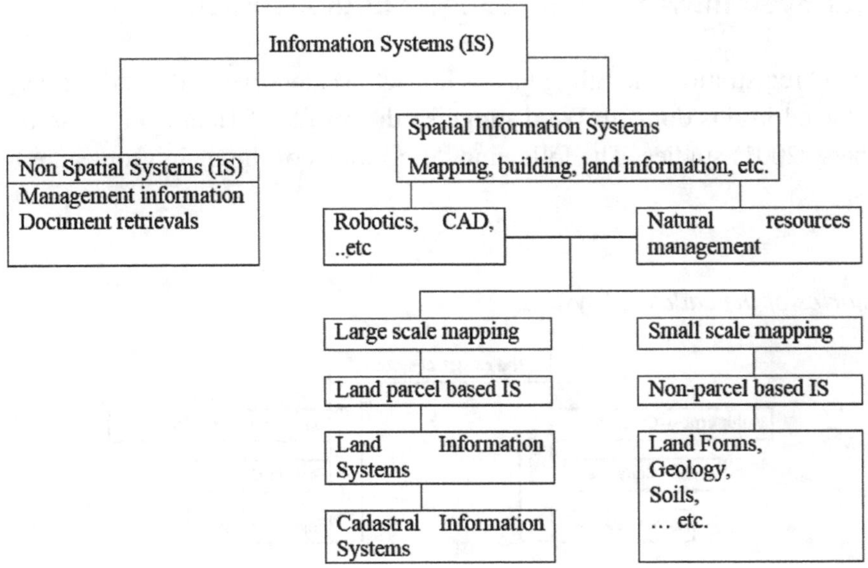

- Environmental information system, related to resources, vegetation, and wildlife;
- Infrastructure information system, focusing on engineering and utility structures (underground services);
- Cadastral information systems, applying to all lands and rights;
- Socio-economic information systems, incorporating census and population data.

The LIS are considered as principal tools to re-engineer the framework of development projects. They enable decision-makers to have an idea on the relationships between humankind and land, infrastructures and business systems, and between the human being and technological issues.

## CATEGORIES OF THE EXISTING CADASLTRAL SYSTEMS

The categorization of cadastral systems is a difficult and ambiguous task. The ambiguity is the result of various factors. Firstly, the cadastre is a very old system with direct relationships to human beings. Secondly, even if it has a static signification that means managing, valuating, and exploring land, it is largely dynamic, as the human knowledge, concepts, and procedures have been changing rapidly and timely. The cadastral systems can be categorized regarding their missions (NRC, 1983; Dale et McLaughlin, 1988), hence they are juridical cadastre that serves legal aspects of land rights, fiscal cadastre that was developed for properties valuation and taxation, and the so-called multi-purpose cadastre that encompasses both legal and fiscal aspects. On the other hand, they can be categorized according to their procedures of land titling and registration. According to Henssen (1995), they can be divided into three groups, which are the English group, the Torrens group, and the Germano/Swiss group (Figure 2).

In addition, all these cadastres may be graphical or digital systems. The former are typically inaccurate and based on graphical plans because they are mainly concerned by land uses. The latter, provide the necessary accuracy to determine analytically the coordinate of land limits and the exact geometrical form of the properties.

### The Cadastral Systems According to their Establishment

The method of land registration and titling has its impact in categorizing the cadastral systems. Several studies are conducted in this domain. We distinguish the results of Henssen (Henssen, 1995) and Effenberg (Effenberg, 2001) studies. The following descriptions of the main groups are based generally on these studies.

*Figure 2. Categories of the cadastral systems*

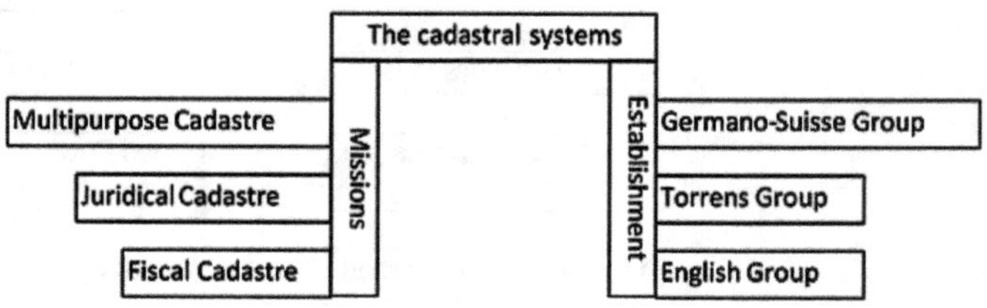

## The English Group

The English group is the first group, which has no formal cadastral map or survey plan to support land titling and registration, but the large-scale topographic maps are used to give accuracy and representation of the real physical features. In this group, we find several countries such as England, Scotland, Wales, Ireland, and Nigeria. In England, the large-scale topographic maps are produced by the UK mapping authority, Ordnance Survey, and their scales vary from 1/1250 in urban zones to 1/2500 in rural areas. In some coastal areas and mountains, the maps at 1/10000 scale are used. The property physical features described in these maps are generally adopted as clear boundaries.

Specifically in England, the land surveyors are not licensed or registered. Therefore, it is not required to involve a surveyor in determining the boundaries because they are produced by the available Ordnance Survey topographic maps. Each property is outlined in red on extracts of the Ordnance Survey topographic maps using a graphical accuracy based on the deed information.

## The Torrens Group

The second group is the Torrens group that uses individual surveys of the parcels or the properties without the specific requirement for a topographic map, but the cadastral map is used for charting and indexing registered or titled land parcels. Various countries are identified within this group such as Australia, New Zealand, some parts of Canada and USA, Morocco, Tunisia, and Syria. The origin of the cadastral concepts used by theses countries can be traced back to the Registrar Robert Torrens act developed in Australia in 27th of January 1858.

For example, in Australia-Canberra, the Australian Capital Territory (ACT), the Land Information Center noted ACTLIC is the Government's agency for surveying, mapping and land information. The center is a section of the Planning and Land Management group (PALM) which represents the dual mission of the local council and the state government within the Department of Urban Services. The ACTLIC maintains digital cadastral maps and all required spatial information. The fundamental spatial data are the land parcel boundaries that make up the ACT cadastre, which has its data daily updated. The reference framework of the accurate positional data is formed of a network of marks positioned every 150 meters along every street in Canberra. This densified system allows the licensed surveyors to fit the parcels to the cadastral maps with a relative precision of 3 centimeters (Effenberg, 2001). All parcels represented in the cadastral maps are accurate and their plan are deposited and registered at the Titles Offices. In Victoria (Australia), The Office of the Surveyor General and the Land Titles Office are grouped into the Office of Geographic Data Coordination to compile digital cadastral maps and maintain spatial data.

## The German/Swiss Group

The third group is the Germanic group, which has the parcel based cadastral map as an inherent component of the cadastral system. This grouped is compound of Germany, Switzerland, Austria, Sweden, Denmark, Alsace-Lorraine, Egypt, Turkey, Thailand, and South Korea. Within the central European countries, the cadastre is used for taxation purposes. However, other aspects such as land use and planning are envisaged in using cadastral maps and records.

The existing cadastres in this group operate mainly as two parts and are held in a separate administrations. The principal administration is generally the Ministry of Justice that is responsible for land titling, registration of rights, ownerships, and mortgages. The second administration that is responsible for national cadastral survey and recording parcel boundaries, can be a Ministry of Finance, Housing, Agriculture, a Ministry of Interior, or a Federal Office such as in Austria.

## THE CADASTRAL SYSTEMS ACCORDING TO THEIR MISSIONS

### The Fiscal Cadastre

The cadastre is defined as a general public inventory of land objects in order to assess values of land properties and to process land taxation. It constitutes an important source of wealth for societies. The essential reference used to investigate this system is from Dale Peter and John McLaughlin (1988). According to the authors (pp.46-62), the earliest fiscal cadastre was developed along the Tigris, Euphrates, and Nile Rivers where revenues of the Pharaohs were generated by assessing land income.

The beneficiaries of land belonging to the Pharaoh had to pay taxes, in form of rent. Greeks and Romans elaborated land records and survey systems to support taxation purposes. However, the modern uses of the principle of fiscal cadastre started from 1720 to 1723, when the Italian provinces of Milan and Mantua were mapped.

After this date, the Austro-Hungarian Empire carried out a cadastral survey to cover its entire territory between 1785 and 1789. In 1807, Napoleon Bonaparte introduced a new approach, which became largely accepted by most countries of Europe. The main philosophy of this system is that the land properties are capable of producing incomes over time, hence they are the basis of wealth.

To develop the new system, Napoleon appointed the Mathematician Delambre to chair a commission whose task was: "To survey ......more than 100 million parcels, to classify these parcels by the fertility of the soil, and to evaluate the productive capacity of each one; to bring together under the name of each owner a list of the separate parcels which he owns to determine, on the basis of their total productive capacity, their total revenue and to make of this assessment a record which should thereafter serve as the basis of future assessment".

The steps needed in executing the fiscal cadastre are:

- Surveying, to discover and identify all parcels that are to be valued;
- Assessment, to classify each parcel and determine its value;
- Collection, to raise revenue from those who exploit, not necessarily the owner, the land property.

The principle of taxation is based on two different methods, property tax and land tax. Firstly, taxes may be recognized by assessing and taxing all uses of lands such as buildings for residential or commercial aims. Secondly, taxes are based on the value of land as determined by the fertility of its soils and its productive capacity. The major purposes of this system are to serve the taxation objectives by administering land tax policy and to provide fiscal information for the expropriation of land for government purposes.

Two essential features characterize the fiscal cadastre. Firstly, it is a general system covering all the territory. Secondly, it is a graphical system that depicting the property framework with a graphical accuracy, sometimes, without ties to local or national control network.

## The Juridical Cadastre

The juridical cadastre is the basis for land titling and land registration. It is concerned with all forms of property and provides security of tenure. It is linked to registration rather than taxation. This system is often the principal source of information about ownership of land rights. The legal institution attached to the juridical cadastre is the registration and titling administration.

Its principal goal is to keep safely registers that list names of owners, the nature of the tenure, and all other ancillary information. Each parcel is identified on a map and referenced to its registers by using the same identification. The system of registration and titling is generally derived from historical and cultural aspects of a society. The main systems presently recognized were created in England and in Australia by Sir Robert Torrens. The major benefits of this system such as certainty of ownership, security of tenure, reduction in land disputes, and security for credit, described by Ian Williamson (1986), lead to greater social cohesion and provide information to respond effectively and efficiency to all needs of land market users.

Although, Moreno (1984) has remarked that many juridical cadastral systems encountered many problems: "too costly, too complicated and too time consuming, the cost of precise boundary surveys, coupled with the operation of the prescribed title registration systems inherited by the countries, is out of proportion to the benefits to be derived therefrom, is beyond the resources of almost every developing country. We might, therefore, have to adjust and adapt our technologies, or alternatively, have to change our recommendations to accord with realities".

The operations conducted in a juridical cadastre are legal, technical, and institutional. Four main processes are distinguished:

1. **Adjudication:** It is an official operation to establish what rights exists, who exercise them and to what restrictions they are subjected. This process may be conducted sporadically such as here and there, now and then, namely whenever or wherever there is a demand or other reasons to determine ownership of a property. Or, they can be conducted systematically in a methodical and orderly sequence in which all properties are brought on to the register and survey area by area.
2. **Demarcation:** This process is also called the boundary delineation, which is normally achieved by the emplacement of the corner monument or pegs in the ground. It enables to mark the limits of each land property on the ground.
3. **Land Surveying:** It follows the legal survey (demarcation) and provides a basis for producing cadastral maps that may help the administration of the land for planning and controlling the development of the territory cadastral system.
4. **Registration:** It records the attributes and a description of the land associated with each parcel or property. Additional information can be provided such as the shape and size of the property, the approximate location of its boundaries, and details about adjoining properties.

## The Multi-Purpose Cadastre

The concept of a multipurpose cadastre was discussed in 1975 by John McLaughlin in his Ph.D. dissertation titled "The nature, design and development of multipurpose cadastre", at the University of Wisconsin. Following this date, the Committee on Geodesy, USA, wrote a report that described the multipurpose cadastre as "a framework that supports continuous, readily available, and comprehensive land related information at the parcel level" (NRC, 1980).

The aim of introducing this new concept was to remedy problems inherent in the existing system such as accessibility, duplication, aggregation, confidentiality, and institutional structure. In addition, the Committee on Geodesy (NRC, 1980) stated, "there is a critical need for a better land-information system in the United States to improve land-conveyance procedures, furnish a basis for equitable taxation, and provide much-needed information for resource management and environmental planning".

In 1983, a Panel on a Multipurpose Cadastre held and reaffirmed the statements of the Committee on Geodesy report regarding the need for a multipurpose cadastre at all levels of government in the USA. The results of the project were approved by the governing board of the National Research Council (NRC, 1983), in a set of procedures and standards. From this date, the concept of a modern cadastre was defined to serve not only the record of interests in land but also to encompass the public land management, infrastructure, and physical records on the land. The report stated that the modern concept of a cadastral system is the concept of a Geographic Information System that uses the property land unit as the basic reference unit for gathering and storing information.

The multipurpose cadastre is proposed to overcome difficulties associated with the traditional system. It provides comprehensive information on land and presents all information at the parcel level. It is built around a reliable and accurate spatial framework: base maps, a cadastral overlay linked to juridical information of properties, and a linkage to land information generated by many offices and users. It supports both the legal and fiscal purposes.

Data are used for facilities management, base mapping, value assessment, land use planning and environmental impact assessment. Recently, many scientists and professional users see the multipurpose cadastre as one of the most revolutionary systems in collecting, managing, analyzing, and delivery land information. Several studies were conducted by different scientists to improve the meaning of a multipurpose cadastral system. Campbell (1994) outlined a GIS-based cadastre for Canada lands, which has the capability to expand into a multipurpose land information management. Williamson (2001) argued that modern cadastres are assuming a more multipurpose role, as the mid-1970 vision of the multipurpose cadastre begins to become a reality. Concurrently, the International Federation of Surveyors (FIG) continues to investigate cadastral issues in its commission 7 on cadastre and land management. El-ayachi (2006) developed a new vision toward designing and implementing the multipurpose cadastre for Morocco.

# MODERN CONCEPTS OF THE CADASTRAL SYSTEMS

## Cadastre 2014: The Origin and Meanings

In 1998, the FIG working group 7 analyzed the existing cadastral systems and outlined the ongoing reforms. Based on the results of this investigation, the working group highlighted the basic aspects of the existing systems; enlarged the existing meaning of cadastre; and presented the characteristics of the new system. Regarding the legal and organizational aspects, the major cadastral systems have the civil law as the basis of their legal systems. The parcel constitutes the basic unit in most jurisdictions. Further, the existing cadastres include both land registration and cadastral mapping, and the land registration deals with interests in land such as land right, restrictions and responsibilities. In about half jurisdictions, the strategic planning and control are realized by the same organization for both registration and mapping, and always kept by public sector. The existing systems served mainly either legal or fiscal purposes. However, many systems are also multi-purposes.

The existing systems present some strengths and weaknesses. The most generally noted strengths are the guarantee of title and the security of legal system. However, the weaknesses are the lack of computerization, the weak links between the legal and technical components of the cadastre, and some deficiencies in financial and administrative matters (Kaufmann and Steudler, 1998).

## The Justifications for Cadastre 2014

The working group outlined the following justifications for cadastre 2014:

1. **Creating Political Stability:** The land rights have an essential influence on the social relationships between individuals and communities. All economic and politic decisions are based on the manner in which societies are dealing with land property matters. Then, the new system is suggested to guarantee the reliability of lands and land markets.
2. **Omit Conflicts:** With its public documentation, the cadastre 2014 has the aim to regulate land use and define what resources are desired and allowed or forbidden to use. Thus an arsenal of laws, in form of restrictions and responsibilities, should be developed to minimize problems in concurrent land use, waste of resources, and environmental damages; if not natural resources societies will be subjected in a non-sustainable development to an over-utilization.
3. **Support of Economy:** The fair and clear legal situation of land matters in cadastre 2014 may diminish the risks of financial loss. With its standardized and complete structure, investors will save time and money. Further, in the process of globalization, the new system has to meet the needs of international companies in their production, marketing, service, research and development transactions.
4. **Flexibility and Effectiveness:** The cadastre 2014, as a fundamental component of a national spatial data infrastructure, will document all legal and physical aspects of lands, and accurately will improve the development of social aspects in the society.
5. **Support of Sustainable Development:** The important items in this point are: guarantee of ownership, security of credit, reducing land disputes, facilitating the land reforms, improving land use, and supporting environmental management.

## The Main Definitions for a Future Vision

Through the currently proposed strategy of the FIG, the new definition of cadastral system is adopted. Then new meaning of cadastral system has been suggested to encompass the new terms, land object and cadastre 2014.

The land object means every piece of land in which homogeneous conditions are defined by law, which determines also all natural objects such as rivers, lakes, forests, and mountains. On the other hand, all phenomena, rights, or restrictions tied to an area or piece on the surface of the earth, constitute land objects that should be defined by law. Then all land objects defined either by a private or public law and are submitted to an identical juridical parameters are called legal land object (Kaufman and Steudler, 1998).

For the future cadastre, the working group enlarges the existing definitions, which are appropriate for the recent situations; and adopted the definition: "Cadastre 2014 is a methodically arranged public inventory of data concerning all legal land objects in a certain country or district, based on a survey of their boundaries. Such legal land objects are systematically identified by means of some separate designation. They are defined either by private of by public law. The outlines, the identifier together with descriptive data, may show for each separate land object the nature, size, value and legal rights or restrictions associated with the land object".

## The Statements of the Cadastre 2014

The basic characteristics of cadastre 2014 are named as the six statements on the development of cadastre in the next twenty years (Kaufman and Steudler, 1998):

1. **Mission and Content:** The circumstances of land resources are exposed to change significantly. The growing world population and the revolution of the new technologies have directly influenced the used of natural resources. Then the main mission of new system is to cover a wider field of interests. The restrictions and responsibilities have to be clearly defined.
2. **Organization:** The future requirements involve the largest technology. The working procedures, the special skills, and the accurate technological systems must be taken in account. Then a new and sophisticated organization is needed.
3. **The Changing Role of Maps:** With the new technology, maps will serve only to represent data and phenomena, the new possibilities allow to efficiently store, manage, and analyze land related data.
4. **Information Technology:** The modern cadastre has to provide the basic data model using information technology. The future managers of cadastre 2014 have to develop sophisticated software systems to accelerate the land information processing.
5. **Privatization:** The public administration concentrates its role in controlling and supervision the cadastral matters, the private sector will gain in producing and furnishing, in an accurate and reliable manner, the land related information handled by this sector.
6. **Cost Recovery:** A new analysis based on cost and benefit must be introduced in the new cadastre, because the future system will need considerable investment.

# THE 3D-CADASTRE

## The Definitions and Meanings

Traditionally, the cadastre is concerned with 2D cadastral information even if parcels or properties are maintained in the modern spatial information systems. Recently, a great interest is observed in using space under and above the surface. The advanced technology and the new requirements of the modern society push governments to encourage an intensive multi use of lands. Several projects are considered as good examples of this new trend.

Van der Molen (2003) presented in his paper various experiences of the multi-use of lands in North America. The underground city of Montreal is a kind of the big projects in using a big subsurface shopping mall. The Post Office Square in Boston that has a six underground car parking is obviously one the most multi use of land. In such projects, an appropriate 3D cadastre should be developed to support the new 3D property rights concepts. The traditional concepts and practical approaches must be updated to encompass the new uses and to analyze their impact on the land tenure, land market, and environmental risks. The cadastral systems as defined in the first chapter serve as a fundamental tool that focuses on mapping and registration of the real properties. We should recognize that even if we introduce the new meanings of 3D cadastre, the concept of 2D property definitions must be unchanged. In several cases, we need to establish a 3D cadastre because the 2D cadastre is not able to deal with these situations. In case of a building that is divided in various apartments and they belong to various owners. Another case is that when somebody owns a home built upon a public road. Other cases can be resumed in the construction of tunnels, pipelines, and cables that are held by various private companies and each company needs to determine its rights on the use of these items.

The main objective of the 3D cadastre is to register rights and limited rights on 2D properties by giving more insights in the juridical situation above and under the surface (Stoter and Salzmann, 2003). The insights are defined trough spatially built constructions on, above, and under the surface as 3D properties. The spatial and juridical relationships between 2D and 3D cadastres must be maintained to smoothly move to the full 3D definition of properties.

## The 3D Cadastral Models Around the World

The 3D cadastral information becomes increasingly dealt with several countries around the world. We distinguish the case of Turkey that, since 1985, the cadastre Law accepted to perform and produce the 3D cadastre (Aydin et al., 2004).

In Denmark, the right of ownership is based on private and public Law (Stoter et al., 2004a). According to Danish Law, the rights are not registered directly in cadastre, but 2D and 3D information on the limited rights are added to the title. Danish Law distinguishes between buildings as a whole apartment complex owned by one person and individual apartments are rented, a cooperative housing association while each member has a right of use, or all units are owned by individual persons as ownership and common parts are held in co-ownership. The public servitudes are registered in the land registry and the spatial extents of the limitations are shown on a cadastral map.

In Norway, a committee that was established in 1995 concluded that it is important to improve the cadastral legislation by providing possibilities to produce 3D properties (Stoter et al., 2004b). A proposal for a law on construction properties was submitted and assumed that in 2006 a 3D cadastre law will be enacted. In this law, the surface property is still the basic property object that includes all the permanently fixed constructions.

In Sweden, a law came into force January 2004 to facilitate the creation of 3D property units (Stoter et al., 2004b). The new defined rules should be in accordance with the existing principles of real property law. The 3D property units have to relate to constructions and they are registered in the land registration without requiring surveying and mapping operations.

In British Columbia, Canada, the government allowed the constitution of a 3D parcel within the existing conventional parcel (Stoter et al., 2004b). A geodetic elevation as an absolute value is needed when a parcel is subdivided as a stratified property. All the survey plans are registered in the Crown Land Registry and the Land Titles' Office.

In Queensland in Australia, it is authorized to create 3D property units since 1997 (Stoter et al., 2004b). However, the insights concerning the rights on volumes are not operating in the case of Queen land. The cadastre includes only the footprint of the 3D parcels on the cadastral maps.

These experiences are presented because of the following reasons. Firstly, in the majority of these countries we find that is recognized by the law to perform the 3D property registration. Secondly, most of these countries have their 2D cadastral system completed and there is a need to deal with the 3D information. Thirdly, these cases are illustrated by an important support of literatures that describe both organizational and technical aspects of the 3D cadastre.

## The Marine Cadastre

### The Concept of Marine Cadastre

In the previous chapters, I described the different categories of cadastre that may be fiscal, juridical, or multipurpose system. It can also represent a third (3D cadastre) or a fourth dimension (3D and time). In the governance of oceans and coastal environments, the rights and restrictions may co-exist in space and time. Then the notion of parcel can be complex to use in defining a marine cadastre.

Several discussions are stated about the concept of marine cadastre. According to Robertson et al. (1999), the marine cadastre can be defined as a system that permits to record the boundaries of maritime rights and interests and to manage them spatially in relationship to the boundaries of other neighboring features. It can be considered as a marine information system, encompassing both the nature and spatial extent of the interests and property rights, with respect to ownership, various rights and responsibilities in the marine jurisdiction (Nichols et al., 2000). Fowler and Treml (2001) added that the most significant objective of creating a marine cadastre is the mapping of the shoreline even if the demarcation process can not be applied to marine boundaries.

### The Technical Framework for Constructing a Marine Cadastre

Ng'ang'a et al. (2004) extended the definition of the multipurpose cadastre to encompass the new requirements of developing a marine cadastre. They argued that the technical framework of a marine cadastre should be similar to that of a land based cadastre (NRC, 1983). The components of a multipurpose cadastre can serve as special requirement of a marine cadastre.

The first component, geodetic reference framework, permits the spatial referencing of data to the identifiable positions on the earth surface. In case of marine cadastre, efficient horizontal and vertical datums need to be defined and harmonized. The International Hydrographic Organization and the International Maritime Organization adopted the lowest astronomical tides as the chart datum (vertical datum) to reduce discrepancies when merging data sets that refer to various datums (Groten, 1999).

Since the GPS reported positions in the WGS84 as a horizontal datum it can be used as a marine cadastre datum. When we are interested in heights relative to the physical mean sea level we can compute at every GPS point the difference between the ellipsoid (WGS84) and the chart datum (Ng'ang'a et al., 2004). The differential GPS (DGPS) service can be used to improve the horizontal and vertical positioning accuracy. The marine DGPS would depend on the shore-based network of DGPS stations that will constitute the spatial reference framework of the marine cadastre.

The second component is the establishment of a series of large map scales. In the context of marine cadastre, we distinguish two sources of base map data, which are the existing nautical charts or the new established maps. The contents of marine maps can be the imagery data obtained from Compact Airborne Spectrographic Imager, Light Detection and Ranging, side scan sonar, single beam, or multi beam added to natural and artificial features in nautical charts. The navigational data, the bathymetric data, the magnetic field values, the water depths, the free air gravity and magnetic anomalies should be the fundamental data sets of natural and artificial features (Monahan and Mayer, 1999). These elements are important in protecting navigation and environment and avoiding conflicting uses and rights.

The third component is the registers of interest that in case of marine cadastre should mention all interests associated to marine parcels and boundaries. They should be categorized according to the type of laws that recognize their existence. Several descriptions can be found for these interests. We distinguish the right of exploring water surface, water column, seabed, living resources, non living resources, and subsurface. The fishing rights, the dragging activities, and the some restrictions of using shoreline can be specified as important interests. The cadastral overlays, the fourth component consist of defining marine boundaries and maintaining a form of offshore cadastre based on different activities such as petroleum drilling, mining, construction of public utilities, and aquaculture (Ng'ang'a et al., 2004). The main mission of marine cadastral overlays is surveying the locations of wells such as oil and gas and providing the evidence of boundaries at surface structure.

## CONCLUSION

By examining the new visions and concepts of the cadastral systems, we recognize that new needs are identified and they should be treated with an effective planning and incremental decision making for administrating all natural resources in a good manner. This means the designing of a conceptual framework that meet the user demands now and in the future.

The multipurpose cadastre as initiated since 1980 in USA constitutes an important experience in the world. It furnished a conceptual framework that enables managing natural and physical resources with unique, homogeneous, and contiguous interests. The main keys of succeeding in this unique experience in USA are linked to various factors and conditions. Firstly, the American society can be considered as the newest modern society in the world. Certainly, the Americans accumulate the worldwide experiences around the world and develop their own integrated cadastral system profiting from the existing land administration and good practices.

Secondly, the availability of the vast resources and space contributes to the gaining of more wealth from these resources. Thirdly, the effective land management is strongly supported and funded by the awarding projects grant to universities and research groups linked to various agencies operating in mapping and surveying. Consequently, the International Federation of Surveyors (FIG) launched a study in 1998 throughout its commission 7 so as to define a new vision aiming to develop a modern cadastral system. The so called cadastre 2014 was developed and described as a multipurpose cadastre that will support the new technology and provide spatial and non spatial information linked to legal land objects. This vision seems to be in general influenced by the USA conception. However it does not discuss profoundly the technical and organizational requirements for establishing the cadastre 2014. Only general trends are presented and various statements concerning financial and technological aspects are declared.

Concerning the 3D and marine cadastre, a considerable amount of interests has been generated about their new concepts and the benefits of their establishment. But, it is obviously necessary to make serious efforts to modernize and update the legal framework of these system and to encourage more research in this field.

## REFERENCES

Aydin, C. C., Deir, O., & Atasoy, M. (2004). Third Dimension (3D) in cadastre and its Integration with 3D GIS in Turky. TS25 Appropriate Technologies for good Land Administration II. FIG Working Week 2004, Athens, Greece.

Campbell, G. J. (1994). *A GIS-based multipurpose cadastre for Canada lands. FIG XX Congress*, Melbourne, Australia.

Cichocinski, P. (1999). Digital cadastral maps in land information systems. LIBER Quarterly, 9(2).

Dale, F. P., & McLaughlin, J. D. (1988). *Land Information Management. An introduction with special reference to cadastral problems in third world countries*. New York: Oxford University Press Inc.

Dale, F. P., & McLaughlin, J. D. (1999). *Land administration*. New York: Oxford University Press Inc.

Dowson, E., & Sheppard, V.L.O. (1952). *Land registration*. Her Majesty's Stationary Office. Colonial office publication, No.13.

Effenberg, W. (2001). *Spatial information Systems, the maintenance of digital cadastral maps* (Ph.D dissertation). University of Melborne.

El Ayachi, M. (2006). *Design of a new approach for developing a Multipurpose cadastral system of Morocco (Organizational and technical aspects)* (PhD thesis). IAV Hassan II, Rabat, Morocco.

El Ayachi, M., & Semlali, E. H. (2001). Digital cadastral map, a multipurpose tool for sustainable development. *Proceeding of the International conference on spatial information for sustainable development*.

FIG. (1995). Statement on the cadastre. Bureau of the FIG.

FIG. (2002). Land information management for sustainable development: best practice guidelines in city-wide Land Information Management. Working Group No.3, Commission 3. FIG publication No.31.

Fowler, C., & Treml, E. (2001). Building a marine cadastral information system for the United States- a case study. *Computers, Environment and Urban Systems*, *25*(4), 493–507. doi:10.1016/S0198-9715(00)00047-8

Groten, C. (1999). coastal boundaries and vertical datums. In *Proceedings of the international conference on technical aspects of maritime boundary delination and delimitation*. International Hydrographic Bureau.

Henssen, J. (1995). Basic principles of the main cadastral system in the world. In *Proceeding of the Seminar Modern Cadastres and cadastral Innovation*. International Federation of Surveyors (FIG).

Kaufmann, J., & Steudler, D. (1998). Cadastre 2014, a vision for a future cadastral system. FIG commission 7. Working Group 1.

Larsson, G. (1991). *Land registration and cadastral systems*. Essex, UK: Longman Scientific and Technical.

Monahan, D., & Mayer, L. (1999). An examination of publicity available bathymetric data sets using digital mapping tools to determine their applicability to article 76 of UNCLOS. In *Proceedings of the international conference on technical aspects of maritime bouandary delineation and delimitation*. International Hydrographic Bureau.

Moreno, R. (1984). Cadastres and land registration. *Proceeding of the International workshop on land tenure administration*.

Ng'ang'a, S., Sutherland, M., Cockburn, S., & Nichols, S. (2004). Toward a 3D marine cadastre in support of good ocean governance. *Computers, Environment and Urban Systems*, *28*(5), 443–470. doi:10.1016/j.compenvurbsys.2003.11.002

Nichols, S., Monahan, D., & Sutherland, M. D. (2000). Good governance of Canada's offshore and coastal zone. towards an understanding of the marine boundary issues. *Geomatica*, *54*(4), 415–424.

NRC. (1980). Need for a multipurpose cadastre. National Research Council National.

NRC. (1983). *Panel on a Multi-Purpose Cadastre. Procedures and standards for a multi-purpose cadastre*. National Academy Press.

Robertson, B., Benwell, G., & Hoogsteden, C. (1999). Marine cadastre. Administration infrastructure requirements. UN-FIG conference on land tenure and cadastral infrastructures for sustainable development, Melbourne, Australia.

Silva, L (2001). Cadastres in the 21st Century (6). A model for the information system of the Portuguese Land Register. *GIM International, the Worldwide Magazine for Geomatics, 15*(6), 61-63.

Silva, M. A., & Stubkjær, E. (2002). A review of methodologies used in research on cadastral development. *Computers, Environment and Urban Systems*, *26*(5), 403–423. doi:10.1016/S0198-9715(02)00011-X

Steudler, D., Williamson, I., Kaufmann, J., & Grant, D. (1997). Benchmarking cadastral systems. *The Australian Surveyor, 42*(3), 87-107.

Stoter, J., Oosterom, P. J. M., Ploeger, H.D., & Aalders, H.J.G.L (2004). Conceptual 3D cadastral models applied in several countries. TS25 Appropriate Technologies for good Land Administration II. FIG Working Week 2004, Athens, Greece.

Stoter, J., Munk, E., & Bodum, L. (2004). 3D registration of real property in Danmark. TS25 Appropriate Technologies for good Land Administration II. FIG Working Week 2004, Athens, Greece.

Stoter, J., & Salzmann, M. (2003). Towards a 3D cadastre. where do cadastral needs and technical possibilities meet? *Computers, Environment and Urban Systems*, *27*(4), 395–410. doi:10.1016/S0198-9715(02)00039-X

Tuomaala, J., & Uimonen, M. (1998). *Introducing the new Object-Oriented Cadastral Information System (JAKO) of Finland.* FIG XXI Congress, commission 7, cadastre and land management, Brighton, UK.

Van der Molen, P. (2002). The dynamic aspects of land administration. an often-forgotten component in system design. *Computers, Environment and Urban Systems*, *26*(5), 361–381. doi:10.1016/S0198-9715(02)00009-1

Van der Molen, P. (2003). Institutional aspects of 3D cadastres. *Computers, Environment and Urban Systems*, *27*(4), 383–394. doi:10.1016/S0198-9715(02)00038-8

Williamson, I (1986). Cadastral and land information systems in developing countries. *The Australian Surveyor, 33*(1), 27-43.

Williamson, I. (1997). The justification of cadastral systems in developing countries. *Geomatica*, *51*(1), 21–36.

# Chapter 14
# Management of the Public Land in Morocco Actors and Logics

**Idrais Abderrahim**
*Institut Agronomique et Vétérinaire Hassan II, Morocco*

**Semlali El Hassan**
*Institut Agronomique et Vétérinaire Hassan II, Morocco*

**Mohamed Ettarid**
*Institut Agronomique et Vétérinaire Hassan II, Morocco*

**El Idrissi Abdelwahed**
*National Institute of Urban Planning, Morocco*

## ABSTRACT

*The land heritage held by the public authorities represents an important part of the Moroccan land system. The management of this land heritage obeys various logics arising from its functional, legal, and institutional complexity. The objective of this study is to identify the numerous logics that drive the management of land assets by various actors and to describe the impact of these logics on the land management. The adopted methodology is to analyze the theoretical aspects and functions of the land tenure, study the regulatory framework of the land and the territorial land, and deal with the institutional framework of the land heritage management. The achieved consequences of this study have enabled the authors to distinguish three logics: the conservative logic, the regulatory logic, and the promoting logic. The goal behind the chapter is to provide the stakeholders with a functional paradigm to take into account for creating the conciliation tools reinforcing the coordination to carry out the urban development of the country.*

DOI: 10.4018/978-1-5225-5939-9.ch014

## INTRODUCTION

The land is the essential element in the definition of economic and social development projects in Morocco. The land heritage is shared between two forms of ownership: private property and the one belonging to the State and to territorial communities

The public ownership of land in Morocco is manifest in the federal lands, private lands, and forestry, and property Habous public. The municipal field, private and public, belongs to the territorial communities. Other statutes in land are appropriate by individuals but subject to the direct management of the State. It is of collective lands, lands Guiches and lands Habous private.

The management of this heritage is of crucial importance; view the important area of public land. This management which is the responsibility of the public authorities is framed by several variants which generate different logics of management. The question which arises: what is the logic of actors in the field of public management of land heritage? What are the factors producing these logics of management?

This study intends to distinguish the different logics of actors in the public land management. It adopts for tool the legal analysis of the concepts and functions of the land, as well as functions and roles of the actors, all the more the theoretical overview on the concepts and functions of land under different disciplines.

## MATERIAL AND METHODOLOGY

In order to be able to distinguish the logic which animate the actors in the management of the heritage of the land in Morocco, we have opted in the first time to a conceptual analysis on the concepts and functions of land; then, we conducted the legal analysis of the land heritage of Morocco for release the definition and functions of each land status, and finally we have made a legal analysis of the institutional framework of the land in Morocco to distinguish the actors and their responsibilities in the management of land heritage

### The Concepts and Functionality of Land

#### The Concepts of Land

The land refers to the set of social relations that have for support the Earth or the territorial space. The concept of the land can be apprehended under four approaches:

- **An Economic Approach:** It considers land as a well-exchangeable and as a commodity. This said, the transactions on the earth as an economic good involve many economic benefits.
- **Approach of Development:** Land is a support of agricultural production or support of urban development.
- **A Social Approach:** Often used by sociologists and geographers in recital the land as an object of desire (the main issue of communities and individuals) who organizes the social relationships between men and between men and the nature.
- **An Environmental Approach:** Land is an absolute environmental fact; the preservation of the land is a condition to the preservation of the environment.

## The Functions of Land

The land has neither the same role nor the same functions for all the actors. Of the small farmer in large proponent, the employee at the politician, the land reveals several functions: economic, development, political, social and environmental ones.

- **The Economic Function:** Land can play important economic roles as well for the public authorities, businesses, households and individuals. For the public authorities, the land is a way to generate tax revenues. For businesses, it presents an asset which allows access to bank credit and to the financing of investments. Whereas for households and individuals, the land is an important tool for the savings because of the high land values due to urbanization.
- **The Function of Development:** In relation to the issues of development, the land can be a set for development projects. In rural areas, the land is the support of the agricultural production for farmers, it presents the raw material for agricultural investment. In urban areas, the field is a base for the policies on urban planning; it is used in the production of urban services (housing, leisure, economic activities).
- **The Social Function:** Land plays a social role. Access to land is often conditioned by the social identities. Belonging to the original families and the status in the family weighs in the question of access to the ground and distribution of rights on the land. This access differs according to the status and age. As well, the land can be seen as a continuity of families, of the tribes through the appropriation in undivided shares of the property but also may redistribute the roles through the sharing of rights on the land heritage.
- **The Environmental Function:** Land has an environmental function. It is regarded as a support of biodiversity and ecosystem services, the land is the physical space of the environmental resources.

## The Legal Framework of Land

The land in Morocco is governed by numerous legal texts dictating the rules of law applicable to each category of land. These texts relate to the State-owned land, collective lands, land Habous and the land of the territorial communities.

## The State-Owned Land

Real Property State-owned come from the land of the Makhzen. This land owned by the State, a legal person in public law, includes the public domain of the State, the private domain of the State and the forest domain.

## The Public Domain of the State

The public domain of the State is governed by the Dahir of 01-07-1914 on the public domain of the State, the Dahir of 01-08-1925 concerning the uses of water and the Dahir of 30-11-1918 on the temporary occupation of the public domain.

The concept of the public domain of the State presents itself as all the parts of the territory, all works and all real property natural (shore of sea, desert areas, water in all its forms) or artificial (communication channels, canals and ports...) . This public domain is divided into public maritime domain, public domain Road, hydraulic public domain and public domain rail.

This public domain has the function to be for the use of all, and cannot be owned privately and its administration belongs to the State guardian of the community. Consequently, it is inalienable, elusive and not perceptible.

## The Private Domain of the State

The private domain of the State is governed by the Dahir of 3 January 1916 on the special procedure of the delimitation of the area of the State.

The concept of the private domain of the State is not clearly defined by the legislation. It presents itself in practice as being the property of which the State is the owner and not part of his public domain.

## The Forest Domain of the State

The forest domain is governed by the Dahir of 10 October 1917 on the exploitation and conservation of forests, as it has been supplemented and amended and the Dahir of 30 September 1976 relating to the organization of the participation of the population in the development of the forest economy.

The concept of the forest area is presented as any portion of the land which includes the state forests, dunes land and maritime dunes, the state-owned land planted or to reforest. As well, this domain includes the annexes to these lands to know the forest homes and their annexes, forest roads and plantations.

The function of this forest area is not explicitly disclosed by the texts in force. This function is present in two forms: an ecological function of conservation of forest resources (soil, water, plants) and a social function and economic (route, clearing, temporary occupation, distraction).

## Collective Lands

Collective lands are governed by a series of legislative texts. They are the Dahir of 27 April 1919 organizing the administrative supervision of the communities "Native" and regulating the management and disposal of collective property, the Dahir of 18 February 1924 on the special regulation for the delimitation of collective lands, Dahir of 19 March 1951 regulating the management and disposal of collective property and of the Dahir No. 1-69-30 of 25 July 1969 relating to collective lands located in the irrigation areas.

The concept of collective lands refers to lands of cultures and of course belonging in full ownership to the tribes, fractions, douars or other ethnic groups. These lands are subject to a use and collective enjoyment according to the traditional rules and customary practices of each ethnic community, under the trusteeship and the control of the authority responsible for the Interior.

Collective Lands have to serve to enhance the incomes of the members of the ethnic communities and to ensure the improvement of the conditions of their lives. They provide the answer to the growing needs in agricultural products for the populations the occupants, and are used for source of income, in the case of exploitation, assignment or rental for the ethnic community.

## The Property "Habous"

Property Habous are governed by the code of the Habous promulgated by the Dahir No. 1.09.236 of 23 February 2010.

The concept of property "Habous" is capital assets of a perpetual way or temporal, intended for the service of a charitable work public or private, and formed by way of contract, command or by the force of the law. This heritage includes public, private, or mixed "Habous".

This land heritage performs social functions. The public "Habous" is intended to serve the public utility, the private Habous serves as a source of income for the concerned to the limit of three generations and the mixed "Habous" fulfils the two functions of public utility and source of income for the concerned according to the dimensions defined shares previously.

## The Area of the Territorial Communities

The field of territorial communities is governed primarily by the Dahir of 19 October 1921 on the municipal field, the Dahir of 12 May 1937 amending and supplementing the Dahir of 19 October 1921 on the municipal field, the Dahir of 14 November 1949 relating to certain authorizations of occupation of the municipal public domain and the viziriel of 31 December 1921 determining the mode of management of the municipal field. The Organic Laws of the territorial communities refer to legal framework of this heritage.

The area of the territorial communities, baptized the municipal field, consists of the public domain and the private domain. The public domain is defined as lands which are formally assigned, whereas the private domain is presented as property of the territorial communities are not assigned to the municipal public domain.

The function of the municipal public domain derives from its components. With regard to the infrastructure (streets, paths, places, public gardens, water, lighting..) and cemeteries, this area fulfils a function of public service.

The private domain municipal is composed of buildings or buildings acquired and construct components have the function to serve the municipality, to be exploited, exchanged or disposed of to increase the income of the municipality or to face extraordinary spending.

# THE INSTITUTIONAL FRAMEWORK OF THE LAND MANAGEMENT

The land management is the responsibility of the State and territorial communities. Each land status is subject to the management of a ministerial department to ensure that this heritage performs its functions in accordance with the provisions of the legal texts in force.

Five ministerial departments ensure the management of the various statutes of the land heritage, either federal lands (public, private, and forestry), collective and the "Habous"; it is respectively the Department of the equipment, of Finance, of the Waters and Forests, of the Habous and of the Interior. While the heritage of territorial communities is managed by organs of these communities.

## The Department of the Equipment and the Transport

Department of equipment and transport is the only authority responsible for the management of the public domain of the State. The responsibilities of this department in the field of land management are defined by Decree No. 2-06-472 of 2 Chaabane 1429 (4 August 2008) setting out the functions and organization of the Ministry of Equipment and Transport.

The Ministry of Equipment and Transport is responsible to establish and implement the conservation policy of the public domain of the State, to ensure its preservation and its protection and the control of its management. It is responsible for the management of the public road, and of the acquisition of the land used for the construction of roads and highways.

The Department of Equipment and Transport thus ensures the delimitation, the management and conservation of the public domain port and maritime. It is in charge of improving the control of the public maritime domain and its protection against any illegal occupation.

## The Department of Finance

The direction of domains of the State under the Department of Finance is the authority responsible for the management of the private domain of the State. The functions of this branch are laid down by the decree n° 2-07-995 of 23 chaoual 1429 (23 October 2008) relating to the powers and to the organization of the Ministry of the Economy and Finances.

The Department of Finance is in charge to carry out the acts of administration and provision on the movable and immovable property constituting the private heritage of the State. His direction is concerned in this case to ensure the management, protection and recovery of the property of the private domain of the State.

The protection of this area is provided by the Direction of areas through the delimitation of the private domain property, the clearance of their legal situation, by the control of certain real estate operations, by its ability to exercise the actions in justice issues affecting the real estate of the State.

The direction of the areas operates for the valorization of the private domain of the State, in ensuring the allocation of the property of his domain for the benefit of the public services, and the transfer of lands falling within the private domain of the State.

## The Department of Water Affairs and Forestry

The government authority responsible for water and forests is the only one empowered to manage the forest domain. The functions of this department are dictated by Decree No. 2-04-503 of 21 Hijja 1425 (1 February 2005) on the powers and organization of the High Administration of waters and forests and of the fight against desertification.

The High Administration of waters and forests ensures, on delegation of the head of government, the administration, the preservation of the forest area of the State and other goods subject to forestry regime, the police and the control of the application of the texts relating to this field.

The Department of Water Affairs and Forestry ensures, in terms of management, the preparation and follow up of programs relating to the achievement and maintenance of infrastructure and equipment in the forest area of the State, the treatment and the follow up of files of land transactions affecting this area.

It carries out the legal studies and prepares the draft laws and regulations relating to the administration of the forest domain of the State.

In the field of protection, the High Administration of Waters ensures the preparation of program's delimitation and registration of the forestry land, oversees the appraisal and monitoring of files relating to litigation in the field of forestry, and ensures the police control and protection of the property of the forestry domain.

## The Department of the Interior

The Ministry of the Interior is the public authority responsible for ensuring the guardianship on the ethnic communities. The functions of the Department of the Interior in the management of collective lands are exercised under the decree n° 2-97-176 of 14 Chaabane 1418 (15 December 1997) relating to the attributions and to the organization of the Ministry of the Interior.

The Ministry of the Interior ensures, through the Department of Rural Affairs, the guardianship on the ethnic communities, the management and the conservation of their heritage, the defense of their interests and the restructuring of collective lands.

The Ministry of the Interior ensures the guardianship on the exercise of ownership's right, the enjoyment and its sharing by the ethnic communities. It ensures the delimitation and the registration of collective property, and authorizes the use in justice and the registration of these lands by ethnic community.

It shall act on all actions relating to the exploitation and transfer of lands right, of alienation, lease and development. It decides on the sharing of the benefits arising from development projects operated on collective lands. The Guardianship Council is the instance of appeal against all decisions of Nouabs of ethnic communities in the field of enjoyment.

The Department of Rural Affairs works for the improvement of the agricultural production of rural structures, and helps to reduce the social and regional disparities by the realization of projects of a socio-economic nature. It acts, in liaison with all the departments, agencies, public, semi-public and private, with a view to promote programs and actions intended to improve the conditions of life of rural populations.

## The Department of the Habous

The Management of "Habous" propriety is the responsibility of the governmental authority of the Habous. The Assignments of the department are laid down by the Dahir No. 1-16-38 of 17 Jumada I 1437 (26 February 2016) laying down the functions and organization of the Ministry of Habous and Islamic Affairs.

The Department of the Habous is responsible to transmit the mission of the Habous's institution, its conservation, its valorization, development of its heritage, the improvement of these revenues spent in the charitable works, including religious actions.

In the management of the heritage, the Department of the Habous establishes the strategies and plans relating to the management of the Habous Public. It ensures the management of the Habous affairs, the classification and the conservation of documents relating to. It put in conscience the benefactors of the importance of the Habous, and encourages them to give and ensures the supervision of the modalities of its constitution.

In the field of protection and conservation, the Department of the Habous works for the conservation and the protection of Habous property. It controls the Habous private, and ensures the Nidarat, and follows all the prénotations and litigation relating to the heritage of the Habous.

In terms of valuation, the Department of the Habous realizes studies, plans and projects aimed at the enhancement and promotion of "Habous" to take advantage of their potential, and take all the necessary measures to guarantee their profitability to serve their missions.

### The Territorial Communities

The territorial communities are responsible for managing their land heritage. The organic laws 111-14 relating to regions, 112-14 related to the provinces and prefectures, and 113-14 relating to the Commons, are legal texts managing these lands.

The territorial communities ensure the management of their heritage land through their elected instances. They are engaged in the constitution, the management, preservation and enhancement of their heritage land.

The Council of the territorial community status on the management of the land heritage of the territorial community, its conservation and maintenance. It examines the constitution of the land heritage necessary for the territorial community for the exercise of its missions, and authorizes the exchange, the allocation and reallocation of property of this field. It accepts donations and bequests to land made to the territorial community.

The President of the Council of the territorial community is in charge to ensure the management and conservation of the heritage land through the holding of the inventories of the assets of the territorial community, the update of bases of consistency of these goods and the clearance of their legal situation. It undertakes the acts of rental, sale, exchange, acquisition as well as any other transaction relating to the private domain of the territorial community. He takes the necessary measures for the management of the public domain of the territorial community and delivers accordingly the permissions for temporary occupation of this area and this in accordance with the legislation and regulations.

## RESULTS

The theoretical analysis of concepts and functions of land combined with the legal analysis of the Land Statutes in Morocco and land missions assigned to public authorities, has allowed us to highlight the logic which animate the public actors in the field of public land management.

In this sense, three major logics can be distinguished: the conservative logic, the regulatory logic and the promoter logic.

## THE CONSERVATIVE LOGIC

The conservative logic attaches to the physical conservation of land. It aims the physical protection of land against any substantial degradation and in area and ensures that the heritage accomplishes its roles and functions.

This logic of land management is linked to social and environmental functions of land. It is in our case the management of public domain of the State, Habous and the domain of forestry.

## The Management of the Public Domain

The management of public domain of the State is subject to the conservative logic. Given that, this heritage fulfils a function of public service; its preservation against any degradation is the ample objective of the Department of the equipment responsible for its management.

The conservative logic of this department is manifested in the actions carried out in terms of protection of this area and those undertaken in the framework of its valuation.

- **The Protection of the Public Domain:** The protection of the public domain of the State against any degradation is ensured in the first time by its delimitation and registration in the name of the State. As well, its protection against any use not consistent with the duties of public services is guaranteed through the administrative police who are in charge of checking all the offenses relating to illegal occupation and use of this heritage.
- **The Enhancement of the Public Domain:** The enhancement of the public domain of the State is done in accordance with its vocation of public service. However, different use and custom can take place there in two cases: the temporary occupation and decommissioning. The first track of recovery is authorized for a duration of 20 years, renewable only once against a fee, and may be interrupted by the Administration without compensation in case of non respect of the use. With regard to the decommissioning, it is begun in the one case when the party to decommission no longer fulfils its role of public service, and is only permitted by the regulatory procedure.

The management of the public domain of the State is then reined by the conservative logic, which tends to preserve its character of public service.

## The Management of Habous Lands

The preservation of the originality of the Habous land and enhancement of this heritage to perform the charitable function and social conditions of the institution of the Wakf, are among the foundations of the conservative logic of management of Habous.

The Department of the Habous ensures the conservation and preservation of the institution of the Wakf, and work for the valorization of this heritage to serve the purposes for which it is immobilized.

- **The Preservation of the Institution of the Wakf:** The preservation of the property of the Wakf is a fundamental objective of the management of Habous. This preservation is manifested in the actions of the clearance and registration of heritage carried out by the Administration of Habous, as well as in the functions of Nidarat ensures the supervision on the Habous public, private and mixed lands.

In addition, the protection of property Habous is exercised in the face of the right of expropriation which is at the origin binding nature of the right to property. This right of expropriation can only be exercised with the agreement of the Habous authorities, which makes this heritage to the shelter of any non-conforming use to its originality.

- **The Valorization of Property Habous:** The enhancement of the land heritage of Habous is operated by the Department of Habous via the terms of the exchange real estate, lease or in the framework of a partnership of investment. These methods of recovery preserve the heritage and fulfill the charitable functions, religious and social ones of Habous. These functions are provided by the financial resources drained by these different modalities of recovery.

The management of Habous is controlled by the conservative logic which aims to preserve the originality and the charitable function of this heritage. The preservation of its sustainability is also reflected in the action of valorization of the land heritage of Habous.

## The Management of the Forest Area

The environmental vocation of the forest area and the need to preserve the environmental resources against any degradation, are at the origin of the conservative management of this heritage property.

The Department of Water and Forests ensures the protection of the forest area against any physical degradation. It also acts for the recovery of the forest heritage via a rationalization of this use in accordance to the preservation of its environmental function.

- **The Protection of the Forest Area:** The Administration of Water and Forests secures the forest area through the actions of delimitation and registration of land. It ensures the protection of this area against any degradation and use not in accordance with its vocation, by the shape of the administrative police of waters and forests. This protection is more evident in the defense of the interests of this area during the litigation raised before the courts.
- **The Valorization of the Forest Heritage:** The meaning conservative and protectionist in the forestry field is manifested even in the modes of exploitation of this heritage. This recovery is done via the rational use of the heritage that is translated by the modes of exploitation to know the distraction, the exchange and the temporary occupation.

The distraction of the forest land is authorized in the one case when the public utility requires it. The exchange with or without payment, allows with guarantee to use forest land without compromising it. The temporary occupation is done in the conditions for the respect of the environmental function of the forest area and may not exceed 20 years renewable once only with the possibility to put an end to by the administration.

The management of the forest area depends on the conservative logic which stems of the missions of the forestry administration, and which is the result of the environmental functions of this area.

# THE REGULATORY LOGIC

The regulatory logic refers to the social function of the land. This logic animates the acts of management having for objective to ensure the regulation and the arbitration of the uses, interests and social relations that tie around the land.

The collective appropriation of the heritage of ethnic communities raises the preservation of the interests of the Community, the management of modes of enjoyment of the rights holders, the conservation and valorization of land for the benefit of the members of community.

The supervision of the Ministry of the Interior on the ethnic communities, aims, through the regulatory logic, the regulation of social relations between the members of the ethnic community and between the ethnic communities who are born around the collective lands.

The logic of regulating is reflected by the actions of guardianship exercised on the ethnic communities and on the management of this heritage for the benefit of the members of community.

- **The Administrative Supervision:** The regulatory logic in the supervision of the Ministry of the Interior on the ethnic communities is manifest in the obligation of the approval and the authorization of all acts of management of the collective heritage.

Guardianship is responsible for managing the sharing of the enjoyment between members of the community, of the registration of the land to the names of the communities, to the approval of the lists of right holders. It ensures the control of the management of community affairs, the designation of Nouabs and the consideration of the objections on the decisions of the Naieb institution.

The authorization of the tutorship is mandatory to carry out any land transaction by the community (rental, sharing and assignment). The guardianship assures the equitable distribution of income of land transactions and development projects, operated on the collective land, on different members of community according to the traditions and customs of each community.

- **The Management of the Heritage for the Benefit of Community:** The Department of Interior ensures management actions of the collective heritage for the benefit of ethnic communities. The actions for the development of this heritage have the objective to guarantee the benefit of rights holders.

The different development projects operated on the collective land improve the productivity of these lands, to ensure the improvement of their profitability, and the enhancement of incomes of community members and ensure the improvement of life conditions of holders.

The administration of Interior opts for partnerships with the different actors of development to ensure the rights holders to benefit from projects.

The conservation of the collective nature of the ownership of collective lands, the control of the internal management of the affairs of community, the preservation of the interests of rights holders and the restriction of the assignment for the legal persons of public law are the traits of a regulatory logic of management of this heritage.

## THE PROMOTER LOGIC

The promoter logic ensures the function of development and the economic function of the land. This logic of management makes the land heritage a media development projects in the urban and rural areas. It guarantees as well the use of land as a factor of production and a source of funding.

The Promoter logic is manifested in the management of the private domain of the State and in the administration of the land heritage of the territorial communities. The private domain of the State is managed under the angle promoter of economic development for the benefit of the State and the private sector. The municipal field is also managed in this promoter optical, but in order to make it a source of financing for the territorial communities.

## The Management of the Private Domain of the State

The Department of Finance is responsible for the management of the private domain of the State under the logic to promote investment and stimulate economic development. This promoter logic is manifested in acts of management of this heritage property. These acts proponents are manifested in three fundamental axes: the constitution and reconstitution of the land domain, the securing of the property and its legal clearance and its mobilization for development of investments.

- **The Constitution and the Reconstitution of the Private Domain:** In order to respond to the needs of development, the department of the areas of the State is responsible to constitute the private land. This reserve is formed through the acquisition to the amicable or by way of expropriation of private properties, and by way of the distraction of the public domain of the State and of the forestry domain.

The administration of finance, in order to perpetuate the land resource development, initiates actions of reconstitution of the reserve land. This reconstitution is made using the resources arising from the disposal of state property.

- **Securing and the Clearance of the Property:** The protection of the private domain of the State is one of the pillars of the intervention promoter of the Department of Finance. Securing is manifested in the census of state-owned property, in their delimitation and by their registration. The department of private domain ensures its protection. It has the power to sue, in the name of the State, against any illegal occupation of the private domain.
- **The Mobilization of Lands for Development Projects:** The domain lands are subject to rules of private law. This allows him to serve its role of base for development projects. The administration of finance takes the management acts necessary to put this domain at the service of development.

It affects, on the one hand, the property of the private domain to the administrations to fulfill their missions, and begins actions of long-term Lease and Assignment of state-owned land for public and private operators of economic development, to carry out projects of development.

The Department of Finance, anxious to stimulation of the economic development of the country, works to put lands in service of development projects. Its actions are then animated by the promoter logic.

## The Management of the Territorial Community's Heritage

The territorial communities, owners of their areas, are responsible for the management of their property portfolio. This heritage land is called to be managed in logic of guarantee to these communities' financial resources to carry out their missions of economic and social development.

The administration of the territorial communities domain, under its two forms, public and private, is ensured by the instances of these territorial communities. This management is manifested in two components: the mastery and the conservation of the heritage and the improvement of its resources through its valorization.

- **The Mastery of the Territorial Community's Heritage:** The mastery of the territorial communities' heritage is ensured by the missions of the elected instances of these communities. The elected councils decide to preserve heritage property, maintain and protect it. For the President, he manages this heritage and ensures its mastery through the updating of lands databases, the clearance of their legal situation and their registration in the name of the territorial community.

These actions concern public and private domain. The municipal public domain is mastered by its delimitation and registration and also by the power of the administrative police granted to the President of the community council. The private domain is mastered by the control actions and the registration undertaken by the President.

- **Enhancement of the Territorial Communities' Heritage:** The territorial communities initiate rental, exchange and transfer of property of their private domain. The aim of these actions is to increase the financial capacity of these communities, and to enable them to acquire the financial resources to meet the needs of operating and investment.

The public domain is valorized by temporary occupation authorized by the President of the Community Council. The resources from this area are used to feed communities, and provide financial resources for it.

The management of the territorial communities' field is done under the promoter logic which privileges the economic function of land. This function is able to allow the Community more financial autonomy in relation to the State and to give it the financial opportunities to develop its activities toward the citizens.

## DISCUSSION

The conservative, regulatory, and promoter logics of the public land management are the results of functions and the legal rules governing the functional framework and institutional framework of this heritage. However, the realization of the public land management focused on these logics is encountering difficulties inherent to several constraints of management.

### The Conservative Logic

The conservation of the public domain of the State, of Habous and of the forest estate is the foundation for the management of land. However, the conservation and valorization of these assets are in front of several difficulties.

The preservation of this type of property is often limited by the difficulties of census of their components, and the determination of their consistencies. This is manifested clearly for the Habous property, and in less degree for the public domain and the domain forest.

The cadence relatively low of the registration of the public domain, including Maritime, and the forest area is one of the impediments to the securing of heritage. The public domain suffers from difficulties of its delimitation, and for the forest property the problem resides on the objections on the delimitation of this heritage.

The valorization of the public domain property is confronted often to the problems of management of temporary occupations which, on the one hand, exceed the durations and on the other hand do not generate significant resources in view of the weakness of the royalties.

## The Regulatory Logic

The regulatory logic of management in the case of collective lands is located in front of several difficulties. It is among others the trend of privatization of this heritage, the need of less guardianship and the issues of preservation and enhancement of the heritage for the owners.

The collective heritage has experienced a trend of privatization. This trend is the result of the different forms of right transfers, usually worn as a right of ownership. The having rights, instead of proceeding to a sharing of enjoyment, initiates transfers on the basis of succession, pledge or assignment to third parties.

The Guardianship concentrated in the hands of the Ministry of Interior is claimed by the ethnic communities that challenge more autonomy for the management of their affairs, which puts in difficulties the preservation of the rights and the regulation of the interests of the collectivists.

The valorization of heritage conditioned by the drainage of profits for the collectivists is in front of the difficulties of investment and cooperation with the actors.

## The Promoter Logic

The promoter logic putting the land in the service of economic development is in the face of the difficulties related to the availability of land and to the enhancement of the land to drain significant revenues for the territorial communities.

The impetus for economic development projects require more and more state-owned land cleared, available and well distributed on the national territory. This involves the effort of census of heritage, its registration and its release of illegal occupations. Also, the availability of the heritage in areas of high economic pressure is limited by the difficulties of the reconstitution of state-owned land in the face of its continuing operation.

The valorization of territorial communities' property for channeling of financial resources faces the low mastery of the consistency of the heritage in the absence of appropriate tools for its management. It is also due to the weakness of the performance of the royalties of the public domain and the price of transfers.

# CONCLUSION

The actors responsible for the public land management in Morocco adopt different logics which are divergent and sometimes contradictory. This is the result of the diversity of concepts and functions of the land on the theoretical plan and legal one.

The conservative, regulatory, and promoter logics are the main traits of the public land management in Moroccan. These logics may not operate in isolation of the socio-economic dynamics of the country. This requires a reconciliation of these logics and coordination between actors to be able to align their actions while maintaining their logic of action and ensuring the response to the needs of socio-economic development of the country.

## REFERENCES

ADEF. (2012). *Demystifying the land: State of the places of the research*. Academic Press.

Anonymous. (2009). *Land Governance and securing rights in the countries of the South. White Book of French actors of cooperation*. Technical Committee: Land and Development.

Barrier, O. (1996). *Management of renewable natural resources and conservation of eco systems: Land-environment*. Thesis in law, France.

CHENESANOGO. (2012). *Land issues and sustainable development in Mali*. Thesis.

Dahir No. 1-09-236 of 23 February 2010 establishing the Code of the Habous.

Dahir No. 1-16-38 of 17 Jumada I 1437 (26 February 2016) laying down the functions and organization of the Ministry of Habous and Islamic Affairs..

Dahir of 06 August 1915 on the Exercise of the actions in justice issues affecting the real estate of the State

Dahir of 1 July 1914 on the public domain.

Dahir of 10 October 1917 on the exploitation and conservation of forests

Dahir of 18 February 1924 on the special regulation for the delimitation of collective lands

Dahir of 19 March 1951 regulating the management and disposal of collective property

Dahir of 19 October 1921 on the municipal field.

Dahir of 24 Ramadan 1333 (6 August 1915) on the Exercise of the actions in justice issues affecting the real estate of the State, as it has been modified

Dahir of 27 April 1919 organizing the administrative supervision of the communities "Native" and regulating the management and disposal of collective property

Dahir of 3 January 1916 on the special regulation on the delimitation of the area of the State.

Decree n° 2-04-503 of 21 Hijja 1425 (1 February 2005) on the powers and organization of the Office of the High Commissioner to the waters and forests and to the fight against desertification

Decree n° 2-06-472 of 2 Chaabane 1429 (4 August 2008) setting out the functions and organization of the Ministry of Equipment and Transport.

Decree n° 2-07-995 of 23 chaoual 1429 (23 October 2008) relating to the powers and to the organization of the Ministry of the Economy and Finance

Decree n° 2-97-176 of 14 Chaabane 1418 (15 December 1997) relating to the powers and to the organization of the Ministry of the Interior

The organic law 111-14 relative to the regions.

The organic law 112-14 relative to the provinces and prefectures.

The organic law 113-14 relative to the Communities.

The Roy E. (1998). From the appropriation to the ownership: A brief introduction to the land terminology. In *What land policies for rural Africa? Reconcile practices, legitimacy and legality*. Paris: Karthala.

Viziriel of 26 December 1920 regulations control the employment and re-employment of funds from the expropriation, leases or disposals of perpetual enjoyment of collective lands

## KEY TERMS AND DEFINITIONS

**Conciliation:** An action of mediating between groups.
**Conservative:** Against change.
**Land:** A part of Earth that is not covered by water.
**Logic:** Reasoning conducted based on values.
**Management:** A way of administration.
**Promoter:** A financer and provider of resources.
**Regulatory:** Serving to regulate interests.

# Chapter 15
# A Testifying Development of a Benchmark Study Approach Applicable to Land in the ECOWAS Area

**Moustapha Gning Tine**
*Institut Agronomique et Vétérinaire Hassan II, Morocco*

**Gbeagblewo Edem**
*Institut Agronomique et Vétérinaire Hassan II, Morocco*

## ABSTRACT

*At independence, the leaders of the member states, in the search for ways of development of their young nations, have encountered the thorny problem of land management which was under the control of the customary regimes. West African land legislation is not sufficiently clarified and suffers from a legal dualism that hinders development and can lead to widespread conflicts. The increasingly visible development of a single regional market and the desire to create conditions for lasting social peace in the region make it imperative to secure the means of subsistence and to harmonize their governance. This work is an attempt to develop a benchmark study approach applicable to West African land. It represents a support for land reforms taking into account present and future land issues in the region. It is a methodology that develops both a set of indicators to evaluate and compare the performance of land practices as an approach to analyzing the results and proposing reforms.*

## INTRODUCTION

Since the sedentarization of man, land has proved to be one of the essential elements for both habitat and production. Ensuring the acquisition, exploitation, taxation and transmission of landed property to avoid conflicts was one of the main concerns of any organized society. A set of rules and land practices were therefore developed by each society according to its specificities. The advent of modern states and capitalism will give these land rules and practices more complex missions. In fact, landed property has

DOI: 10.4018/978-1-5225-5939-9.ch015

become a commodity with the development of capitalism. This gives rise to more elaborate rules for securing land transactions. In addition, territory development concerns impose provisions to ensure land and land information requirements for development projects at the national and regional levels.

In the ECOWAS region, made up of 15 member states, land issue is the most important concern of regional policy makers, especially the non-homogeneous status of land in the sub-region. A challenge is related to the update of the land situation in different countries and how to better understand good practices in governance and equitable access to land. The issue of inheritance and de facto discrimination of women in relation to land and habitat are among shortcomings in ECOWAS region. There is a need to harmonize texts relating to the management of natural resources, development projects (ECOWAS, 2014). Since the colonial period, land practices have varied from one ECOWAS state to another. While much of it was governed by the land practices enacted by France, some were under English or Portuguese practices. This difference in practice continued after independence, thus giving countries more advanced than others in terms of resolving land problems in the region. An exchange of experiences would therefore be welcome for continuous improvement and harmonization of land practices in the region by taking into account local specificities in order to achieve good performance. Benchmarking is a good technique to achieve this goal of improving practices.

The aim of this chapter is to propose an approach for conducting a benchmark study linked to land governance in the ECOWAS area in order to have indicators that will enable us to highlight the best practices that can be used in the implementation of future land practices. This will be done by defining relevant and measurable axes and indicators for the study of land as well as the development of an approach for analyzing the results and proposing reforms.

## ESTABLISHMENT OF THE BENCHMARK STUDY APPROACH

The benchmark is a methodology of constantly seeking best practices in order to adopt or adapt their positive aspects and to implement them in order to progress. The objective of the benchmark is to change a current situation, which can be improved to a more competitive situation, with the originality of comparing companies, administrations, services or offices operating in other fields of activity (Loroy, 2005). It is a technique that must be applied to the land field domain with great caution especially in the ECOWAS area. Trying to turn the study towards legislation could not be of great relevance because it is common to find countries with well-elaborated legislation and disappointing results. The land benchmark will rather focus on the impact of land practices. Indeed, an impact-practice analysis will allow us to more effectively grasp the best practices and to facilitate the achievement of the final objective of developing best practices. Land management is a complex field that directly or indirectly influences many sectors, especially the social and economic ones. A benchmark model of land tenure must therefore take into account this complexity of matter and the domains that depend on it. The approach of a land-based benchmark should therefore be oriented towards the search for practices that ensure good governance of land tenure and land administration in ECOWAS region.

### Process for Establishing the Approach

Although many writers have written about the land situation, West Africa is characterized by a great lack of data on the impacts of practices in land governance. To remedy this situation, which constitutes a brake on the practice of the benchmark technique, the authors integrate into the process the definition of the methods of measuring the indicators while taking into account the practical realities of these measures in the field. The proposal of an analysis of results and formulation of reforms will make the best of the existing practices. This analysis approach consists of proposing a set of elements to analyze and understand the results of each indicator and to highlight the best practices that are the basis of the best results.

## Existing Study on Land Monitoring and Comparison

In land matters, a set of indicators exists for specific uses, especially for the internal monitoring of a country's land situation. Thus, the Land Observatory of Madagascar uses as an indicator for the monitoring of land policy (GLTN, 2014):

- Number of land certificates delivered ;
- Average time to issue land documents;
- Percentage of Deeds delivered to women;
- Percentage of land certificates delivered to women;
- Rate deteriorated land records and
- Percentage of land disputes in the Antananarivo court of justice.

Similarly, for the monitoring of land governance in Georgia, the following indicators are used (GLTN, 2014):

- Communal mapped / registered lands,
- Plots registered to women,
- Number and cost of Transactions,
- Collection of the actual land taxes,
- Dispossessed zone and cost.

However, key indicators for comparing several countries on land governance issues, such as tenure security measures and the distribution of land access, are rare or non-existent (GLTN, 2014). In 2012, collaboration between UN-Habitat, the Millennium Challenge Corporation, and the World Bank led to the establishment of the Global Land Indicators Initiative (GLII) to develop indicators for a worldwide comparative study in terms of land Governance and security. The GLII expert group proposed a set of indicators for reflection (GLTN, 2015):

- **Land Regime Security:**
  - **The Guarantee of Rights to Land and Properties:** Percentage of men, women, community and Companies with recognized proof of occupation.
  - **Land Security Perceived:** Percentage of men, women and businesses, which perceive recognized and protected land rights.
  - **Legal Recognition of a Continuum of Land Rights:** The level to which the legal framework recognizes and protects the legitimate land rights and their use, either by custom or legal land regimes.

  - **The Equal Right of Women:** Level at which women and men have the same rights to possess, inherit and hand down landed properties.
  - **The Surface of Mapped Land:** Percentage of area of land mapped on the Map of the legal recognized regime .
- **Land Disputes and Their Regulation:**
  - **Effectiveness of the Regulation of Land Disputes:** time to solve a land dispute / property.
  - **Efficiency of the Regulation of Land Disputes:** Percentage reported to land disputes which have been regulated.
- **Services of Landed Properties Administration:**
  - **Income as a Percentage of Land Tax:** The tax on the land properties related to GDP.
  - **Efficiency of Land Administration:** Time to conduct a land operation / property

Although these indicators proposed by the group of experts touch on several important points of the land issue, these indicators do not cover agrarian land development and urbanization issues.

## Focus and Indicators

The axes must allow, on the one hand, the study of the performance of land administration and on the other hand, the efficiency of land practices. They are defined in such a way that the broad lines of good land governance as defined in the FAO study are covered. Thus the focus will be:

- **Land Administration:** Investigating how the administrative organization and legal administrative approaches contribute to the land administration performance.
- **The Securing of Goods and Land Transactions:** Investigates to what extent the legal provisions of land security and their practice contribute to solve land security problems.
- **The Land in the Development:** Investigates how land legislation and land status in particular contribute to effectively lay land requirements when talking of development policies and preservation of the environment.

These three axes will enable us to highlight the economic, social and environmental impact of the various land practices to identify the inadequacies of legislation and practices with the aim of continuous improvement.

## Development of Indicators

An indicator is a value that quantifies or qualifies phenomena and assesses both the current situation and the evolution of a territory. Indicators are tools for decision-making that synthesize and summarize information, assess the impacts or outcomes of a policy, plan, strategy or physical achievement for a given period. The indicators also allow to locate / identify the changes of situations in time or space, facilitate monitoring and help to prioritize actions. Thus an indicator must be:

- Quantifiable;
- Sensitive and reliable over time and
- Adapted to Target group.

A good choice of indicators is based on explicit causality, so that a change in the indicator is directly related to a variation in the phenomenon studied (FAO, 2007). In order to make comparable and relevant measurements of the results of land practices at the level of each axis, several indicators are defined. These indicators are defined in order to satisfy three objectives:

- Allow to measure the impact of land practices on a territory in the ECOWAS region,
- Allow to monitor the evolution in a same country
- Allow comparison of impacts among the countries of the region in spittheir different sizes of areas, economies and population .

To meet these three requirements, indicators will be defined on the basis of social or environmental economic factors directly correlated with the evolution of land practices and mostly defined as a ratio enabling to avoid biases due to variations in parameters (population, surface area ...) of each country.

## INDICATORS OF LAND ADMINISTRATION

Land administration is the set of government structures implemented to meet the organizational and technical needs in land matters. The performance of the land administration will directly affect the level of awareness of the population on land issues, the availability of land information, the duration of land tenure and the confidence of the costumers; who are the owners; in the services provided by these administrations. Thus, the performance of land administration will be measured through six indicators, which are:

- **Population Awareness Rate:** According to FAO (2003), anyone who knows his/her rights and knows what to do if they are not respected is better able to protect them than a less well-informed person;
- **Availability of Land Information:** "The dissemination of land information, the concerned parts and their rights are fundamental for an effective land administration " (FAO, 2003);
- Duration of registration; Recording, authorization and expropriation (GLTN, 2015);
- Owners satisfaction rate about government services;
- Completeness of land information ;
- Coverage rate of land registration.

The methods for measuring these indicators are defined in Table 1.

### Indicators of Rights Security and Land Transactions

Land tenure security is the promotion of access to and ownership of land by all sections of the population without distinction of group, gender or religion and in an equitable manner. It also consists in guaranteeing each owner the freedom to use or transfer his property and to ensure fair land transactions. The level of security of tenure will be linked to the number and nature of land disputes and to the set of mechanisms for resolving these conflicts. The indicators that will measure land security will be:

*Table 1. Indicators of land administration*

| Indicators | Measuring Methods |
|---|---|
| Population awareness rate on land issues | • Ask questions about the services or agents to check for a primary land operation sample-based of 1000 (more for populations passing 10 million) people of less than 18 years old, taken in a Random manner throughout the country<br>• The number of satisfactory responses to the total number of interviewed is the awareness rate |
| The availability of land information | • Calculate the average time necessary for individuals, professionals or structures of the State to get a response to land information requests.<br>• There will be a day for any computerized system and accessible via the Internet without displacement for any information. |
| Duration of the procedures | • highlight the legal duration of the procedures.<br>• Conduct a survey on a sample of 100 people minimum for each step to highlight the practical durations .<br>• calculate the gap between averages of pratical and legal durations. |
| Satisfaction rate | • Take a sample in a Random manner 100 people minimum among the recipients of land administration services for a given period .<br>• Ask them to criticize the quality of the service and the rate it over 10 to evaluate the service .<br>• The arithmetic average of all marks is the satisfaction rate |
| Completeness of land information | • Take a sample of archived land records in land services<br>• Confronting the information listed to the reality after an investigation<br>• the updated files report on the total size of the sample is the land information completeness rate. |
| The coverage rate of the land registration | • summon the capacity of all documented plots are in land service archives,<br>• Make the ratio between the total capacity obtained and the area of the country |

- Sensation rate of land security (GLTN, 2015);
- Litigation rate for land reason: realizes the weight of land disputes on the courts and their contributions to the social divide (GLTN, 2014);
- The frequency accordingly to the type of land disputes: highlights the different types of conflict (inheritance; among communities, encroachment, transaction fraud, expropriation ... ..) and their magnitudes;
- The acceptance rate of land conflicts judgments reflects the quality of regulation rules adopted by the courts and their acceptance by the population;
- The percentage of women have right to property ed ty: reports from the land of e held by women in a country (GLTN, 2014).

The methods for measuring these indicators are defined in Table 2.

## Indicators of Land Tenure in Development

Land is the basis for the development of all human activities. From agriculture to finance to mining, crafts and the processing industry, all sectors of the economy are based on land. The harmonious development of cities and the preservation of the environment are highly dependent on them. Thus the indicators of this last axis will be:

- Part of investments on mortgage collateral;
- Share of land revenue in national revenue (GLTN, 2014);

*Table 2. Rights Security and Land Transactions Indicators*

| Indicators | Measuring Methods |
|---|---|
| The sensation of security of tenure | • Take a sample Random land rights holders (bare properties ty, usufruct, pasture ...), made up of holders whose rights are registered and holders whose rights are not recorded (customary)<br>• Ask them questions about their feelings about the security of their land rights<br>• Making the ratio between the number of people claiming to feel in security on the size of the sample |
| Land appeal rate | • It is the ratio of the number of land-related complaints to the total number of complaints received by the courts for a given period |
| Frequency by type of conflict | • Classify all land disputes by Categories<br>• Count the number of conflict of each category,<br>• Make the ratio of the number of conflicts by category and the total number of land conflicts. |
| Acceptance rate of judgments | • Take a sample e judgments by the courts of a given period,<br>• Investigation about concerned parts and evaluate their satisfaction of judgment on the scale of 10,<br>• A mark higher than 5 is considered as "satisfaction".<br>• Find the proportion of "satisfaction" with respect to the entire sample. |
| The percentage of women with property rights | • Take a random sample of properties (registered as well as non-registered<br>• Investigate to find the holders (in the cadastral land recorder on the field)<br>• Highlighting the relationship between the properties held by women and all the investigated properties (this report must be compared to the percentage of women in the population in order to be exploitable) |

- Share of land costs in development projects (equipment, environment);
- Duration of land proceedings over development project duration;
- Coverage of urban master plans;
- Execution rate of urban master plans;
- Retention rate of natural areas (forest, sea, river);
- Agri-land development rate;
- Average size of agricultural parcels.

The methods for measuring these indicators are defined in Table 3.

The Table 4 provides an overview of the axes, indicators and their direction of positive change. The direction of positive change is the direction of variation of the indicator (+ increase or - decrease) which can be perceived as an improvement.

## Approach to Analyzing Results and Formulating Reforms

The measurement of indicators in all the States under study will give a large number of figures to illuminate the impact of practices in each State. However, a simple comparison of these figures will be of no use and may even lead to harmful reform decisions rather than improvements.

To really draw relevant lessons from these figures, a comparative analysis must be conducted. This requires comparing not only the figures but also the practices that can justify these results at the level of each country.

Thus, the authors propose for each indicator a set of (practical) elements that can be exploited to highlight the practices that have given the results obtained. To do this, the authors rely on the correlations between each indicator and the practices.

*Table 3. Indicators of land tenure in development*

| Indicators | Measuring Methods |
|---|---|
| Share of investments on mortgage collateral | • This is the ratio of loans granted by banks on mortgage collateral to total loans over a given period |
| Share of land receipts | • It is the ratio of receipts from land transactions and taxes to total national revenues for a given period |
| Share of land costs in development projects | • Take a sample of development projects executed over a given period<br>• Evaluate land costs (expropriation cost, change of status ..) for all projects<br>• The ratio of land cost and all the project cost, is the part of co land u ts |
| Duration of land procedures in development projects | • Take a sample of development projects executed over a given period<br>• Assess the duration of land procedures (expropriation duration, change of status ....)For each project<br>• Make the ratio between land procedures duration and the total duration for each project<br>• Take the average of all reports |
| Coverage of urban master plans | • Find towns effective surface by analyzing remote sensing images<br>• Find the surface covered by the urban master plans<br>• Make a ratio between the two areas |
| Execution rate of urban master plans | • Find the complete area of buildings conform to urban master plan<br>• Make the ratio of the total area consistent and the area covered by the urban master plan |
| Conservation rate of natural areas | • Assess area covered by forests; cost right-of-river between two given period of time using remote sensing<br>• Make the ratio between the surfaces of both period of time |
| Agro-land development rate | • Find the total capacity of developed agricultural surfaces.<br>• Make the ratio between developed surface sand the agricultural area (UAA) |
| Average size of agricultural parcels | • Take a random sample of agricultural plots<br>• Averaging their capacity |

## RESULTS OF LAND ADMINISTRATION INDICATORS

### Rate of Awareness on Land Issues

The results of this indicator will be explained by analyzing the following elements:

- The communication strategies used by administrations to inform the public on land issues. In fact the type of media (radio, tele, press ...), support type (written, audio, video ...), the language of communication as well as the quality and legitimacy of people in charge of information are factors that directly influence the level of awareness.
- The proximity of administration services to the population.

### Availability of Land Information

It is clear that jurisdictions with well-operated, computerized systems could more rapidly formulate requests and achieve results. However, computerization is not the only way to achieve good performances especially for remote areas where a good mastery of techniques based on the paper format can prove to be very suitable. Thus the availability of information can be explained by:

- The quality of the data bases (land information system) used;
- The nature of the data bases used (numeric or paper);

*Table 4. Summary table of axes and indicators*

| Axes | Title of Indicator | Unit of Measure | Sense of Positive Development |
|---|---|---|---|
| Land Administration | The rate of awareness of the population | % | + |
| | The availability of land information | % | + |
| | The duration of the registration, recording, authorization and expropriation | day | - |
| | The satisfaction rate of landowners with land administration services | % | + |
| | Completeness of land information; | % | + |
| | The coverage rate of the land registration. | % | + |
| Property Security and Transaction | The rate of sensation of security of tenure (GLTN, 2015); | % | + |
| | The rate of legal action for land | % | - |
| | The frequency by type of land disputes | % | - |
| | The rate of acceptance of judgments of land disputes | % | + |
| | The percentage of women with property rights (GLTN, 2014). | % | + |
| Land in development | Investment share on mortgage collateral; | % | + |
| | Share of land receipts in national revenues (GLTN, 2014); | % | + |
| | Share of land costs in development projects (equipment, environment); | % | - / + |
| | Duration of the procedure over the duration of development projects; | % | - |
| | Coverage of urban master plans; | % | + |
| | Implementation rates of urban master plans; | % | + |
| | Conservation rates of natural areas (forest, maritime, river); | % | - |
| | Agro-land management rate; | % | + |
| | Average size of agricultural parcels. | Ha | + |

- The competence of agents in charge of archiving;
- The nature of the diffusion means of land information ;
- The nature of the communication of information among the various services.

## Duration of the Proceedings

To understand the results of the length of the registration process, registration, obtaining authorizations, expropriation ..., it would be useful to analyze:

- Legal provisions namely:
    - The *periods* laid down by the legislation: this will make it possible to know to what extent the legal durations influence the practical durations;
    - penalties for non-compliance with legal periods;
    - The number of services involved;
- And practical effects such as:
- The number of available agents and their Competence;

- The system of communication between agents;
- The proximity of services.

## The Satisfaction Rate

This indicator is strongly correlated positively with the first three. Thus, in addition to the results of previous indicators, the satisfaction rate will be explained in:

- The reception quality of agents;
- The equity measures in the management of files;
- Operations costs.

## Completeness of Land Information

The level of completeness of land information will find its elements of analysis in:

- The records update frequencies and other documents;
- The nature of the update tools;
- The number and qualification of agents in charge of setting to date.

## The Coverage Rate of Land Registration

The land registration coverage ratio requires for its justification in a country analysis:

- The character of land registration in the country: voluntary or mandatory;
- Costs and registration proceedings;
- Projects in favor of the of recording large sets.

## Litigation Rate, Frequency by Type of Conflict and Land Tenure Sensation

The rate of use and frequency by type of conflict can find their explanation in the same type of practices. Actually, if the use rate is the weight of land records at the court level, frequency spring types of the most common conflicts, to establish a Pareto chart. So to understand the results of these three indicators, the analysis elements are:

- Administration indicators results;
- The regimes and land statutes in force (Who owns the land how to own? and how to transmit properties?);
- Expropriation procedures and compensation value;
- The inclusion of customary rules in land regulations;
- The quality of property delimiting properties and legal investigation (tools, standards, techniques, agents;
- Setting of legal land transactions;
- The level of corruption in land government.

## Judgments Acceptance Rate

For a better land security in a country, not only practices should prevent up conflicts but also to allow a fair settlement of conflicts. Understanding the results of this indicator will return to the analysis of the elements:

- Adequate of land legislation with the country realities (customs);
- The quality of land disputes judgment;
- The competence of judges;
- The level of corruption in the courts;
- The quality of private judiciary expertise (topographical and real estate expertise).

## Percentage of Women With Property Rights

Ensuring women's access to land is essential to the assurance of social equity. The values of this indicator find their explanation by analyzing two key elements:

- Customary practices of land goods transfer: these practices can grant or deny women the right to access to properties;
- Women education;
- Sources and women's income level;
- The family law.

## Land Indicators Results in Development

All of these axes indicators find an explanation for their results with the analysis of the same elements namely:

- The results of the administration;
- Land safety results;
- Housing policies of the population segments with limited incomes;
- Public and private land investments on a large scale;
- Laws on public land management on government organs activities in terms of land management (FAO, 2003);
- The laws on the management of resources;
- Planning for land use and the control of the subdivisions;
- Laws about land taxes.

It should be noted that all the results are dependent on each other. Indeed, the results of an axis to understanding those of other axes. This requires a comprehensive analysis of the results to better understand. In addition, this dependence results to better understand a simple comparison will be meaningless.

## Reform Formulation Process

After the data collection and the analysis results, the next step of the technique of the benchmark is that of the formulation of reforms. This step is very important because it is the culmination of the whole process. Indeed, it is learned from comparative analysis, a set of practices so performance is revealed through the indicator values of certain countries. These good practices highlighted thus replace the old practices whose failure proved through the indicators measured values. These good practices, once set up, will determine the future of the land situation in the coming years. It is therefore essential to pay particular attention to this step will determine the likely success or failure of the benchmark project land practices.

To complete this stage of the benchmarking project in land area, it is important to keep in mind that practices the best results are not always the best set up for another country. Indeed successful practices to remember to be realistic about the realities of the new country. It will take into account:

- The availability of funding for the establishment of reforms
- The availability of qualified human resources necessary to meet the needs of reforms
- The cultural level of the population to new r e alit e s that will be introduced by the red form. For example: ask for a mostly illiterate population è computer you to make proceed hard internet would be a great failure.

To give more chances to the success of the reforms, the benchmark results should be communicated massively. In addition, all parties involved in land issues should be involved in action plans development process. Once defined actions to be taken, a monitoring team must be established to monitor the performance of actions and report any problems encountered ground.

## CONCLUSION

The approach proposed in the context of this work is a support for the establishment of new reforms in addition to some in the region. It will allow states to:

- Follow improvements led by practices set up,
- To compare themselves to neighbors in the region, and
- To get inspiration from neighboring practices in order to make more certain reforms.

It will even allow leaders of regional institutions to have land management tools to the West African scale. The results of comparisons allow a classification of states according to their performance and encourage competition, source of innovation.

## REFERENCES

ECOWAS. (2014). *New regime for community acts*. Retrieved from http://www.ecowas.int/ecowas-law/regulations-directives-and-other-acts/

FAO. (2003). Le régime foncier et le développement rural. *FAO Etudes sur les régimes fonciers 3*. Retrieved from http://www.fao.org/3/a-y4307f.pdf

GLTN. (2014). *Expert Group Meeting on Land Indicators*. Retrieved from http://sdg.iisd.org/events/expert-group-meeting-on-land-indicators/

GLTN. (2015). *Options for global reporting on GLTN / GLII land indicators in the context of the sustainable development goals*. United Nations Human Settlements Programme (UN-Habitat). Report 9/2015.

Lepoivre, F. (2004). Benchmarking concept et méthodologie. *Petite entreprise et partenariat*. Retrieved from https://documentfr.com/category/Small-Business-Entrepreneurship

## KEY TERMS AND DEFINITIONS

**ECOWAS:** It stands for the Economic Community of West African States. Established on May 28th, 1975 via the treaty of Lagos, ECOWAS is a 15-member regional group with a mandate of promoting economic integration in all fields of activity of the constituting countries. Member countries making up ECOWAS are Benin, Burkina Faso, Cape Verde, Cote d'Ivoire, The Gambia, Ghana, Guinea, Guinea Bissau, Liberia, Mali, Niger, Nigeria, Sierra Leone, Senegal and Togo. Considered one of the pillars of the African Economic Community, ECOWAS was set up to foster the ideal of collective self-sufficiency for its member states.

# Compilation of References

Achard, F., Eva, H. D., Stibig, H.-J., Mayaux, P., Gallego, J., Richards, T., & Malingreau, J.-P. (2002). Determination of deforestation rates of the world's humid tropical forests. *Science, 297*(5583), 999–1002. doi:10.1126cience.1070656 PMID:12169731

Adad, M. A., & Semlali, E. H. (2015). *Etude d'un modèle pour les données foncières et cadastrales marocaines à travers le modèle du domaine d'administration des terres*. Istanbul, Turkey: WCS-CE - The World Cadastre Summit, Congress & Exhibition.

ADEF. (2012). *Demystifying the land: State of the places of the research*. Academic Press.

African Development Bank (AfDB). (2014). *Morocco's Growth Diagnostic: Identifying Morocco's Binding Constraints to Broad-Based Growth*. Author.

African Land Policy Initiative. (2016). *Guidelines for the development of the curricula on land governance in Africa*. Addis Ababa: Land Policy Initiative.

African Union, African Development Bank, Economic Commission for Africa. (2010). *Framework and Guidelines on Land policy in Africa*. Addis Ababa: ECA.

Akesbi, N. (2001). Les exploitations agricoles au Maroc: Diagnostic à la lumière du recensement Général Agricole. *MEDIT, 4*.

Al Baouaba & Al Watania. (2015). *Portail national des collectivités territoriales*. Ministère de l'Intérieur.

ALBACHARIA. (2017). Retrieved from http://www.albacharia.ma/mapController?op=show

Aldosari, A. (2006). Middle East, Western Asia, and North Africa. In C. Carpenter (Ed.), *World and its Peoples* (p. 1258). New York: Marshall Cavendish Corporation.

Alemu, Z. G. (2012). *Livelihood strategies in rural South Africa: Implications for poverty reduction*. Paper prepared for presentation at the International Association of Agricultural Economists (IAAE), *Triennial Conference*, Foz do Iguacu, Brazil.

Allan, R. P., & Soden, B. J. (2008). Atmospheric warming and the amplification of precipitation extremes. *Science, 321*(5895), 1481–1483. doi:10.1126cience.1160787 PMID:18687921

Allen, R. G., Pereira, L. S., Raes, D., & Smith, M. (1998). Crop evapotranspiration —guidelines for computing crop water requirements. FAO Irrigation and drainage paper 56. Food and Agriculture Organization, Rome. Rome.

Allen, R. G. (2000). Using the FAO-56 dual crop coefficient method over an irrigated region as part of an evapotranspiration intercomparison study. *Journal of Hydrology (Amsterdam)*, *229*(1), 27–41. doi:10.1016/S0022-1694(99)00194-8

Allen, R. G., Pereira, L. S., Raes, D., & Smith, M. (1998). *Crop Evapotranspiration-Guidelines for Computing Crop Water Requirements* (Vol. 56). Rome: FAO.

Amalina, N. A. (2014). *Adoption of land administration domain model for land administration in Malaysia*. (Thesis).

Amalina, N. Z., Abdul Rahman, A., Hassan, M. I., & Tan, L. C. (2015). Conceptual Modelling of 3D Cadastre and LADM. WCS-CE - The World Cadastre Summit, Congress & Exhibition, Istanbul, Turkey.

Amalina, N. Z., Abdul Rahman, A., Jamil, H., Teng, C. H., Tan, L. C., & Looi, K. S. … van Oosterom, P. (2014). *Development of a Prototype for the Assessment of the Malaysian LADM Country Profile*. FIG Congress 2014 Engaging the Challenges – Enhancing the Relevance, Kuala Lumpur, Malaysia.

Amsalu, A., & de Graaff, J. (2006). Farmers' views of soil erosion problems and their conservation knowledge at Beressa watershed, central highlands of Ethiopia. *Agriculture and Human Values*, *23*(1), 99–108. doi:10.100710460-005-5872-4

Amsalu, T., & Addisu, S. (2014). *Assessment of Grazing Land and Livestock Feed Balance in Gummara-Rib Watershed, Ethiopia*. Academic Press.

ANCFCC. (2017). Retrieved from www.ancfcc.gov.ma

Andarzian, B., Bannayan, M., Steduto, P., Mazraeh, H., Barati, M. E., Barati, M. A., & Rahnama, A. (2011). Validation and testing of the AquaCrop model under full and deficit irrigated wheat production in Iran. *Agricultural Water Management*, *100*(1), 1–8. doi:10.1016/j.agwat.2011.08.023

Andreas, E. L. (1989). Two-wavelength method of measuring path-averaged turbulent surface heat fluxes. *Journal of Atmospheric and Oceanic Technology*, *6*(2), 280–292. doi:10.1175/1520-0426(1989)006<0280:TWMOMP>2.0.CO;2

Andréassian, V, (2004). *Couvert forestier et comportement hydrologique des bassins versants*. La houille blanche/N° 2-2004-31.

Andréassian, V. (2002). *Impact de l'évolution du couvert forestier sur le comportement hydrologique des bassins versants. Tome 1*. Thèse de Doctorat de l'Université Paris 6. Ecole doctorale: Géosciences et Ressources Naturelles Spécialité: Hydrologie.

Andréassian, V., Tangara, M., & Muraz J. (2001). Evaluer l'impact de l'évolution du couvert forestier sur le comportement hydrologique des bassins versants: méthodologie et premiers résultats fondés sur les données de l'IFN. *Rev. for. fr. liii -3-4-2001*, 475-480.

Anonymous. (2009). *Land Governance and securing rights in the countries of the South. White Book of French actors of cooperation*. Technical Committee: Land and Development.

Aouragh, M. (2012). *Dynamique du « paysage-arganeraie » dans le Sud-ouest marocain. Apport des données de télédétection et perspectives de les intégrer dans un SIG* (Thèse de Doctorat). Géographie, l'université Paris IV Sorbonne.

Arvor, D., Dubreuil, V., & Ronchail, J., & S., P. (2008). Apport des données TRMM 3B42 à l'étude des précipitations au Mato Grosso. *Climatologie*, *5*, 49–69.

Asiama, K. O., Bennett, R., & Zevenbergen, J. (2017). Land consolidation for sub-Saharan Africa's customary lands—The Need for Responsible Approaches. *Am. J. Rural Dev*, *5*, 39–45.

Asiama, K. O., Bennett, R., Zevenbergen, J., & Asiama, S. O. (2018). Land valuation in support of responsible land consolidation on Ghana's rural customary lands. *Survey Review*, 1–13. doi:10.1080/00396265.2018.1467672

Asrar, G., Fuchs, M., Kanemasu, E. T., & Hatfield, J. L. (1984). Estimating Absorbed Photosynthetic Radiation and Leaf Area Index from Spectral Reflectance in Wheat1. *Agronomy Journal*, *76*(2), 300–306. doi:10.2134/agronj1984.00021962007600020029x

Astrup, R., & Larson, B. C. (2006). Regional variability of species-specific crown openness for aspen and spruce in western boreal Canada. *Forest Ecology and Management*, *228*(1), 241–250. doi:10.1016/j.foreco.2006.02.048

Atnafu, Dejen, & Vijverberg. (2011). Assessment of the Ecological Status and Threats of Welala and Shesher Wetlands, Lake Tana Sub-Basin (Ethiopia). *Journal of Water Resource and Protection*, *3*(7), 540–547.

AUC-ECA-AfDB Consortium. (2010). Framework and guidelines on land policy in Africa: A framework to strengthen land rights, enhance productivity and secure livelihoods. AUC-ECA-AfDB.

AUC-ECA-AfDB Consortium. (2011). Land policy in Africa: West Africa regional assessment. AUC-ECA-AfDB.

AUC-ECA-AfDB Consortium. (2012). Tracking progress in land policy formulation and implementation in Africa. AUC-ECA-AfDB.

AUC-ECA-AfDB Consortium. (2014). Guiding principles on large-scale land-based investments in Africa. AUC-ECA-AfDB.

Aydin, C. C., Deir, O., & Atasoy, M. (2004). Third Dimension (3D) in cadastre and its Integration with 3D GIS in Turky. TS25 Appropriate Technologies for good Land Administration II. FIG Working Week 2004, Athens, Greece.

Babqiqi, A., & Messouli, M. (2013). Simulation of climate and its implication on agriculture in Morocco using Statistical DownScaling. *International Journal of Latest Research in Science and Technology*, *2*(5), 83-96.

Bach, H., Clausen, T. J., Trang, D. T., Emerton, L., Facon, T., Hofer, T., ... Whiting, L. (2011). *From local watershed management to integrated river basin management at National and Transboundary levels. Lao PDR*. Mekong River Commission.

Balaghi, R., Jlibene, M., Tychon, B., & Eerens, H. (2013). Agrometeorological cereal yield forecasting in Morocco (2013 ed.). National Institute of Agronomic Research (INRA): Devision of information and communication (INRA).

Baldwin, V. C., Burkhart, H. E., Westfall, J. A., & Peterson, K. D. (2001). Linking Growth and Yield and Process Models to Estimate Impact of Environmental Changes on Growth of Loblolly Pine. *Forest Science*, *47*(1), 77–82.

Bandyopadhyay, P. K., & Mallick, S. (2003). Actual evapotranspiration and crop coefficients of wheat (Triticum aestivum) under varying moisture levels of humid tropical canal command area. *Agricultural Water Management*, *59*(1), 33–47. doi:10.1016/S0378-3774(02)00112-9

Banque Mondiale. (2008). *Marchés fonciers pour la croissance économique au Maroc*. Washington, DC: Word Bank.

Barandela,, R.M., & Valdovinos, J.S. (2003). New applications of ensembles of classifiers. *Pattern Analysis & Applications*, *6*, 245–256. doi:10.100710044-003-0192-z

Baret, F., Guyot, G., & Major, D. J. (1989). Crop biomass evaluation using radiometric measurements. *Photogrammetria*, *43*(5), 241–256. doi:10.1016/0031-8663(89)90001-X

Barreiro, S., Schelhaas, M.-J., Kändler, G., Antón-Fernández, C., Colin, A., Bontemps, J.-D., ... Ferezliev, A. (2016). Overview of methods and tools for evaluating future woody biomass availability in European countries. *Annals of Forest Science*, *73*(4), 823–837. doi:10.100713595-016-0564-3

Barrier, O. (1996). *Management of renewable natural resources and conservation of eco systems: Land-environment*. Thesis in law, France.

Bathust Declaration. (1999). *The BucharestDeclaration on land administration for sustainable development. FIG Publication no. 21*. Copenhagen, Denmark: FIG.

Beerech, H.V., Mrslatha, B.M., Thimmaraja Yadava, G., & Dandur, N. (2014). An approcach for identification and classification of crops using multispectral images. *IJERT, 3*(5).

Benabdelouahab, T., Balaghi, R., Hadria, R., Lionboui, H., Djaby, B., & Tychon, B. (2016). Testing Aquacrop to Simulate Durum Wheat Yield and Schedule Irrigation in a Semi-Arid Irrigated Perimeter in Morocco. *Irrigation and Drainage*, *65*(5), 631–643. doi:10.1002/ird.1977

Benabdelouahab, T., Balaghi, R., Hadria, R., Lionboui, H., Minet, J., & Tychon, B. (2015). Monitoring surface water content using visible and short-wave infrared SPOT-5 data of wheat plots in irrigated semi-arid regions. *International Journal of Remote Sensing*, *36*(15), 4018–4036. doi:10.1080/01431161.2015.1072650

Benabdelouahab, T., Balaghi, R., Hadria, R., Lionboui, H., & Tychon, B. (2016). Assessment of vegetation water content in wheat using near and shortwave infrared SPOT-5 Data in an irrigated area. *Revue des Sciences de l'Eau*, *29*(2), 97–107. doi:10.7202/1036542ar

Benediktsson, J. A., Chanussot, J., & Fauvel, M. (2007). Multiple classifiers in remote sensing: from basics to recent developments. *7th International Workshop on Multiple Classifier Systems (MCS)*, 501-512. 10.1007/978-3-540-72523-7_50

Ben-Gal, A., Kool, D., Agam, N., van Halsema, G. E., Yermiyahu, U., Yafe, A., ... Dag, A. (2010). Whole-tree water balance and indicators for short-term drought stress in non-bearing 'Barnea' olives. *Agricultural Water Management*, *98*(1), 124–133. doi:10.1016/j.agwat.2010.08.008

Béranger, S., Blanchard, F., Archambault, A., & Allier, D. (2006). *Utilisation des outils d'aide à la décision dans la gestion des méga sites* [Rapport final]. Retrieved from http://infoterre.brgm.fr/rapports/RP-55223-FR.pdf

Bettinger, P., Boston, K., Siry, J., & Grebner, D. (2009). *Forest management and planning*. Amsterdam: Academic.

Bharathkumar, L., & Mohammed-Aslam, M. (2015). A Crop Pattern Mapping of Tumkur Taluk Using NDVI Technique" A Remote Sensing and GIS Approach. *Aquat. Procedia*, *4*, 1397–1404. doi:10.1016/j.aqpro.2015.02.181

Bi, H., Li, X., Liu, X., Guo, M., & Li, J. (2009). A case study of spatial heterogeneity of soil moisture in the Loess Plateau, western China: A geostatistical approach. *International Journal of Sediment Research*, *24*(1), 63–73. doi:10.1016/S1001-6279(09)60016-0

Bouchaou, L., Michelot, J. L., Qurtobi, M., Zine, N., Gaye, C. B., Aggarwal, P. K., ... Vengosh, A. (2009). Origin and residence time of groundwater in the Tadla basin (Morocco) using multiple isotopic and geochemical tools. *Journal of Hydrology (Amsterdam)*, *379*(3–4), 323–338. doi:10.1016/j.jhydrol.2009.10.019

Bouderbala, N. (1990). *Les systèmes de propriété foncière au Maghreb: le cas du Maroc*. Montpelier: CIHEAM.

Bowen, I. S. (1926). The ratio of heat losses by conduction and by evaporation from any water surface. *Physical Review*, *27*(6), 779–787. doi:10.1103/PhysRev.27.779

Breidenbach, J., & Astrup, R. (2012). Small area estimation of forest attri- butes in the Norwegian National Forest Inventory. *European Journal of Forest Research*, *131*(4), 1–13. doi:10.100710342-012-0596-7

Briem, G. J., Benediktsson, J. A., & Sveinsson, J. R. (2002). Multiple Classifiers Applied to Multisource Remote Sensing Data. *IEEE Transactions on Geoscience and Remote Sensing*, *40*(10), 2291–2300. doi:10.1109/TGRS.2002.802476

Brisson, N., Mary, B., Ripoche, D., Jeuffroy, M. H., Ruget, F., Nicoullaud, B., ... Delécolle, R. (1998). STICS: A generic model for the simulation of crops and their water and nitrogen balances. I. Theory and parameterization applied to wheat and corn. *Agronomie*, *18*(5-6), 311–346. doi:10.1051/agro:19980501

Bryceson, D. F. (1999). African rural labour, income diversification & livelihood approaches a long-term development perspective. *Review of African Political Economy*, *26*(80), 171–189. doi:10.1080/03056249908704377

Burba, G. (2013). Eddy Covariance Method for Scientific, Industrial, Agricultural and Regulatory Applications: a Field Book on Measuring Ecosystem Gas Exchange and Areal Emission Rates. Lincoln, NE: Academic Press.

Burkhart, H. E., & Tomé, M. (2012). *Modeling forest trees and stands*. Springer Science & Business Media. doi:10.1007/978-90-481-3170-9

C.M. (2011). *Constitution Marocaine. Royaume du Maroc*. Secrétariat Général du Gouvernement, Direction de l'Imprimerie Officielle.

C.T. (2009). *Gouvernance foncière et sécurisation des droits dans les pays du Sud*. Comité Technique du Foncier et Développement, Agence Française de Développement internationale.

Campbell, G. J. (1994). *A GIS-based multipurpose cadastre for Canada lands. FIG XX Congress*, Melbourne, Australia.

Cantiani, P., De Meo, I., Ferretti, F., & Paletto, A. (2010). Forest functions evaluation to support forest landscape management planning. *Forestry Ideas*, *16*(1), 39.

Carter, M. R. (2002). Soil quality for sustainable land management. *Agronomy Journal*, *94*(1), 38–47. doi:10.2134/agronj2002.0038

Cawley, G. C., & Talbot, N. L. C. (2003). Efficient leave-one-out cross-validation of kernel fisher discriminant classifiers. *Pattern Recognition*, *36*(11), 2585–2592. doi:10.1016/S0031-3203(03)00136-5

Cawley, G. C., & Talbot, N. L. C. (2004). Fast exact leave-one-out cross-validation of sparse least-squares support vector machines. *Neural Networks*, *17*(10), 1467–1475. doi:10.1016/j.neunet.2004.07.002 PMID:15541948

Ceccato, P., Flasse, S., & Grégoire, J. M. (2002). Designing a spectral index to estimate vegetation water content from remote sensing data: Part 2. Validation and applications. *Remote Sensing of Environment*, *82*(2–3), 198–207. doi:10.1016/S0034-4257(02)00036-6

Ceccato, P., Flasse, S., Tarantola, S., Jacquemoud, S., & Grégoire, J. M. (2001). Detecting vegetation leaf water content using reflectance in the optical domain. *Remote Sensing of Environment*, *77*(1), 22–33. doi:10.1016/S0034-4257(01)00191-2

Ceccato, P., Gobron, N., Flasse, S., Pinty, B., & Tarantola, S. (2002). Designing a spectral index to estimate vegetation water content from remote sensing data: Part 1: Theoretical approach. *Remote Sensing of Environment*, *82*(2–3), 188–197. doi:10.1016/S0034-4257(02)00037-8

CESE. (2012). *Le système fiscal marocain: développement économique et cohésion sociale*. Conseil Economique, Social et Environnemental, rapport du conseil économique et social, Auto-Saisine n° 9/2012.

Chapman, L. (2007). Potential applications of near infrared hemispherical imagery in forest environments. *Agricultural and Forest Meteorology*, *143*(1), 151–156. doi:10.1016/j.agrformet.2006.12.006

Chaponnière, A., Maisongrande, P., Duchemin, B., Hanich, L., Boulet, G., Escadafal, R., & Elouaddat, S. (2005). A combined high and low spatial resolution approach for mapping snow covered areas in the Atlas mountains. *International Journal of Remote Sensing*, *26*(13), 2755–2777. doi:10.1080/01431160500117758

Chehbouni, A., Escadafal, R., Duchemin, B., Boulet, G., Simonneaux, V., Dedieu, G., ... Sobrino, J. A. (2008). An integrated modelling and remote sensing approach for hydrological study in arid and semi-arid regions: The SUDMED programme. *International Journal of Remote Sensing*, *29*(17-18), 5161–5181. doi:10.1080/01431160802036417

CHENESANOGO. (2012). *Land issues and sustainable development in Mali*. Thesis.

Cheng, J., Bo, Y., Zhu, Y., & Ji, X. (2014). A novel method for assessing the segmentation quality of high-spatial resolution remote-sensing images. *International Journal of Remote Sensing*, *35*(10), 3816–3839. doi:10.1080/01431161.2014.919678

Cheng, T., Rivard, B., Sánchez-Azofeifa, A. G., Féret, J. B., Jacquemoud, S., & Ustin, S. L. (2012). Predicting leaf gravimetric water content from foliar reflectance across a range of plant species using continuous wavelet analysis. *Journal of Plant Physiology*, *169*(12), 1134–1142. doi:10.1016/j.jplph.2012.04.006 PMID:22608180

Chia, J., & Lee, B. (2017). An exploratory methodology to incorporate transfer location in transit spatial coverage quantification. In *Proceedings of the Australasian Transport Research Forum (ATRF) 2017*. Australasian Transport Research Forum (ATRF).

Chigbu, U. E. (2013b). *Territorial development: suggestions for a new approach to rural development in Nigeria* (PhD thesis). Technical University of Munich, Germany.

Chigbu, U. E., & Klaus, M. (2013, May 6). Insecurity-Generating System of Land Tenure and its Impact on Rural Development: Evidence from Uturu, Nigeria. FIG Working Week.

Chigbu, U. E. (2012). Village renewal as an instrument of rural development: Evidence from Weyarn, Germany. *Community Development (Columbus, Ohio)*, *43*(2), 209–224. doi:10.1080/15575330.2011.575231

Chigbu, U. E. (2013a). Rurality as a choice: Towards ruralizing rural areas in sub-Saharan African countries. *Development Southern Africa*, *30*(6), 812–825. doi:10.1080/0376835X.2013.859067

Chigbu, U. E. (2013c). Fostering rural sense of place: The missing piece in Uturu, Nigeria. *Development in Practice*, *23*(2), 264–277. doi:10.1080/09614524.2013.772120

Chigbu, U. E. (2015). Repositioning culture for development: Women and development in a Nigerian rural community. *Community Work & Family*, *18*(3), 336–352. doi:10.1080/13668803.2014.981506

Chigbu, U. E. (2015). Ruralisation: A tool for rural transformation. *Development in Practice*, *25*(7), 1067–1073. doi:10.1080/09614524.2015.1071783

Chigbu, U. E., Izugbara, C. O., & de Vries, W. T. (2018). Land, culture, culture loss and community: rural insights from sub-Saharan Africa. In S. Kenny, B. McGrath, & R. Phillips (Eds.), *The Routledge Handbook of Community Development: Perspectives from Around the Globe* (pp. 98–114). London: Routledge.

Chigbu, U. E., Schopf, A., de Vries, W. T., Masum, F., Mabikke, S., Antonio, D., & Espinoza, J. (2017). Combining land-use planning and tenure security: A tenure responsive land-use planning approach for developing countries. *Journal of Environmental Planning and Management*, *60*(9), 1622–1639. doi:10.1080/09640568.2016.1245655

Chigbu, U. E., Tenadu, K., & Mwasumbi, A. (2017). Curriculum Reform in Land Governance Education: The Need for Transforming Existing Curricula in Africa. *World Bank Conference on Land and Poverty*, Washington, DC.

Choi, W., & Deal, B. (2008). Assessing hydrological impact of potential land use change through hydrological and land use change modeling for the Kishwaukee River basin (USA). *Journal of Environmental Management*, *88*(4), 1119–1130. doi:10.1016/j.jenvman.2007.06.001 PMID:17689854

Cichocinski, P. (1999). Digital cadastral maps in land information systems. LIBER Quarterly, 9(2).

Clark, C. O. (1945). Storage and the unit hydrograph. *Transactions of the American Society of Civil Engineers*, *110*, 1419–1488.

Cochran, W. G. (1977). *Sampling techniques* (3rd ed.). Academic Press.

Cohen, J. M., & Uphoff, N. T. (1980). Participation's place in rural development: Seeking clarity through specificity. *World Development*, *8*(3), 213–235. doi:10.1016/0305-750X(80)90011-X

Congalton, R. (1991). Areview of Assessing the Accuracy of Classifications of Remotly Sensed Data. *Remote Sensing of Environment*, *37*(1), 35–46. doi:10.1016/0034-4257(91)90048-B

Conseil Economique Social et Environnemental. (2013). *Open data. La libération des données publiques au service de la croissance et de la connaissance*. Rabat: CESE.

Conseil Economique Social et Environnemental. (2014). *Cohérence des politiques sectorielles et accords de libre échange*. Rabat: CESE.

Constable, J. V. H., & Friend, A. L. (2000). Suitability of process-based tree growth models for addressing tree response to climate change. *Environmental Pollution*, *110*(1), 47–59. doi:10.1016/S0269-7491(99)00289-4 PMID:15092855

Coppin, P. R., & Bauer, M. E. (1996). Digital change detection in forest ecosystems with remote sensing imagery. *Remote Sensing Reviews*, *13*(3–4), 207–234. doi:10.1080/02757259609532305

Coppin, P., Jonckheere, I., Nackaerts, K., Muys, B., & Lambin, E. (2004). Review ArticleDigital change detection methods in ecosystem monitoring: A review. *International Journal of Remote Sensing*, *25*(9), 1565–1596. doi:10.1080/0143116031000101675

Côté, J.-F., Fournier, R. A., Frazer, G. W., & Olaf Niemann, K. (2012). A fine-scale architectural model of trees to enhance LiDAR-derived measurements of forest canopy structure. *Agricultural and Forest Meteorology, 166–167*, 72–85. doi:10.1016/j.agrformet.2012.06.007

Cour des comptes. (2014). *Evaluation de la stratégie Maroc Numeric 2013*. Rabat: Cour des Comptes.

Cour des comptes. (2015). *Annual report of the Public Finance Court for the year 2015*. Rabat: Cour des Comptes.

Craglia, M. (2010). *Building INSPIRE: The Spatial Data Infrastructure for Europe*. Retrieved from http://www.esri.com/news/arcnews/spring10articles/building-inspire.html

Czechowski, P. O., Owczarek, T., Badyda, A., Majewski, G., Rogulski, M., & Ogrodnik, P. (2018). Preliminary comparative assessment of PM10 hourly measurement results from new monitoring stations type using stochastic and exploratory methodology and models. *E3S Web of Conferences, 28.*

Dahir No. 1-09-236 of 23 February 2010 establishing the Code of the Habous.

Dahir No. 1-16-38 of 17 Jumada I 1437 (26 February 2016) laying down the functions and organization of the Ministry of Habous and Islamic Affairs..

Dahir of 06 August 1915 on the Exercise of the actions in justice issues affecting the real estate of the State

Dahir of 1 July 1914 on the public domain.

Dahir of 10 October 1917 on the exploitation and conservation of forests

Dahir of 18 February 1924 on the special regulation for the delimitation of collective lands

Dahir of 19 March 1951 regulating the management and disposal of collective property

Dahir of 19 October 1921 on the municipal field.

Dahir of 24 Ramadan 1333 (6 August 1915) on the Exercise of the actions in justice issues affecting the real estate of the State, as it has been modified

Dahir of 27 April 1919 organizing the administrative supervision of the communities "Native" and regulating the management and disposal of collective property

Dahir of 3 January 1916 on the special regulation on the delimitation of the area of the State.

Dale, F. P., & McLaughlin, J. D. (1988). *Land Information Management. An introduction with special reference to cadastral problems in third world countries*. New York: Oxford University Press Inc.

Dale, F. P., & McLaughlin, J. D. (1999). *Land administration*. New York: Oxford University Press Inc.

Dalmau, Alarcón, & Oliva. (2017). Crop Classification in Satellite Images through Probabilistic Segmentation Based on Multiple Sources. *Sensors (Basel), 17*(6). PMID:28608825

Dassot, M., Constant, T., & Fournier, M. (2011). The use of terrestrial LiDAR technology in forest science: Application fields, benefits and challenges. *Annals of Forest Science, 68*(5), 959–974. doi:10.100713595-011-0102-2

de Vries W. & Chigbu U. E. (2017). Responsible land management – concept and application in a territorial rural context. *Fub –Flächenmanagement und Bodenordnung, 79*(2), 65-73.

De Vries, P. G. (2012). *Sampling theory for forest inventory: a teach-yourself course*. Springer Science & Business Media.

de Wit, A. J. W., Boogaard, H. L., & van Diepen, C. A. (2004). Using NOAA–AVHRR estimates of land surface temperature for regional agrometeorogical modelling. *International Journal of Applied Earth Observation and Geoinformation, 5*(3), 187–204. doi:10.1016/j.jag.2004.03.003

DeBarry, P. A. (2004). *Watersheds: Processes, Assessment and Management*. John Wiley & Sons, Inc.

Decree n° 2-04-503 of 21 Hijja 1425 (1 February 2005) on the powers and organization of the Office of the High Commissioner to the waters and forests and to the fight against desertification

Decree n° 2-06-472 of 2 Chaabane 1429 (4 August 2008) setting out the functions and organization of the Ministry of Equipment and Transport.

Decree n° 2-07-995 of 23 chaoual 1429 (23 October 2008) relating to the powers and to the organization of the Ministry of the Economy and Finance

Decree n° 2-97-176 of 14 Chaabane 1418 (15 December 1997) relating to the powers and to the organization of the Ministry of the Interior

Decroux, . (2007). *Ouvrage Droit Foncier Marocain*. Editions la Porte.

Demarty, J., Ottlé, C., Braud, I., Olioso, A., Frangi, J. P., Bastidas, L. A., & Gupta, H. V. (2004). Using a multiobjective approach to retrieve information on surface properties used in a SVAT model. *Journal of Hydrology (Amsterdam), 287*(1), 214–236. doi:10.1016/j.jhydrol.2003.10.003

Department of the Interior U.S. (2016). *Geological survey product guidelandsat surface reflectance-derived Spectral indices version 3.3 December 2016*. Author.

Desclée, B., Bogaert, P., & Defourny, P. (2006). Forest change detection by statistical object-based method. *Remote Sensing of Environment, 102*(1), 1–11. doi:10.1016/j.rse.2006.01.013

Desta, L., Kassie, M., Benin, S., & Pender, J. (2000). *Land degradation and strategies for sustainable development in the Ethiopian highlands: Amhara Region*. Socio-economic and Policy Research working paper 32. International Livestock Research Institute, Nairobi, Kenya.

DGUA. (2007). *Rapport final de l'étude portant sur l'analyse de la problématique foncière sous le régime du DAHIR khalifien de 1914, et son impact sur le développement urbain dans les provinces du nord, Direction générale de l'urbanisme et de l'architecture*. Retrieved from www.muat.gov.ma

Di Gregorio, A. (2016). *Land Cover Classification System: Classification concepts*. Software version 3. Available at: http://www.fao.org/3/a-i5232e.pdf

Diatta, M. (2007). *Application de la télédétection et des Systèmes d'Information Géographique à la gestion des forêts classées du SENEGAL: cas de la forêt de Thiès. Mémoire de DESS, RECTAS*. Ilé-Ifè, Nigéria: OAU Campus.

Dinku, T., Chidzambwa, S., Ceccato, P., Connor, S. J., & Ropelewski, C. F. (2008). Validation of high-resolution satellite rainfall products over complex terrain. *International Journal of Remote Sensing*, *29*(14), 4097–4110. doi:10.1080/01431160701772526

Donia, N. (2012). Application of remotely sensed imagery to watershed analysis a case study of Lake Karoun, Egypt. *Arabian Journal of Geosciences*, *6*(9), 3217–3228. doi:10.100712517-012-0621-7

Doorenbos, J., & Kassam, A. H. (1979). Yield response to water. FAO Irrigation and Drainage, 33, 193.

Doorenbos, J., & Pruitt, W. O. (1977). *Guidelines for predicting crop water requirements* (Vol. 24). Rome: FAO.

Dowson, E., & Sheppard, V.L.O. (1952). *Land registration*. Her Majesty's Stationary Office. Colonial office publication, No.13.

Driouch. (2003). *Ouvrage « les origines du système d'immatriculation foncière »*. Les publications de la série de la connaissance juridique. DECRET n° 2-13-424..

Driouech, F. (2010). *Distribution des précipitations hivernales sur le Maroc dans le cadre d'un changement climatique: descente d'échelle et incertitudes*. Thèse de Doctorat. Université de Toulouse.

Dube, K. (2016). Implications of rural irrigation schemes on household economy. A case of Lower Gweru Irrigation Scheme, Zimbabwe. *South African Journal of Agricultural Extension*, *44*(1), 75–90. doi:10.17159/2413-3221/2016/v44n1a371

Dubreuil, P. (1974). *Introduction à l'analyse hydrologique*. Paris: Ed. Masson-ORSTOM.

Duchemin, B., Hadria, R., Erraki, S., Boulet, G., Maisongrande, P., Chehbouni, A., ... Simonneaux, V. (2006). Monitoring wheat phenology and irrigation in Central Morocco: On the use of relationships between evapotranspiration, crops coefficients, leaf area index and remotely-sensed vegetation indices. *Agricultural Water Management*, *79*(1), 1–27. doi:10.1016/j.agwat.2005.02.013

Duchemin, B., Hagolle, O., Mougenot, B., Benhadj, I., Hadria, R., Simonneaux, V., ... Chehbouni, A. G. (2008). Agrometerological study of semi-arid areas: An experiment for analysing the potential of time series of FORMOSAT-2 images (Tensift-Marrakech plain). *International Journal of Remote Sensing*, *29*(17-18), 5291–5299. doi:10.1080/01431160802036482

Duchemin, B., Maisongrande, P., Boulet, G., & Benhadj, I. (2008). A simple algorithm for yield estimates: Evaluation for semi-arid irrigated winter wheat monitored with green leaf area index. *Environmental Modelling & Software*, *23*(7), 876–892. doi:10.1016/j.envsoft.2007.10.003

Ducrey, M., & Bartoli, M. (1974). Utilisation des photographies hémisphériques pour le cubage des arbres sur pied. *Annales des Sciences Forestieres*, *31*(4), 229–248. doi:10.1051/forest/19740403

Du, L., Tian, Q., Yu, T., Meng, Q., Jancso, T., Udvardy, P., & Huang, Y. (2013). A comprehensive drought monitoring method integrating MODIS and TRMM data. *International Journal of Applied Earth Observation and Geoinformation*, *23*, 245–253. doi:10.1016/j.jag.2012.09.010

Economic Commission for Africa. (2007). *e-Strategies - Best Practices and Lessons Learnt.* Addis Ababa: ECA.

ECOWAS. (2014). *New regime for community acts.* Retrieved from http://www.ecowas.int/ecowas-law/regulations-directives-and-other-acts/

Eddaif, N. (2015). *Suivi spatio-temporel des surfaces de l'arganier par imagerie satellitaire optique à moyenne résolution spatiale: Application à la forêt Mseguina-Sud Ouest du Maroc.* Mémoire de Master: Sciences et Technologie de l'Espace, Facultés des sciences de Rabat.

Effenberg, W. (2001). *Spatial information Systems, the maintenance of digital cadastral maps* (Ph.D dissertation). University of Melborne.

E-GOV Program. (2014). Retrieved from http://www.egov.ma/sites/default/files/Programme%20eGov%20Morocco.pdf

Ehleringer, J. R., Roden, J., & Dawson, T. E. (2000). Assessing ecosystem-level water relations trough stable isotope ratio analysis. In Methods in Ecosystem Science. Springer.

El Abboudi, A. (2015). *Cartographie de la densité du couvert arboré de l'arganier de la province d'Ait Baha.* Résumé des communications, Congrès international de l'arganier, 3éme édition.

El Alaoui El Fels, A., & Saidi, M. E. (2014). Simulation et spatialisation du risque d'inondation dans une vallée anthropisée, le cas de la vallée de l'Ourika (Haut Atlas, Maroc). European Scientific Journal, 10(17).

El Ayachi, M. (2006). *Design of a new approach for developing a Multipurpose cadastral system of Morocco (Organizational and technical aspects)* (PhD thesis). IAV Hassan II, Rabat, Morocco.

El Ayachi, M., & Semlali, E. H. (2001). Digital cadastral map, a multipurpose tool for sustainable development. *Proceeding of the International conference on spatial information for sustainable development.*

El Bouanani. (2011). *Immobilier d'Entreprise au Maroc.* ICH Rabat INAU.

El Ghrib, A. (2016). *Privatization of collective agricultural Land in Morocco. World Bank Conference on Land and poverty Conference*, Washington, DC.

El Hafid, R., El Mourid, M., Samir, K., & Bakoulou, B. (1996). Caractérisation de certaines variétés de blé, orge et du triticale sous différentes situations hydriques en conditions de champ et simulées. *Al awamia, 93*, 7-26.

El Mansouri, L. (2014). Une approche basée objet combinée avec les classificateurs avancés (SVM, RF, Extra Trees) pour la détection des changements des bâtiments. La revue internationale de Géomatique, Février 2014.

El Mansouri, L. (2013). *Object-based approach and Tree-based ensemble classifications for mapping Building changes.* IARIA.

El-Ayachi, M. (2017). A paradigm of improving land information management. In S. Faiz & K. Mahmoudi (Eds.), *Handbook of research on geographic information systems applications and advancements*. Hershey, PA: IGI Global. doi:10.4018/978-1-5225-0937-0.ch022

Eldho, P. (2012). *Module 4 – (L12 -L18) Watershed modeling*. Available at http://nptel.ac.in/courses/105101010/downloads/Lecture12.pdf

Elia, E. A., Zevenbergen, J. A., Lemmen, C. H. J., & van Oosterom, P. J. M. (2013). The land administration domain model (LADM) as the reference model for the Cyprus land information system (CLIS). Survey Review, 45(329).

Enemark, S., Clifford Bell, K., Lemmen, C., & McLaren, R. (2014). *Fit-For-Purpose Land Administration. International Federation of Surveyors*. Copenhagen: FIG.

Er-Raki, S., Chehbouni, A., Boulet, G., & Williams, D. G. (2010). Using the dual approach of FAO-56 for partitioning ET into soil and plant components for olive orchards in a semi-arid region. *Agricultural Water Management*, *97*(11), 1769–1778. doi:10.1016/j.agwat.2010.06.009

Er-Raki, S., Chehbouni, A., Guemouria, N., Duchemin, B., Ezzahar, J., & Hadria, R. (2007). Combining FAO-56 model and ground-based remote sensing to estimate water consumptions of wheat crops in a semi-arid region. *Agricultural Water Management*, *87*(1), 41–54. doi:10.1016/j.agwat.2006.02.004

Er-Raki, S., Chehbouni, A., Hoedjes, J., Ezzahar, J., Duchemin, B., & Jacob, F. (2008). Improvement of FAO-56 method for olive orchards through sequential assimilation of thermal infrared-based estimates of ET. *Agricultural Water Management*, *95*(3), 309–321. doi:10.1016/j.agwat.2007.10.013

Er-Raki, S., Ezzahar, J., Khabba, S., Jarlan, L., Kharrou, M. H., & Chehbouni, G. (2013). *Micrometeorology tools for measuring evapotranspiration from the leaf to the region. In Evapotranspiration: processes, sources and environmental implications* (pp. 1–22). New York: Nova Science.

Espinoza, J. (2011). What is required to promote Good Land Governance in the 21st Century. *Word Bank Conference on Land and Poverty*, Washington, DC.

Estifanos Abera. (2014). *Assessment of micro- watershed vulnerability for soil erosion in Ribb watershed using GIS and Remote Sensing*. Author.

Ethiopian Forestry Action Plan (EFAP). (1994). Final Report Volume II –The Challenges for Development and Volume III – Issues and Action. EFAP Secretariat.

Ettarid, M., El-ayachi, M., Ettaqy, E., & Elyamani, M. (2003). L'Orthophotographie numérique au service du cadastre national. *Proceeding of the 2nd FIG Regional Conference. Session TS14 Cadastre in the Mediterranean*. Available from: https://www.researchgate.net/publication/238073176_L'Orthophotographie_numerique_au_service_du_cadastre_national

Ezzine, H., Bouziane, A., & Ouazar, D. (2014). Seasonal comparisons of meteorological and agricultural drought indices in Morocco using open short time-series data. *International Journal of Applied Earth Observation and Geoinformation*, *26*, 36–48. doi:10.1016/j.jag.2013.05.005

Fabiola Flores, P., & Mario Lillo, S. (2010). Simple Air Temperature Estimation Method from MODIS Satellite Images on a Regional Scale. *Chilean Journal of Agricultural Research*, *70*, 436–445.

Falkowski, M. J., Wulder, M. A., White, J. C., & Gillis, M. D. (2009). Supporting large-area, sample-based forest inventories with very high spatial resolution satellite imagery. *Progress in Physical Geography*, *33*(3), 403–423. doi:10.1177/0309133309342643

FAO. (2003). Le régime foncier et le développement rural. *FAO Etudes sur les régimes fonciers 3*. Retrieved from http://www.fao.org/3/a-y4307f.pdf

Farahani, H. J., Izzi, G., & Oweis, T. Y. (2009). Parameterization and Evaluation of the AquaCrop Model for Full and Deficit Irrigated Cotton All rights reserved. *Agronomy Journal*, *101*(3), 469–476. doi:10.2134/agronj2008.0182s

Farcy, C., & Devillez, F. (2005). New orientations of forest management planning from an historical perspective of the relations between man and nature2. *Forest Policy and Economics*, *7*(1), 85–95. doi:10.1016/S1389-9341(03)00013-3

Féderation Internationale des Géomètres. (2012). *Spatially Enabled Society*. Copenhagen: FIG.

Féderation Internationale des Géomètres. (2015). Fit for purpose land administration. Copenhagen: International Federation of Surveyors (FIG).

Fekkoul, H. (2013). Groundwater contamination by nitrates, salinity and pesticides: Case of the unconfined aquifer of triffa plain (Eastern Morocco). *Revue Marocaine des Sciences Agronomiques et Vétérinaires*, *2*, 12–36.

Feng, H., Chen, C., Dong, H., Wang, J., & Meng, Q. (2013). Modified Shortwave Infrared Perpendicular Water Stress Index: A Farmland Water Stress Monitoring Method. *Journal of Applied Meteorology and Climatology*, *52*(9), 2024–2032. doi:10.1175/JAMC-D-12-0164.1

Fernand, J. (1995). Multiple-use forestry-a concept of communication. *Multiple-Use Forestry in the Nordic Countries*, 67–80.

FGCC. (1989). Geometric geodetic accuracy standards and specifications for using GPS relative positioning techniques. Version 5.0. preliminary document. National Ocean Service. NOAA.

FIG. (1995). Statement on the cadastre. Bureau of the FIG.

FIG. (2002). Land information management for sustainable development: best practice guidelines in city-wide Land Information Management. Working Group No.3, Commission 3. FIG publication No.31.

Fitzgerald, A., Hooper, N., & Cook, J. (2013). Implementing open licensing in government open data initiatives: a review of Australian government. practice. *Proceedings of the 9th International Symposium on Open Collaboration*. 10.1145/2491055.2491094

Florian, S. (2009). *Potential of a Post-Classification Change Detection Analysis to Identify Land Use and Land Cover Changes. Bachelor's Degree, Physical Geography and Ecosystem Analysis*. Lund University.

Fouad M. (2015). *Analyse basée télédétection pour la révision de la carte de répartition de l'écosystème à arganeraie et l'étude diachronique de sa dynamique spatio-temporelle*. Résumé des communications, Congrès international de l'arganier, 3éme édition.

Fowler, C., & Treml, E. (2001). Building a marine cadastral information system for the United States- a case study. *Computers, Environment and Urban Systems*, *25*(4), 493–507. doi:10.1016/S0198-9715(00)00047-8

Franklin, S. E. (2001). *Remote sensing for sustainable forest management*. CRC Press. doi:10.1201/9781420032857

Freeze, F. (1962). *Elementary Forest Sampling*. USDA.

Gangkofner, U. G., Pradhan, P. S., & Holcomb, D. W. (2008). Optimizing the High-Pass Filter Addition Technique for Image Fusion. *Photogrammetric Engineering and Remote Sensing*, *74*(9), 1107–1118. doi:10.14358/PERS.74.9.1107

Garen, D. C., & Moore, D. S. (2005). Curve number hydrology in water quality modeling: Uses, abuses, and future directions. *Journal of the American Water Resources Association*, *41*(2), 377–388. doi:10.1111/j.1752-1688.2005.tb03742.x

Geerts, S., & Raes, D. (2009). Deficit irrigation as on-farm strategy to maximize crop water productivity in dry areas. *Agricultural Water Management*, *96*(9), 1275–1284. doi:10.1016/j.agwat.2009.04.009

Gerrits, A. M. J. (2010). The role of interception in the hydrological cycle. Academic Press.

Geurts, P., Irrthum, A., & Wehenkel, L. (2009). Supervised learning with decision tree-based methods in computational and systems biology. *Molecular BioSystems*, *5*, 1593–1605.

Ghulam, A., Qin, Q., Teyip, T., & Li, Z. L. (2007). Modified perpendicular drought index (MPDI): A real-time drought monitoring method. *ISPRS Journal of Photogrammetry and Remote Sensing*, *62*(2), 150–164. doi:10.1016/j.isprsjprs.2007.03.002

GLTN. (2014). *Expert Group Meeting on Land Indicators*. Retrieved from http://sdg.iisd.org/events/expert-group-meeting-on-land-indicators/

GLTN. (2015). *Options for global reporting on GLTN / GLII land indicators in the context of the sustainable development goals*. United Nations Human Settlements Programme (UN-Habitat). Report 9/2015.

Gorgoso, J. J., Rojo, A., Cámara-Obregón, A., & Diéguez-Aranda, U. (2012). A comparison of estimation methods for fitting Weibull, Johnson's SB and beta functions to Pinus pinaster, Pinus radiate and Pinus sylvestris stands in northwest Spain. *Forest Systems*, *21*(3), 446. doi:10.5424/fs/2012213-02736

Gregoire, T. G., & Valentine, H. T. (2007). *Sampling strategies for natural resources and the environment*. CRC Press.

Groot, A., Cortini, F., & Wulder, M. A. (2015). Crown-fibre attribute relationships for enhanced forest inventory: Progress and prospects. *Forestry Chronicle*, *91*(3), 266–279. doi:10.5558/tfc2015-048

Groten, C. (1999). coastal boundaries and vertical datums. In *Proceedings of the international conference on technical aspects of maritime boundary delination and delimitation*. International Hydrographic Bureau.

Guérif, M., & Duke, C. (1998). Calibration of the SUCROS emergence and early growth module for sugar beet using optical remote sensing data assimilation. *European Journal of Agronomy*, *9*(2), 127–136. doi:10.1016/S1161-0301(98)00031-8

Guerraoui. (1986). *Les Problèmes Fonciers et l'Agence Foncière Nationale*. Actes du Symposium sur la Gestion Foncière au Maroc et en Tunisie, Barcelone.

Gu, H.-h., Yu, Z.-b., Yang, C.-g., Ju, Q., Lu, B.-h., & Liang, C. (2010). Hydrological assessment of TRMM rainfall data over Yangtze River Basin. *Water Science and Engineering*, *3*(4), 418–430. doi:10.3882/j.issn.1674-2370.2010.04.005

Guigoz, Y., Giuliani, G., & Ray, N. (2013). *State of the art in Open Source GIS and SDI*, ARPEGEO project Symposium on Environmental Data Sharing for the Benefit of the South Caucasus Region.

Gumma, M. K., Thenkabail, P. S., Maunahan, A., Islam, S., & Nelson, A. (2010). Mapping seasonal rice cropland extent and area in the high cropping intensity environment of Bangladesh using MODIS 500m data for the year 2010. *ISPRS Journal of Photogrammetry and Remote Sensing*, *91*, 98–113.

Hachem, S., Duguay, C. R., & Allard, M. (2012). Comparison of MODIS-derived land surface temperatures with ground surface and air temperature measurements in continuous permafrost terrain. *The Cryosphere*, *6*(1), 51–69. doi:10.5194/tc-6-51-2012

Hadria, R., Benabdelouahab, T., Mahyou, H., Balaghi, R., Bydekerke, L., El Hairech, T., & Ceccato, P. (2018). Relationships between the three components of air temperature and remotely sensed land surface temperature of agricultural areas in Morocco. *International Journal of Remote Sensing*, *39*(2), 356–373. doi:10.1080/01431161.2017.1385108

Hadria, R., Duchemin, B., Jarlan, L., Dedieu, G., Baup, F., Khabba, S., ... Le Toan, T. (2010). Potentiality of optical and radar satellite data at high spatio-temporal resolutions for the monitoring of irrigated wheat crops in Morocco. *International Journal of Applied Earth Observation and Geoinformation*, *12*(1), S32–S37. doi:10.1016/j.jag.2009.09.003

Hadria, R., Duchemin, B., Lahrouni, A., Khabba, S., Er-raki, S., Dedieu, G., ... Olioso, A. (2006). Monitoring of irrigated wheat in a semi-arid climate using crop modelling and remote sensing data: Impact of satellite revisit time frequency. *International Journal of Remote Sensing*, *27*(6), 1093–1117. doi:10.1080/01431160500382980

Hadria, R., Khabba, S., Lahrouni, A., Duchemin, B., Chehbouni, A., Carriou, J., & Ouzine, L. (2007). Calibration and Validation of the Shoot Growth Module of STICS Crop Model: Application to Manage Irrigation Water in the Marrakesch/Al Haouz Plain. *Arabian Journal for Science and Engineering*, *32*, 87–101.

Halwatura, D., & Najim, M. M. M. (2013). Application of the HEC-HMS model for runoff simulation in a tropical catchment. *Environmental Modelling & Software*, *46*, 155–162. doi:10.1016/j.envsoft.2013.03.006

Haralick, K. (1973). *Textural Features for image Classification*. Institute of Electrical and Electronics Engineers.

Haregeweyn, N., Poesen, J., Nyssen, J., De Wit, J., Haile, M., Govers, G., & Deckers, S. (2006). Reservoirs in Tigray (Northern Ethiopia): Characteristics and sediment deposition problems. *Land Degradation & Development*, *17*(2), 211–230. doi:10.1002/ldr.698

Hartikainen, M., Eyvindson, K., Miettinen, K., & Kangas, A. (2016). Data-Based Forest Management with Uncertainties and Multiple Objectives. In *International Workshop on Machine Learning, Optimization and Big Data* (pp. 16–29). Springer. 10.1007/978-3-319-51469-7_2

Haut Commissariat au Plan. (2008). *Monographie régionale GHARB – CHRADA – BENI HSSEN*. Author.

Hazewinkel, M. (2001). Encyclopedia of Mathematics. Springer.

HCEFLCD. (2017). *Inventaire Forestier National. Haut-Commissariat aux Eaux et Forêts et à la Lutte Contre la Désertification*. Retrieved from http://www.eauxetforets.gov.ma

HCEFLCD. (2017). *Mesguina Forest management plan* (Draft document). Author.

HCP. (2016). *High Commission for Planning (Morocco)*. Retrieved from https://www.hcp.ma/region-rabat/attachment/775960/

He, C. (2003). Integration of geographic information systems and simulation model for watershed management. *Environmental Modelling & Software*, *18*(8-9), 809–813. doi:10.1016/S1364-8152(03)00080-X

Hengl, T., Heuvelink, G. B. M., Perčec Tadić, M., & Pebesma, E. (2012). Spatio-temporal prediction of daily temperatures using time-series of MODIS LST images. *Theoretical and Applied Climatology*, *107*(1-2), 265–277. doi:10.100700704-011-0464-2

Henssen, J. (1995). Basic principles of the main cadastral system in the world. In *Proceeding of the Seminar Modern Cadastres and cadastral Innovation*. International Federation of Surveyors (FIG).

Hespanha. (2012). *Development Methodology for an Integrated Legal Cadastre Deriving Portugal Country Model from the Land Administration Domain Model* (PhD thesis). Geboren te Lisboa, Portugal.

Hintz, D. (2012). *Data Harmonization Principles and Development Approaches as Applied to INSPIRE SDIs*. INSPIERE - GMES Information Brochre. Technische Universität München.

Hsiao, T. C., Heng, L., Steduto, P., Rojas-Lara, B., Raes, D., & Fereres, E. (2009). AquaCrop: The FAO crop model to simulate yield response to water. III. Parameterization and testing for maize. *Agronomy Journal*, *101*(3), 448–459. doi:10.2134/agronj2008.0218s

Hsu, K., Gao, X., Sorooshian, S., & Gupta, H. V. (1997). Precipitation Estimation from Remotely Sensed Information Using Artificial Neural Networks. *Journal of Applied Meteorology*, *36*(9), 1176–1190. doi:10.1175/1520-0450(1997)036<1176:PEFRSI>2.0.CO;2

Hudak, A. T., Evans, J. S., & Stuart Smith, A. M. (2009). LiDAR Utility for Natural Resource Managers. *Remote Sensing*, *1*(4), 934–951. doi:10.3390/rs1040934

Huete, A. R., Liu, H., Batchily, K., & Van Leeuwen, W. (1997). A comparison of vegetation indices over a global set of TM images for EOS-MODIS. *Remote Sensing of Environment*, *59*(3), 440–451. doi:10.1016/S0034-4257(96)00112-5

Hunt, E. R. Jr, & Rock, B. N. (1989). Detection of changes in leaf water content using Near- and Middle-Infrared reflectances. *Remote Sensing of Environment*, *30*(1), 43–54. doi:10.1016/0034-4257(89)90046-1

Hussein, F., Janat, M., & Yakoub, A. (2011). Simulating cotton yield response to deficit irrigation with the FAO AquaCrop model. *Spanish Journal of Agricultural Research*, *9*(4), 1319–1330. doi:10.5424jar/20110904-358-10

Ibannain, F., & Donnay, J.P. (2012). *Prior analysis of a data infrastructure in Morocco*. Retrieved from http://orbi.ulg.ac.be/handle/2268/142431

Idrees, M., Vahideh, Y. A., & Shariff, R. (2012). Comparing Approaches and Strategies for NSDI Implementation Between the Developed and Developing World. *The International Journal of Spatial Data Infrastructures*.

International Telecommunication Union. (2017). *WSIS-SDG Matrix-linking WSIS Action Lines with Sustainable development Goals*. Geneva: ITU.

Iqbal, M. A., Shen, Y., Stricevic, R., Pei, H., Sun, H., Amiri, E., ... del Rio, S. (2014). Evaluation of the FAO AquaCrop model for winter wheat on the North China Plain under deficit irrigation from field experiment to regional yield simulation. *Agricultural Water Management*, *135*(0), 61–72. doi:10.1016/j.agwat.2013.12.012

ISO. (2012). Modèle du Domaine d'administration des Terres (LADM). ISO 19152: 2012(F), Information géographique - première édition 2012-12-01, Projet de Norme Marocaine 2016 (Indice de classement IC 17.8.417).

Jacquemoud, S., Bacour, C., Poilvé, H., & Frangi, J. P. (2000). Comparison of Four Radiative Transfer Models to Simulate Plant Canopies Reflectance: Direct and Inverse Mode. *Remote Sensing of Environment*, *74*(3), 471–481. doi:10.1016/S0034-4257(00)00139-5

Jakobsson, A., & Lysandros Tsoulos, L. (2007). The Role of Quality in Spatial Data Infrastructures. *Proceedings of the 23 rd International Cartographic Conference*.

Jamieson, P. D., Porter, J. R., & Wilson, D. R. (1991). A test of computer simulation model ARC-WHEAT1 on wheat crop grown in New Zealand. *Field Crops Research*, *27*(4), 337–350. doi:10.1016/0378-4290(91)90040-3

Jeníček, M. (2007). Effects of land cover on runoff process using SCS CN method in the upper Chomutovka catchment. In *Proceedings of the first Scientific Conference on Integrated catchment management for hazard mitigation*. Remote Sensing Department, University of Trier.

Jensen, J. R. (2007). *Remote Sensing of the Environment: An Earth resource perspective*. Prentice Hall.

Johnson, E. W. (2000). *Forest sampling desk reference*. CRC Press. doi:10.1201/9781420042498

Johnston, M., Lindner, M., Parrotta, J., & Giessen, L. (2012). Adaptation and mitigation options for forests and forest management in a changing climate. *Forest Policy and Economics*, *24*(Supplement C), 1–2. doi:10.1016/j.forpol.2012.09.007

Joliveau, T. (2004). *Géomatique et gestion environnementale du territoire: Recherche sur un usage géographique des SIG. Mémoire d'Habilitation à Diriger des Recherches en Sciences Humaines*. Université de Rouen.

Jonckheere, I., Fleck, S., Nackaerts, K., Muys, B., Coppin, P., Weiss, M., & Baret, F. (2004). Review of methods for in situ leaf area index determination: Part I. Theories, sensors and hemispherical photography. *Agricultural and Forest Meteorology*, *121*(1), 19–35. doi:10.1016/j.agrformet.2003.08.027

Jones, C. A., Kiniry, J. R., & Dyke, P. T. (1986). *CERES-Maize: a simulation model of maize growth and development*. College Station, TX: Texas A&M University Press.

Joyce, R. J., Janowiak, J. E., Arkin, P. A., & Xie, P. (2004). CMORPH: A Method that Produces Global Precipitation Estimates from Passive Microwave and Infrared Data at High Spatial and Temporal Resolution. *Journal of Hydrometeorology*, *5*(3), 487–503. doi:10.1175/1525-7541(2004)005<0487:CAMTPG>2.0.CO;2

Kangas, A., & Maltamo, M. (2006). *Forest inventory methodology and applications*. Dordrecht: Springer. doi:10.1007/1-4020-4381-3

Kang, H. J., & Doermann, D. (2005). Selection of Classifiers for the Construction of Multiple Classifier Systems. *Proceedings of the Eight International Conference on Document Analysis and Recognition*, 263-268.

Kaufmann & Steudler. (1998). Cadastre 2014. A vision for a future cadastral system. FIG Commission, Brighton, UK.

Kaufmann, J., & Steudler, D. (1998). Cadastre 2014, a vision for a future cadastral system. FIG commission 7. Working Group 1.

Keesstra, S., Nunes, J., Novara, A., Finger, D., Avelar, D., Kalantari, Z., & Cerdà, A. (2018). The superior effect of nature based solutions in land management for enhancing ecosystem services. *The Science of the Total Environment*, *610*, 997–1009. doi:10.1016/j.scitotenv.2017.08.077 PMID:28838037

Kershaw, J. A., Ducey, M. J., Beers, T. W., & Husch, B. (2016). *Forest mensuration* (5th ed.). Chichester, UK: John Wiley & Sons. doi:10.1002/9781118902028

Khabba, S., Duchemin, B., Hadria, R., Ezzahar, J., Chehbouni, A., Lahrouni, A., & Hanich, L. (2009). Evaluation of digital hemispherical photography and plant canopy analyser for measuring Vegetation area index of orange orchards. *Journal of Agronomy*, *8*(2), 67–72. doi:10.3923/ja.2009.67.72

Kleinn, C. (2003). New technologies and methodologies for national forest inventories. *UNASYLVA-FAO*, 10–15.

Klinghoffer, A. J. (1973). Modernisation and political development in Africa. *The Journal of Modern African Studies*, *11*(1), 1–19. doi:10.1017/S0022278X00008053

Knebl, M. R., Yanga, Z. L., Hutchisonb, K., & Maidment, D. R. (2005). Regional scale flood modeling using NEXRAD rainfall, GIS, and HEC-HMS/RAS: A case study for the San Antonio River Basin Summer 2002 storm event. *Journal of Environmental Management*, *75*(4), 325–336. doi:10.1016/j.jenvman.2004.11.024 PMID:15854726

Koch, B., Heyder, U., & Weinacker, H. (2006). Detection of Individual Tree Crowns in Airborne Lidar Data. *Photogrammetric Engineering and Remote Sensing*, *72*(4), 357–363. doi:10.14358/PERS.72.4.357

Kogan, F. N. (2000). *Contribution of Remote Sensing to Drought Early Warning*. Paper presented at the Early Warning Systems for Drought Preparedness and Drought Management.

Köhl, M., Magnussen, S., & Marchetti, M. (2006). *Sampling methods, remote sensing and GIS multiresource forest inventory*. Berlin: Springer. doi:10.1007/978-3-540-32572-7

Laaroussi. (2017). *Expert-comptable, commissaire aux comptes*. Cabinet EurO-CAF Consulting. Retrieved from oec-casablanca.ma/img/uploads/IR_agricole_10_01_17.pptx

Lafuente. (1999). *La politique berbère de la France et le nationalisme marocain*. Academic Press.

Lahmer, I. (2017). Extraction stratifiée par âge des agrumes à partir d'une série temporelle des images Sentinel 2 dans la plaine de triffa, région de l'oriental du Maroc. In Rapport de projet de fin d'étude. Institut Agronomique et Veterinaire Hassan II.

Laliberte, A. S., Fredrickson, E. L., & Rango, A. (2007). Combining decision trees with hierarchical object-oriented image analysis for mapping arid rangelands. *Photogrammetric Engineering and Remote Sensing*, *73*(2), 197–207. doi:10.14358/PERS.73.2.197

Land Policy Initiative. (2016). Needs assessment and mapping of training institutions on land governance in North Africa. Consultancy report UNECA.

Lardon, S., & Roche, S. (2008). Représentations spatiales dans les démarches participatives. Production et usages. *Revue Internationale de Géomatique*, *18*(4).

Larsson, G. (1991). *Land registration and cadastral systems*. Essex, UK: Longman Scientific and Technical.

Lasaponara, R., & Masini, N. (2012). *Satellite Remote Sensing: A New Tool for Archaeology*. Springer. doi:10.1007/978-90-481-8801-7

Leckie, D. G., Tinis, S., Nelson, T., Burnett, C., Gougeon, F. A., Cloney, E., & Paradine, D. (2005). Issues in species classification of trees in old growth conifer stands. *Canadian Journal of Remote Sensing*, *31*(2), 175–190. doi:10.5589/m05-004

Legrouri, A., Kettani, D., & Kalpakian, J. (2012). *Using Demand Side Management to Adapt to Water Scarcity and Climate Change in the Saïss Basin, Morocco*. Ifrane, Morocco: Al Akhawayn University.

Lemmen, C. (2012). *A Domain Model for Land Administration* (PhD thesis). Faculty of ITC, The Nederland.

Lemmen, C., van Oosterom, P., Uitermark, H., & de Zeeuw, K. (2013). *Land administration domain model is an ISO standard now*. The World Bank.

Lepoivre, F. (2004). Benchmarking concept et méthodologie. *Petite entreprise et partenariat*. Retrieved from https://documentfr.com/category/Small-Business-Entrepreneurship

Letcher, R. A., & Gioppini, C. (2005). Policies and tools for sustainable water management in the European Union. *Environmental Modelling & Software*, *20*(2), 93–98. doi:10.1016/j.envsoft.2004.01.003

Li, J., Hu, B., & Woods, M. (2015). A two-level approach for species identification of coniferous trees in central Ontario forests based on multispectral images. *IEEE Journal of Selected Topics in Applied Earth Observations and Remote Sensing*, *8*(4), 1487–1497. doi:10.1109/JSTARS.2015.2423272

Li, L., Hong, Y., Wang, J., Adler, R. F., Policelli, F. S., Habib, S., ... Okello, L. (2009). Evaluation of the real-time TRMM-based multi-satellite precipitation analysis for an operational flood prediction system in Nzoia Basin, Lake Victoria, Africa. *Natural Hazards*, *50*(1), 109–123. doi:10.100711069-008-9324-5

Lionboui, H., Benabdelouahab, T., Elame, F., Hasib, A., & Boulli, A. (2016). Multi-year agro-economic modelling for predicting changes in irrigation water management indicators in the Tadla sub-basin. *International Journal of Agricultural Management*, *5*(4), 96–105. doi:10.5836/ijam/2016-05-96

Lionboui, H., Benabdelouahab, T., Hasib, A., & Boulli, A. (2016). Analysis of Farms Performance Using Different Sources of Irrigation Water: A Case Study in a Semi-Arid Area. *International Journal of Agricultural Management and Development*, *6*(2), 145–154.

Lionboui, H., Fadlaoui, A., Elame, F., & Benabdelouahab, T. (2014). Water pricing impact on the economic valuation of water resources. *International Journal of Education and Research*, *2*(6), 147–166.

Liu, Y. (2018). Introduction to land use and rural sustainability in China. *Land Use Policy*, *74*, 1–4. doi:10.1016/j.landusepol.2018.01.032

Loague, K., & Green, R. E. (1991). Statistical and graphical methods for evaluating solute transport models: Overview and application. *Journal of Contaminant Hydrology*, *7*(1-2), 51–73. doi:10.1016/0169-7722(91)90038-3

Lobell, D. B., Asner, G. P., Ortiz-Monasterio, J. I., & Benning, T. L. (2003). Remote sensing of regional crop production in the Yaqui Valley, Mexico: Estimates and uncertainties. *Agriculture, Ecosystems & Environment*, *94*(2), 205–220. doi:10.1016/S0167-8809(02)00021-X

Lu, D., Mausel, P., Brondizio, E., & Moran, E. (2004). Change detection techniques. *International Journal of Remote Sensing*, *25*(12), 2365–2401. doi:10.1080/0143116031000139863

Lush, V., Bastin, L., & Lumsden, J. (2012). *Geospatial data quality indicators*. International Spatial Accuracy Research Association.

M.C. (2011). *Moroccan Constitution.* Kingdom of Morocco, General Secretariat of the Government, Directorate of Official Printing. Retrieved from http://adala.justice.gov.ma/production/constitution/fr/Constitution.pdf

Mader, M., Matijevi'cb, H., & Roi'ca, M. (2015). Analysis of possibilities for linking land registers and other official registers in the Republic of Croatia based on LADM. *Land Use Policy*, *49*, 606–616. doi:10.1016/j.landusepol.2014.10.025

Magel, H. (2014). *Rural-urban partnership: key for a better land and water management?* International Symposium on occasion of IFAT 2014 on Intelligent Peri-Urbanisation through Decentralisation, Munich, Germany.

Mailly, D. (2017). Hemispherical Photography in Support of Forest Inventory and Silviculture. In *Hemispherical Photography in Forest Science: Theory, Methods, Applications* (pp. 227–252). Dordrecht: Springer. doi:10.1007/978-94-024-1098-3_8

Mäkelä, A., Landsberg, J., Ek, A. R., Burk, T. E., Ter-Mikaelian, M., Ågren, G. I., ... Puttonen, P. (2000). Process-based models for forest ecosystem management: Current state of the art and challenges for practical implementation. *Tree Physiology*, *20*(5–6), 289–298. doi:10.1093/treephys/20.5-6.289 PMID:12651445

Malthus, T. R. (1989). On Population (First Essay on Population, 1798, and Second Essay on Population, 1803). Modern Library, for Random House.

Mandallaz, D. (2008). *Sampling Techniques for Forest*. Academic Press.

MAPM. (2016). *Le recensement agricole, une étape essentielle pour la création du Registre National Agricole*. Retrieved from http://www.agriculture.gov.ma

Maree R. (2005). *Decision trees for automatic images classification* (PhD Thesese). Ulg University.

Marée, R., Geurts, P., & Wehenkel, L. (2016). Towards Generic Image Classification using Tree-based Learning: An Extensive Empirical Study. *Pattern Recognition Letters*, *74*, 17–23. doi:10.1016/j.patrec.2016.01.006

MCC. (2018). *Millennium Challenge Corporation of the United State of America*. Retrieved from www.mcc.gov

McCullagh, A., & Nieuwenhuis, M. (2015). *Review of tools for growth forecasting and productivity assessment in forestry in Ireland*. IrIsh Forestry.

McElhinny, C., Gibbons, P., Brack, C., & Bauhus, J. (2005). Forest and woodland stand structural complexity: Its definition and measurement. *Forest Ecology and Management*, *218*(1), 1–24. doi:10.1016/j.foreco.2005.08.034

McLaughlin, J. D. (1977). *Maritime accuracy study, Land Registration and Information Service technical report*. Fredericton, Canada: University of New Brunswick.

Meng, X., Currit, N., & Zhao, K. (2010). Ground filtering algorithms for airborne LiDAR data: A review of critical issues. *Remote Sensing*, *2*(3), 833–860. doi:10.3390/rs2030833

Merganič, J., & Sterba, H. (2006). Characterisation of diameter distribution using the Weibull function: Method of moments. *European Journal of Forest Research*, *125*(4), 427–439. doi:10.100710342-006-0138-2

Michael, T., & Millward, L. (2014). *Accessing Information is Our Right" A Practical Guide to Promoting Access to Public Information in Morocco*. Retrieved from http://www.uneskit.org/sites/default/files/dai_illus_int.pdf

Milewski, A., Elkadiri, R., & Durham, M. (2015). Assessment and Comparison of TMPA Satellite Precipitation Products in Varying Climatic and Topographic Regimes in Morocco. *Remote Sensing*, *7*(5), 5697–5717. doi:10.3390/rs70505697

Ministry of Agriculture, Fisheries, Rural Development and Water and Forests (MAFRDWF). (2010). *Study for the development of Information System Master Plan of the Department of Agriculture, carried out by the Financial Directorate*. Author.

Ministry of Agriculture, Fisheries, Rural Development and Water and Forests (MAFRDWF). (2011). *Study of the Establishment of a Geo-Information System for the Accompanying of the Green Morocco Plan by the Information Systems Department*. Author.

Ministry of Agriculture. (2013). *Fisheries, Rural Development and Water and Forests*. MAFRDWF-ISD.

Ministry of Agriculture. Fisheries, Rural Development and Water and Forest (MAFRDWF). (2017). *Le fond de développement Agricole*. Retrieved from http://www.agriculture.gov.ma/sites/default/files/fda2017/FDA_2017_VF.pdf

Mkhabela, M. S., & Bullock, P. R. (2012). Performance of the FAO AquaCrop model for wheat grain yield and soil moisture simulation in Western Canada. *Agricultural Water Management*, *110*(0), 16–24. doi:10.1016/j.agwat.2012.03.009

Moffitt, C. B., Hossain, F., Adler, R. F., Yilmaz, K. K., & Pierce, H. F. (2011). Validation of a TRMM-based global Flood Detection System in Bangladesh. *International Journal of Applied Earth Observation and Geoinformation*, *13*(2), 165–177. doi:10.1016/j.jag.2010.11.003

Moges Kidane, B., Amare Sewnet, M., & Alemu Bezie, D. (2015). Multi-temporal Land Use Land Cover Change and Dynamics of Blue Nile Basin by Using GIS and Remote Sensing Techniques, North-Western Ethiopia. *International Journal of Environmental Sciences*, *4*(2), 81–88.

Monahan, D., & Mayer, L. (1999). An examination of publicity available bathymetric data sets using digital mapping tools to determine their applicability to article 76 of UNCLOS. In *Proceedings of the international conference on technical aspects of maritime bouandary delineation and delimitation*. International Hydrographic Bureau.

Monserud, R. A. (2003). Evaluating forest models in a sustainable forest management context. *Forest Biometry. Modelling and Information Sciences*, *1*(1), 35–47.

Montealegre, A. L., Lamelas, M. T., & de la Riva, J. (2015). A comparison of open-source LiDAR filtering algorithms in a mediterranean forest environment. *IEEE Journal of Selected Topics in Applied Earth Observations and Remote Sensing*, *8*(8), 4072–4085. doi:10.1109/JSTARS.2015.2436974

Mora, B., Wulder, M. A., & White, J. C. (2010). Identifying leading species using tree crown metrics derived from very high spatial resolution imagery in a boreal forest environment. *Canadian Journal of Remote Sensing*, *36*(4), 332–344. doi:10.5589/m10-052

Moreno, A., Maselli, F., Chiesi, M., Genesio, L., Vaccari, F., Seufert, G., & Gilabert, M. A. (2014). Monitoring water stress in Mediterranean semi-natural vegetation with satellite and meteorological data. *International Journal of Applied Earth Observation and Geoinformation*, *26*(0), 246–255. doi:10.1016/j.jag.2013.08.003

Moreno, R. (1984). Cadastres and land registration. *Proceeding of the International workshop on land tenure administration.*

Morgan, J. L., & Gergel, S. E. (2013). Automated analysis of aerial photographs and potential for historic forest mapping. *Canadian Journal of Forest Research*, *43*(8), 699–710. doi:10.1139/cjfr-2012-0492

Moroccan Government. (2013, March 18). Official Bulletin n° 6135 (p. 2590).

Motevalli, S., Hosseinzadeh, M., Esmaili, R., Derafshi, K., & Gharehchahi, S. (2012). Assessing the Effects of Land use Change on Hydrologic Balance of Kan Watershed using SCS and HEC-HMS Hydrological Models – Tehran, IRAN. *Australian Journal of Basic and Applied Sciences, 6*(8), 510-519.

Moukrim, S., Lahssini, S., Mharzi Alaoui, H., Rifai, N., Arahou, M., & Rhazi, L. (2018). Modélisation de la distribution spatiale des espèces endémiques pour leur conservation: Cas de l'Argania spinosa (L.) Skeels. *Revue d'Écologie*, *73*(2), 153–166.

Moulin, S. (1999). Impacts of model parameter uncertainties on crop reflectance estimates: A regional case study on wheat. *International Journal of Remote Sensing*, *20*(1), 213–218. doi:10.1080/014311699213730

Moutaib, O., & Oujamaa, A. (2017). Cartoghrahie agrumes dans la plaine de Tiffa. In Rapport de projet de fin d'étude. Institut Agronomique et Veterinaire Hassan II.

MoWR (Ministry of Water Resources). (2002). *Water Sector Development Program.* Addis Ababa, Ethiopia: Author.

Muller, E., & Décamps, H. (2000). Modeling soil moisture-reflectance. *Remote Sensing of Environment*, *76*(2), 173–180. doi:10.1016/S0034-4257(00)00198-X

Musy, A., & Higy, C. (1998). Hydrologie appliquée. Edition *H*G*A*.

Naesset, E. (2002). Determination of mean tree height of forest stands by digital photogrammetry. *Scandinavian Journal of Forest Research*, *17*(5), 446–459. doi:10.1080/028275802320435469

Næsset, E. (2002). Predicting forest stand characteristics with airborne scanning laser using a practical two-stage procedure and field data. *Remote Sensing of Environment*, *80*(1), 88–99. doi:10.1016/S0034-4257(01)00290-5

Næsset, E., Gobakken, T., Solberg, S., Gregoire, T. G., Nelson, R., Ståhl, G., & Weydahl, D. (2011). Model-assisted regional forest biomass estimation using LiDAR and InSAR as auxiliary data: A case study from a boreal forest area. *Remote Sensing of Environment*, *115*(12), 3599–3614. doi:10.1016/j.rse.2011.08.021

Najine, A., Jaffal, M., Khammari, K. E., Aïfa, T., Khattach, D., Himi, M., ... Aqil, H. (2006). Contribution de la gravimétrie à l'étude de la structure du bassin de Tadla (Maroc): Implications hydrogéologiques. *Comptes Rendus Geoscience*, *338*(10), 676–682. doi:10.1016/j.crte.2006.04.015

Nelson, G. C., Bennett, E., Berhe, A. A., Cassman, K., DeFries, R., Dietz, T., ... Zurek, M. (2006). Anthropogenic Drivers of Ecosystem Change: An Overview. *Ecology and Society*, *11*(2), 29p. doi:10.5751/ES-01826-110229

Ng'ang'a, S., Sutherland, M., Cockburn, S., & Nichols, S. (2004). Toward a 3D marine cadastre in support of good ocean governance. *Computers, Environment and Urban Systems*, *28*(5), 443–470. doi:10.1016/j.compenvurbsys.2003.11.002

Nichols, S., Monahan, D., & Sutherland, M. D. (2000). Good governance of Canada's offshore and coastal zone. towards an understanding of the marine boundary issues. *Geomatica*, *54*(4), 415–424.

Nilson, T., & Kuusk, A. (2004). Improved algorithm for estimating canopy indices from gap fraction data in forest canopies. *Agricultural and Forest Meteorology*, *124*(3), 157–169. doi:10.1016/j.agrformet.2004.01.008

Ning, Z., Yang, H., Qiming, Q., & Lu, L. (2013). VSDI: A visible and shortwave infrared drought index for monitoring soil and vegetation moisture based on optical remote sensing. *International Journal of Remote Sensing*, *34*(13), 4585–4609. doi:10.1080/01431161.2013.779046

NRC. (1980). Need for a multipurpose cadastre. National Research Council National.

NRC. (1983). *Panel on a Multi-Purpose Cadastre. Procedures and standards for a multi-purpose cadastre*. National Academy Press.

Nurelegn Mekuriaw, G., & Amare Sewnet, M. (2014). Land Use/Cover Dynamics in Ribb Watershed, North Western Ethiopia. *Journal of Natural Sciences Research, 4*.

Ohana-Levi, N., Karnieli, A., Egozi, R., Givati, A., & Peeters, A. (2015). Modeling the Effects of Land-Cover Change on Rainfall-Runoff Relationships in a Semiarid, Eastern Mediterranean Watershed. Hindawi Publishing Corporation. *Advances in Meteorology*.

Ojike, M. (1946). *My Africa*. New York: John Day.

Oleyiblo, J. O., & Li, Z. J. (2010). Application of HEC-HMS for flood forecasting in Misai and Wan'an catchments in China. *Water Science and Engineering, 3*(1), 14-22. 10.3882/j.issn.1674-2370.2010.01.002

Olioso, A., Chauki, H., Courault, D., & Wigneron, J.-P. (1999). Estimation of Evapotranspiration and Photosynthesis by Assimilation of Remote Sensing Data into SVAT Models. *Remote Sensing of Environment*, *68*(3), 341–356. doi:10.1016/S0034-4257(98)00121-7

Oliver, C. D., & Larson, B. C. (1996). *Forest stand dynamics: updated edition*. John Wiley and sons.

Opeyemi, Z.A. (2006). *Change detection in land use land cover using Remote sensing data and GIS: A case study of Ilorin and its environs in Kwara State*. Academic Press.

ORMVAM. (2017). Potentialités agricoles de la région de l'Oriental. Berkane: Office Régional de Mise en Valeur Agricole de la Moulouya.

ORMVAT. (2009). *Rapport annuel de l'Office de la mise en valeur agricole de Tadla*. ORMVAT.

Ouatiki, H., Boudhar, A., Tramblay, Y., Jarlan, L., Benabdelouhab, T., Hanich, L., ... Chehbouni, A. (2017). Evaluation of TRMM 3B42 V7 Rainfall Product over the Oum Er Rbia Watershed in Morocco. *Climate (Basel)*, *5*(1), 1. doi:10.3390/cli5010001

P.M. (2018). Letter to the head of the Economic, Social and Environmental Council of Morocco. No. 159. January 22, 2018.

Palahí, M., Pukkala, T., Blasco, E., & Trasobares, A. (2007). Comparison of beta, Johnson's SB, Weibull and truncated Weibull functions for modeling the diameter distribution of forest stands in Catalonia (north-east of Spain). *European Journal of Forest Research*, *126*(4), 563–571. doi:10.100710342-007-0177-3

Paletto, A., Ferretti, F., Cantiani, P., & De Meo, I. (2012). Multi-functional approach in forest landscape management planning: An application in Southern Italy. *Forest Systems*, *21*(1), 68–80. doi:10.5424/fs/2112211-11066

Palmer, I. (1977). Rural women and the basic-needs approach to development. *Int'l Lab. Rev.*, *115*, 97.

Panda, R. K., Behera, S. K., & Kashyap, P. S. (2003). Effective management of irrigation water for wheat under stressed conditions. *Agricultural Water Management*, *63*(1), 37–56. doi:10.1016/S0378-3774(03)00099-4

Paré. (2008). *Land Use Dynamics, Tree Diversity, and Local Perception of Dry Forest Decline in Southern Burkina Faso, West Africa* (Doctoral Thesis). Department of Forest Genetics and Plant Physiology, Umeå, Sweden.

Pašakarnis, G., & Maliene, V. (2010). Towards sustainable rural development in Central and Eastern Europe: Applying land consolidation. *Land Use Policy*, *27*(2), 545–549. doi:10.1016/j.landusepol.2009.07.008

Paul, M., & wa Gĩthĩnji, M. (2018). Small farms, smaller plots: Land size, fragmentation, and productivity in Ethiopia. *The Journal of Peasant Studies*, *45*(4), 757–775. doi:10.1080/03066150.2016.1278365

Penuelas, J., Pinol, J., Ogaya, R., & Filella, I. (1997). Estimation of plant water concentration by the reflectance Water Index WI (R900/R970). *International Journal of Remote Sensing*, *18*(13), 2869–2875. doi:10.1080/014311697217396

Percivall, G. (2006). OpenGIS International Standards for GEOSS Interoperability Arrangements. *IEEE International Symposium on Geoscience and Remote Sensing*.

Peters, L. (2014). *Statistical analysis of geospatial information relevance of SDI's and international standards.* ESRI.

Phuhlisani, S. (2009). *International and Local Approaches to Rural Development: Key Issues and Questions: A Review of the Literature for the Drakenstein Municipality*. Phuhlisani Solutions.

Plan Maroc Numeric. (2013). Retrieved from http://www.egov.ma/sites/default/files/Maroc%20Numeric%202013.pdf

Plumjeaud, C. (2011). *Modèles et méthodes pour l'information spatio-temporelle évolutive* (Unpublished doctoral dissertation). Université de Grenoble, France.

PNCT. (2017). *Portail National Des Collectivités Territoriales*. Ministère de l'Intérieur.

Ponce, V. M., & Hawkins, R. H. (1996). Runoff curve number: Has it reached maturity? *Journal of Hydrologic Engineering*, *11*(1), 11–19. doi:10.1061/(ASCE)1084-0699(1996)1:1(11)

Porté, A., & Bartelink, H. H. (2002). Modelling mixed forest growth: A review of models for forest management. *Ecological Modelling*, *150*(1), 141–188. doi:10.1016/S0304-3800(01)00476-8

Puissant, A., Rougier, S., & Stumpf, A. (2014). Object-oriented mapping of urban trees using Random Forest classifiers. *International Journal of Applied Earth Observation and Geoinformation*, *26*, 235–245. doi:10.1016/j.jag.2013.07.002

Pulliti, S., Gobakken, T., Ørka, H. O., & Næsset, E. (2016). Assessing 3D point clouds from aerial photographs for species-specific forest inventories. *Scandinavian Journal of Forest Research*, *32*(1), 68–79. doi:10.1080/02827581.2016.1186727

Radoux, J., & Defourny, P. (2007). A quantitative assessment of boundaries in automated forest stand delineation using very high resolution imagery. *Remote Sensing of Environment*, *110*(4), 468–475. doi:10.1016/j.rse.2007.02.031

Raes, D., Steduto, P., Hsiao, T. C., & Fereres, E. (2009b). *AquaCrop:The FAO Crop Model to Simulate Yield Response to Water: Reference Manual Annexes*. Academic Press.

Raes, D., Steduto, P., Hsiao, T. C., & Fereres, E. (2009a). AquaCrop:The FAO Crop Model to Simulate Yield Response to Water: II. Main Algorithms and Software Description. *Agronomy Journal*, *101*(3), 438–447. doi:10.2134/agronj2008.0140s

Rahlf, J., Breidenbach, J., Solberg, S., Næsset, E., & Astrup, R. (2017). Digital aerial photogrammetry can efficiently support large-area forest inventories in Norway. *Forestry*, *00*, 1–19.

Rajabifard, A., Feeney, M.-E. F., & Williamson, I. (2003). *Spatial data infrastructures: concept, nature and SDI hierarchy*. Academic Press.

Rennolls, K., Tomé, M., McRoberts, R. E., Vanclay, J. K., LeMay, V., Guan, B. T., & Gertner, G. Z. (2007). Potential contributions of statistics and modelling to sustainable forest management: review and synthesis. *Sustainable Forestry: From Monitoring and Modelling to Knowledge Management and Policy Science*, 314–341.

Rennolls, K. (1995). Forest height growth modelling. *Forest Ecology and Management*, *71*(3), 217–225. doi:10.1016/0378-1127(94)06102-O

Reyer, C., Lasch-Born, P., Suckow, F., Gutsch, M., Murawski, A., & Pilz, T. (2014). Projections of regional changes in forest net primary productivity for different tree species in Europe driven by climate change and carbon dioxide. *Annals of Forest Science*, *71*(2), 211–225. doi:10.100713595-013-0306-8

Richard, R. (2014). *Geospatial Land Information Management: Addressing the Data and information needs*. Addis Ababa: UNGGIM HLF.

Richardson, A. J., Wiegand, C. L., Wanjura, D. F., Dusek, D., & Steiner, J. L. (1992). Multisite analyses of spectral-biophysical data for Sorghum. *Remote Sensing of Environment*, *41*(1), 71–82. doi:10.1016/0034-4257(92)90062-O

Rihane, R. (2015). *Contribution à l'évaluation des services hydrologiques de la forêt et des aménagements antiérosifs dans le bassin versant de l'Ourika (Mémoire de 3ème cycle)*. Ecole Nationale Forestière d'Ingénieurs, Salé.

Rihane, R. (2015). *Contribution à l'évaluation des services hydrologiques de la forêt et des aménagements antiérosifs dans le bassin versant de l'Ourika. Mémoire de 3ème cycle*. Maroc: Ecole Nationale Forestière d'Ingénieurs.

Robertson, B., Benwell, G., & Hoogsteden, C. (1999). Marine cadastre. Administration infrastructure requirements. UN-FIG conference on land tenure and cadastral infrastructures for sustainable development, Melbourne, Australia.

Rochdane, S., Bounoua, L., Zhang, P., Imhoff, M. L., Messouli, M., & Yacoubi-Khebiza, M. (2014). Combining Satellite Data and Models to Assess Vulnerability to Climate Change and Its Impact on Food Security in Morocco. *Sustainability*, *6*(4), 1729–1746. doi:10.3390u6041729

Rodríguez-García, C., Montes, F., Ruiz, F., Cañellas, I., & Pita, P. (2014). Stem mapping and estimating standing volume from stereoscopic hemispherical images. *European Journal of Forest Research*, *133*(5), 895–904. doi:10.100710342-014-0806-6

Roeckner, E., Bäuml, G., Bonaventura, L., Brokopf, R., Esch, M., Giorgetta, M., . . . Tompkins, A. (2003). *The atmospheric general circulation model ECHAM5. Part I: Model description*. Max Planck Institute for Meteorology Report n° 349.

Rogers, A. S., & Kearney, M. S. (2004). Reducing signature variability in unmixing coastal marsh Thematic Mapper scenes using spectral indices. *International Journal of Remote Sensing*, *25*(12), 2317–2335. doi:10.1080/01431160310001618103

Rokni, K., Ahmad, A., Selamat, A., & Hazini, S. (2014). Water Feature Extraction and Change Detection Using Multitemporal Landsat Imagery. *Remote Sensing*, *6*(5), 4173–4189. doi:10.3390/rs6054173

Rondeux, J. (1999). *La mesure des arbres et des peuplements forestiers*. Les presses agronomiques de Gembloux.

Rondinelli, D. A. (1979). Administration of integrated rural development policy: The politics of agrarian reform in developing countries. *World Politics*, *31*(3), 389–416. doi:10.2307/2009995

Roth, E., Donohue, G., Sack, M., Wallace, R., & Buckingham, M. (2015). *A Process for Keeping Pace with Evolving Web Mapping Technologies*. University of Wisconsin–Madison.

Rouse, J. W., Haas, R. H., Schell, J. A., Deering, D. W., & Harlan, J. C. (1974). Monitoring the vernal advancement and retrogradation (green wave effect) of natural vegetation. Texas A&M Univ.

Roy, T., Dinar, A., Tsur, Y., & Diao, X, (2005). Feedback links between economy-wide and farm-level policies: With application to irrigation water management in Morocco. *Journal of Policy Modeling, 27*, 905–928.

Sader, S. A., Hayes, D. J., Hepinstall, J. A., Coan, M., & Soza, C. (2001). Forest change monitoring of a remote biosphere reserve. *International Journal of Remote Sensing, 22*(10), 1937–1950. doi:10.1080/01431160117141

Sadler, T. D. (2004). Informal reasoning regarding socioscientific issues: A critical review of research. *Journal of Research in Science Teaching, 41*(5), 513–536. doi:10.1002/tea.20009

Saidi, M. E., Daoudi, L., Aresmouk, M. E., & Blali, A. (2003). *Rôle du milieu physique dans l'amplification des crues en milieu montagnard,exemple de la crue du 17 août 1995 dans la vallée de l'Ourika* (Vol. 14-2). Haut Atlas, Maroc: Sécheresse.

Saidi, M. E., Daoudi, L., Aresmouk, M. E., Fniguire, F., & Boukrim, S. (2010). Les crues de l'oued Ourika (Haut Atlas,Maroc): Événements extrêmes en contexte montagnard semi-aride. Comunicações Geológicas, t. 97. *Lisbonne, 2010*, 113–128.

Salama, H., & Tahiri, M. (2010). La gestion des ressources en eau face aux changements climatiques. Cas du bassin Tensift (Maroc). *Larhyss Journal, 8*, 127-138.

Salvemini, M. (2009). *The Infrastructure for Spatial Information in The European Community Vs Regional SDI: The shortest Way for reaching Economic and Social Development. Ninth United Nations Regional Cartographic Conference for the Americas*, New York, NY.

Schelhaas, M.-J., Nabuurs, G.-J., Hengeveld, G., Reyer, C., Hanewinkel, M., Zimmermann, N. E., & Cullmann, D. (2015). Alternative forest management strategies to account for climate change-induced productivity and species suitability changes in Europe. *Regional Environmental Change, 15*(8), 1581–1594. doi:10.100710113-015-0788-z

Schreuder, H. T., Gregoire, T. G., & Wood, G. B. (1993). *Sampling methods for multiresource forest inventory*. John Wiley & Sons.

Schuler, M., Thévoz, L., Walser, O., Joost, S., Debarbieux, B., & Dao, H. (2011). Les visages et usages de l'information géographique dans le processus de décision territoriale. In F. Joerin (Ed.), *Les SIG au service du développement territorial*. Presses Polytechniques et Universitaires Romandes.

Seckler, D., Barker, R., & Amarasinghe, U. (1999). Water scarcity in the twenty-first century. *Water: Research and Development, 15*, 29–42. doi:10.1080/07900629948916

Secrétariat d'Etat Chargé de l'Etat de l'Environnement (SEEE). (2010). *Seconde communication nationale à la convention cadre des nations unies sur les changements climatiques*. Retrieved from http://www.abhatoo.net.ma/content/download/20617/375478/version/1/file/seconde_communication_nationale_convention_cadre_NU_chang_clim.pdf

Seddik. (2010). *Expert-Comptable DPLE.* Fiscalité Immobilière. ICH Rabat INAU.

Servigne, S., Lesage, N., & Libourel, T. (2006). *Quality Components, Standards, and Metadata. Rodolphe Devillers.* London: Robert Jeansoulin. Fundamentals of Spatial Data Quality.

Shen, S., & Leptoukh, G. G. (2011). Estimation of surface air temperature over central and eastern Eurasia from MODIS land surface temperature. *Environmental Research Letters, 6*(4), 045206. doi:10.1088/1748-9326/6/4/045206

Shepherd, K. D., Shepherd, G., & Walsh, M. G. (2015). Land health surveillance and response: A framework for evidence-informed land management. *Agricultural Systems, 132*, 93–106. doi:10.1016/j.agsy.2014.09.002

Sibari, T., Haidi, S., & Ait Foura, A. (2001). Typologie des crues et érosion mécanique dans un bassin versant de zone semi-aride: Bassin versant de I'Inaouene, Maroc. Science et changements planétaires. *Sécheresse, 12*(3), 187–193.

Silva, L (2001). Cadastres in the 21st Century (6). A model for the information system of the Portuguese Land Register. *GIM International, the Worldwide Magazine for Geomatics, 15*(6), 61-63.

Silva, M. A., & Stubkjær, E. (2002). A review of methodologies used in research on cadastral development. *Computers, Environment and Urban Systems, 26*(5), 403–423. doi:10.1016/S0198-9715(02)00011-X

Simonneaux, V., Lepage, M., Helson, D., Metral, J., Thomas, S., Duchemin, B., ... Chehbouni, A. (2009). Estimation spatialisée de l'évapotranspiration des cultures irriguées par télédétection: Application à la gestion de l'irrigation dans la plaine du Haouz (Marrakech, Maroc). *Sécheresse, 20*(1), 123–130.

Sithole, G., & Vosselman, G. (2003). Comparison of filtering algorithms. In *Proceedings of the ISPRS working group III/3 workshop* (pp. 71–78). Academic Press.

Skidmore, E. L., Dickerson, J. D., & Shimmelpfennig, H. (1975). Evaluating surface-soil water content by measuring reflectance. *Soil Science Society of American Proceedings, 39*(2), 238–242. doi:10.2136 ssaj1975.03615995003900020009x

Skovsgaard, J. P., & Vanclay, J. K. (2008). Forest site productivity: A review of the evolution of dendrometric concepts for even-aged stands. *Forestry, 81*(1), 13–31. doi:10.1093/forestry/cpm041

Skovsgaard, J. P., & Vanclay, J. K. (2013). Forest site productivity: A review of spatial and temporal variability in natural site conditions. *Forestry, 86*(3), 305–315. doi:10.1093/forestry/cpt010

Söderbergh, I., & Ledermann, T. (2003). Algorithms for simulating thinning and harvesting in five European individual-tree growth simulators: A review. *Computers and Electronics in Agriculture, 39*(2), 115–140. doi:10.1016/S0168-1699(03)00022-X

Song, X., Zhou, X., Li, X., & Li, X. (2009). *Spatial Heterogeneity Analysis of Soil Moisture Based on Geostatistics.* Paper presented at the Information Science and Engineering (ICISE), 2009 1st International Conference on.

Steduto, P., Hsiao, T. C., Raes, D., & Fereres, E. (2009). AquaCrop—The FAO Crop Model to Simulate Yield Response to Water: I. Concepts and Underlying Principles All rights reserved. No part of this periodical may be reproduced or transmitted in any form or by any means, electronic or mechanical, including photocopying, recording, or any information storage and retrieval system, without permission in writing from the publisher. *Agronomy Journal*, *101*(3), 426–437. doi:10.2134/agronj2008.0139s

Stepper, C., Straub, C., & Pretzsch, H. (2015). Assessing height changes in a highly structured forest using regularly acquired aerial image data. *Forestry*, *88*(3), 304–316. doi:10.1093/forestry/cpu050

Steudler, D., Williamson, I., Kaufmann, J., & Grant, D. (1997). Benchmarking cadastral systems. *The Australian Surveyor, 42*(3), 87-107.

Stone, M. (1974). Cross-Validatory Choice and Assessment of Statistical Predictions. *Journal of the Royal Statistical Society. Series B (Methodological), 36*(2), 111-147. doi: 10.2307/2984809

Stoter, J., Munk, E., & Bodum, L. (2004). 3D registration of real property in Danmark. TS25 Appropriate Technologies for good Land Administration II. FIG Working Week 2004, Athens, Greece.

Stoter, J., Oosterom, P. J. M., Ploeger, H.D., & Aalders, H.J.G.L (2004). Conceptual 3D cadastral models applied in several countries. TS25 Appropriate Technologies for good Land Administration II. FIG Working Week 2004, Athens, Greece.

Stoter, J., & Salzmann, M. (2003). Towards a 3D cadastre. where do cadastral needs and technical possibilities meet? *Computers, Environment and Urban Systems*, *27*(4), 395–410. doi:10.1016/S0198-9715(02)00039-X

Strimbu, V., & Strimbu, B. (2015). *A graph-based segmentation algorithm for tree crown extraction using airborne LiDAR data* (Vol. 104). Academic Press. 10.1016/j.isprsjprs.2015.01.018

Su, F., Hong, Y., & Lettenmaier, D. P. (2008). Evaluation of TRMM Multisatellite Precipitation Analysis (TMPA) and Its Utility in Hydrologic Prediction in the La Plata Basin. *Journal of Hydrometeorology*, *9*(4), 622–640. doi:10.1175/2007JHM944.1

Taha, A., & Mualla, W. (2003). *La prédiction du comportement hydrologique du bassin Awage en utilisant le modèle WMS (Watershed Modelling System).* Publi. N° 278, 2003.

Talhofer, V., Hošková-Mayerová, Š., & Hofmann, A. (2012). Improvement of digital geographic data quality. *International Journal of Production Research*, *50*(17), 4846 - 4859.

Tensay Getnet. (2011). *Sedimentation Modeling for Ribb Dam.* Author.

Teshome, A., Graaff, J., Ritsema, C., & Kassie, M. (2016). Farmers' perceptions about the influence of land quality, land fragmentation and tenure systems on sustainable land management in the north western Ethiopian Highlands. *Land Degradation & Development*, *27*(4), 884–898. doi:10.1002/ldr.2298

The organic law 111-14 relative to the regions.

The organic law 112-14 relative to the provinces and prefectures.

The organic law 113-14 relative to the Communities.

The Roy E. (1998). From the appropriation to the ownership: A brief introduction to the land terminology. In *What land policies for rural Africa? Reconcile practices, legitimacy and legality*. Paris: Karthala.

Thies, M., & Spiecker, H. (2004). Evaluation and future prospects of terrestrial laser scanning for standardized forest inventories. *Forest, 2*(2), 1.

Tian, Q., Tong, Q., Pu, R., Guo, X., & Zhao, C. (2001). Spectroscopic determination of wheat water status using 1650-1850 nm spectral absorption features. *International Journal of Remote Sensing*, *22*(12), 2329–2338. doi:10.1080/01431160118199

Tingting, L., & Chuang, L. (2008). Crop classification by support vector machine with intelligently selected training data for an operational application. *International Journal of Remote Sensing*, *29*(8), 2227–2240. doi:10.1080/01431160701395203

Tingting, L., & Chuang, L. (2010). Study on extraction of crop information using time-series MODIS data in the Chao Phraya Basin of Thailand. *Advances in Space Research*, *45*(6), 775–784. doi:10.1016/j.asr.2009.11.013

Tramblay, Y., Badi, W., Driouech, F., Al Adlouni, S., Neppel, L., & Servat, E. (2012). Modélisation des crues dans le bassin du barrage Makhazine. Rapport interne de l'ABH de Loukous.

Tramblay, Y., Thiemig, V., Dezetter, A., & Hanich, L. (2016). Evaluation of satellite-based rainfall products for hydrological modelling in Morocco. *Hydrological Sciences Journal*, *61*(14), 2509–2519. doi:10.1080/02626667.2016.1154149

Trombetti, M., Riaño, D., Rubio, M. A., Cheng, Y. B., & Ustin, S. L. (2008). Multi-temporal vegetation canopy water content retrieval and interpretation using artificial neural networks for the continental USA. *Remote Sensing of Environment*, *112*(1), 203–215. doi:10.1016/j.rse.2007.04.013

Tuomaala, J., & Uimonen, M. (1998). *Introducing the new Object-Oriented Cadastral Information System (JAKO) of Finland.* FIG XXI Congress, commission 7, cadastre and land management, Brighton, UK.

U.S. Army Corps of Engineers (USACE). (2000). *Hydrologic Modeling System: Technical Reference Manual.* Davis, CA: Hydrologic Engineering Center.

UN-Habitat. (2014). *Good Land governance for the 2030 Agenda.* UNGGIM HLF.

United Nations Committee of Experts on Global Geospatial Information Management. (2017). Compendium of good practices for national institutional arrangements. New York: UNGGIM.

United Nations Committee of Experts on Global Geospatial Information Management. (2017). *National Institutional Arrangements: Instruments, Principles and Guidelines*. New York: UNGGIM.

Valancogne, C., & Nasr, Z. (1989). Une méthode de mesure du débit de sève brute dans de petits arbres par bilan de chaleur. *Agronomie*, *9*(6), 609–617. doi:10.1051/agro:19890609

Van der Molen, P. (2002). The dynamic aspects of land administration. an often-forgotten component in system design. *Computers, Environment and Urban Systems*, *26*(5), 361–381. doi:10.1016/S0198-9715(02)00009-1

Van der Molen, P. (2003). Institutional aspects of 3D cadastres. *Computers, Environment and Urban Systems*, *27*(4), 383–394. doi:10.1016/S0198-9715(02)00038-8

Vanclay, J. K. (1994). *Modelling forest growth and yield: applications to mixed tropical forests*. Wallingford, UK: CAB International.

Vancutsem, C., Ceccato, P., Dinku, T., & Connor, S. J. (2010). Evaluation of MODIS land surface temperature data to estimate air temperature in different ecosystems over Africa. *Remote Sensing of Environment*, *114*(2), 449–465. doi:10.1016/j.rse.2009.10.002

Vapnik, V. N. (1995). *The nature of statistical learning theory*. Springer-Verlag New York, Inc. doi:10.1007/978-1-4757-2440-0

Vastaranta, M., Wulder, M. A., White, J. C., Pekkarinen, A., Tuominen, S., Ginzler, C., ... Hyyppä, H. (2013). Airborne laser scanning and digital stereo imagery measures of forest structure: Comparative results and implications to forest mapping and inventory update. *Canadian Journal of Remote Sensing*, *39*(5), 382–395. doi:10.5589/m13-046

Véga, C., & Durrieu, S. (2011). Multi-level filtering segmentation to measure individual tree parameters based on Lidar data: Application to a mountainous forest with heterogeneous stands. *International Journal of Applied Earth Observation and Geoinformation*, *13*(4), 646–656. doi:10.1016/j.jag.2011.04.002

Verhoef, W., & Bach, H. (2003). Remote sensing data assimilation using coupled radiative transfer models. *Physics and Chemistry of the Earth Parts A/B/C*, *28*(1), 3–13. doi:10.1016/S1474-7065(03)00003-2

Viziriel of 26 December 1920 regulations control the employment and re-employment of funds from the expropriation, leases or disposals of perpetual enjoyment of collective lands

Vogt, J. V., Viau, A. A., & Paquet, F. (1997). Mapping regional air temperature fields using satellite-derived surface skin temperatures. *International Journal of Climatology*, *17*(14), 1559–1579. doi:10.1002/(SICI)1097-0088(19971130)17:14<1559::AID-JOC211>3.0.CO;2-5

Vos, W. (1996). *Changing forest functions in NW Europe: From alienation to involvement*. Academic Press.

Waldhoff, G., & Lussem, U. (2016). *Enhanced Land Use Classification 2015 of the Rur Catchment – Update*. CRC/TR32 Database (TR32DB). .10.5880/TR32DB.19

Wang, Y. G., Zhu, H., & Li, Y. (2013). Spatial heterogeneity of soil moisture, microbial biomass carbon and soil respiration at stand scale of an arid scrubland. *Environmental Earth Sciences*, *70*(7), 3217–3224. doi:10.100712665-013-2386-z

Wehenkel, L., Bastin, C., Phillips, C., & Geurts, P. (2017). Tree Ensemble Methods and Parcelling to Identify Brain Areas Related to Alzheimer's Disease. In *2017 International Workshop on Pattern Recognition in Neuroimaging (PRNI), Proceedings*. IEEE.

Weir, A. H., Bragg, P. L., Porter, J. R., & Rayner, J. H. (1984). A winter wheat crop simulation model without water or nutrient limitations. *The Journal of Agricultural Science*, *102*(2), 371–382. doi:10.1017/S0021859600042702

Weiskittel, A. R. (2011). *Forest growth and yield modeling*. Hoboken, NJ: Wiley. doi:10.1002/9781119998518

Weiss, M., Baret, F., Smith, G. J., Jonckheere, I., & Coppin, P. (2004). Review of methods for in situ leaf area index (LAI) determination: Part II. Estimation of LAI, errors and sampling. *Agricultural and Forest Meteorology*, *121*(1), 37–53. doi:10.1016/j.agrformet.2003.08.001

West, P. W., & West, P. W. (2009). *Tree and forest measurement*. Springer. Retrieved from http://link.springer.com/content/pdf/10.1007/978-3-540-95966-3.pdf

White, J. C., Coops, N. C., Wulder, M. A., Vastaranta, M., Hilker, T., & Tompalski, P. (2016). Remote Sensing Technologies for Enhancing Forest Inventories: A Review. *Canadian Journal of Remote Sensing*, *42*(5), 619–641. doi:10.1080/07038992.2016.1207484

White, J. C., Stepper, C., Tompalski, P., Coops, N. C., & Wulder, M. A. (2015). Comparing ALS and image-based point cloud metrics and modelled forest inventory attributes in a complex coastal forest environment. *Forests*, *6*(10), 3704–3732. doi:10.3390/f6103704

Whitlock, C. (2004). Land management: Forests, fires and climate. *Nature*, *432*(7013), 28–29. doi:10.1038/432028a PMID:15525968

Williamson, I (1986). Cadastral and land information systems in developing countries. *The Australian Surveyor, 33*(1), 27-43.

Williamson, I. E., Wallace, J., & Rajabifard, A. (2009). Land administration for sustainable development. ESRI Academic Press.

Williamson, I. (1997). The justification of cadastral systems in developing countries. *Geomatica*, *51*(1), 21–36.

Williams, R. (1989). The EPIC Crop Growth Model. *Transactions of the ASAE. American Society of Agricultural Engineers*, *32*(2), 497. doi:10.13031/2013.31032

Wilson, O. J. (1998). Village renewal and rural development in the former German Democratic Republic. *GeoJournal*, *46*(3), 247–255. doi:10.1023/A:1007028409123

Woods, M., Pitt, D., Penner, M., Lim, K., Nesbitt, D., Etheridge, D., & Treitz, P. (2011). Operational implementation of a LiDAR inventory in Boreal Ontario. *Forestry Chronicle*, *87*(04), 512–528. doi:10.5558/tfc2011-050

World Bank. (2013). *Climate-Induced Spatio-Temporal Shifts in Natural and Agro-ecosystems in the Middle East and North Africa Region*. A synthesis report. World Bank.

Wulder, M. A., Hall, R. J., Coops, N. C., & Franklin, S. E. (2004). High spatial resolution remotely sensed data for ecosystem characterization. *A.I.B.S. Bulletin*, *54*(6), 511–521.

Wulder, M. A., White, J. C., Hay, G. J., & Castilla, G. (2008). Towards automated segmentation of forest inventory polygons on high spatial resolution satellite imagery. *Forestry Chronicle*, *84*(2), 221–230. doi:10.5558/tfc84221-2

Wulder, M. A., White, J. C., Nelson, R. F., Næsset, E., Ørka, H. O., Coops, N. C., ... Gobakken, T. (2012). Lidar sampling for large-area forest characterization: A review. *Remote Sensing of Environment, 121*(Supplement C), 196–209. doi:10.1016/j.rse.2012.02.001

Xie, Y., Sha, Z., & Yu, M. (2008). Remote sensing imagery in vegetation mapping: A review. *Journal of Plant Ecology, 1*(1), 9–23. doi:10.1093/jpe/rtm005

Yaduvanshi, A., Srivastava, P. K., & Pandey, A. C. (2015). Integrating TRMM and MODIS satellite with socio-economic vulnerability for monitoring drought risk over a tropical region of India. *Physics and Chemistry of the Earth Parts A/B/C, 83*, 14–27. doi:10.1016/j.pce.2015.01.006

Yan, G., Mas, J.-F., Maathuis, B. H. P., Xiangmin, Z., & Dijk, P. M. V. (2006). Comparison of pixel-based and object-oriented image classification approaches—a case study in a coal fire area, Wuda, Inner Mongolia, China. *International Journal of Remote Sensing, 27*(18), 4039–4055. doi:10.1080/01431160600702632

Yener, M. K., Şorman, A. Ü., Şorman, A. A., Şensoy, A., & Gezgin, T. (2007). Modeling studies with HEC-HMS and runoff scenarios in Yuvacik basin, *Turkey. International Congress on River Basin Management.*

Zeleke, K. T., Luckett, D., & Cowley, R. (2011). Calibration and Testing of the FAO AquaCrop Model for Canola. *Agronomy Journal, 103*(6), 1610–1618. doi:10.2134/agronj2011.0150

Zhang, A., & Jia, G. (2013). Monitoring meteorological drought in semiarid regions using multi-sensor microwave remote sensing data. *Remote Sensing of Environment, 134*, 12–23. doi:10.1016/j.rse.2013.02.023

Zhang, K., & Whitman, D. (2005). Comparison of three algorithms for filtering airborne lidar data. *Photogrammetric Engineering and Remote Sensing, 71*(3), 313–324. doi:10.14358/PERS.71.3.313

Zhou, W., & Hong-cha, M. A. (2006). Combining methods and diversity measures in multiples classifier systems for thematic classification of landsat TM image. *Journal of Heilongjiang Institute of Technilogy, 20*(4), 14–21.

# About the Contributors

**Moha El-Ayachi**, Professor, School of Geomatic Sciences and Surveying Engineering, holds a Doctorate of Sciences in a Joint Supervision Program of the University of Minnesota, USA and IAV Hassan II, Morocco on Land and Cadastral Information Management. He served as Head of Department of Geodetic Sciences from 2005 to 2011. Enrolled as Licensed Surveyor at the Order of Licensed Surveying Engineers in Morocco (ONIGT) and served as its Vice President during 2018-2020 and as General Secretary during the period 2012-2016. He is leading currently the Network of Excellence on Land Governance in Africa as Coordinator of the Northern Africa Node. He is Chief Editor of African Journal of Geospatial Sciences and Land Governance. He published more than 30 peer and non-peer review papers in the field of land information, land governance, global positioning system, and GIS development. He's involved in many partnership searching programs at the national and international levels. He is currently consultant Educator and Trainer in Land governance enhancement using the new technology at the African Level. He fluently conducts studies in Arabic, English, and French.

**Loubna El Mansouri** is Professor at the School of Geomatic Sciences and Surveying Engineering within the Institute of Agronomy and Veterinary Medicine (IAV) Hassan II, Rabat, Morocco, where she has been since 2004. From 2014 to 2017 she served as Head of Department of Geodetic Sciences. She conducted her Ph.D studies during 2009-2013 at "SURAFCE" Laboratory in the College of Science, University of LIEGE (ULG) in Belgium. From 2010 to 2012 she was a visiting Ph.D. staff at "MONTEFIORE" Institute of Computer Science at ULG. She received an engineering degree in Geomatic and Geodetic Sciences from IAV Hassan 2 in 2004 and enrolled as licensed Surveyor. Holds a Doctorate of Sciences in Remote Sensing from the ULG in 2013. Her research interests cover both digital image processing and remote sensing. Much of her work has been on improving application of machine learning technique in land use/ land cover data producing for territory management, more over in agriculture and forest domain. Professor El Mansouri published more than 30 papers, coordinator of many R&D projects, and co-advisor of Doctorate studies.

* * *

**Idrais Abderrahim** graduated on Survey Engineering from Hassan II Agronomy and Veterinary Institute Bachelor on policy sciences from Mohammed V faculty of Law Student in the Hight School of administration PhD Student in the Hassan II Agronomy and Veterinary Institute.

**Tarik Benabdelouahab** holds a Ph.D. degree in Remote Sensing and modeling from Liège University, Belgium, a soil sciences Engineer's degree from the Agronomic and veterinarian Institute, Morocco. He joined the National Institute of Agricultural Research (INRA) in January 2008. He is experienced in monitoring and management of natural resources using remote sensing and modeling. He has published more ten papers in refereed journals.

**Abdelghani Boudhar** is Professor of Hydrology and Remote Sensing in Sultan Moulay Slimane University and received his PhD degree in water resources modelling and remote sensing from Caddi Ayyad University in Morocco in 2009. His research focuses on the use of remote sensing datasets to simulate hydrological processes. He has published more than twenty papers (principal author or coauthor) in several refereed journals.

**Uchendu E. Chigbu** is a Nigerian. He is a lecturer and researcher at the Faculty of Civil Geo and Environmental Engineering, at the Technical University of Munich. His primary research areas are best described as Social Geodesy, focusing on how to ensure social transformations through land management, land-use planning, land tenure and land policy/governance improvements. His recent works involve activities in Africa, Asia, and South America. He is also the Coordinator of the Ph.D. Program in Land Management at the Technical University of Munich, Germany.

**Nadia Eddaif** is a Ph.D. student at Med V university in Rabat. She works on Lidar data applied to forest inventory. Sher received her Master degree in Remote sensing and GIS in 2015.

**Rachid Hadria** received his PhD degree in crop modelling and remote sensing from Caddi Ayyad University in Morocco in 2006 and is now working as a research scientist at the National Institute of Agronomic Research. His research includes the combination of remote sensing, GIS and crop modeling to manage natural resources in arid zones. He has published more than twenty papers (principal author or coauthor) in refereed journals.

**Yidnekachew Jember** is a Lecturer in Woldia University. During his 8-year tenure, Yidnekachew get good experience in land administration and management and environmental issues. Before this role, Yidnekachew was an expert in Land Administration and Use Department., where he supervises the activity of ten District offices. Yidnekachew holds a master's degree in Land Administration and Management specialization in Land Information Management System from Bahir Dar University, Ethiopia.

**Abdellatif Khattabi** is an agricultural engineer of the Agronomic and Veterinary Institute Hassan II, Rabat and Ecole Supérieure du Bois, Paris France. Holds a Master of Science in Economics and a Ph.D. in Forestry from the University of Idaho, USA, and a Master of Science in NTIC from the University of Louis Pasteur, Strasbourg, France. Professor at the National Forestry School of Engineering since 1994, and visiting professor at Hassan II University, Casablanca, Mohamed V University, Rabat, Moulay Ismail University, Meknes and Abdelmalek Saadi University, Tetouan. Associate Researcher at the Royal Institute for Strategic Studies (IRES), Rabat, in his program on climate change, and associate researcher at the Institute of Labor (IZA) in Germany. From 1981 to 1994, researcher at the National

Center for Forest Research, Rabat. Author of numerous publications, recipient of numerous research grants and consultant for public and private, national and international organizations (UNESCO, ISESCO, UNDP, UNEP, FAO, World Bank, GIZ, Overseas Development Institute London [ODI]), member of the Fulbright Alumni Association and several professional associations, and lead author of Chapter 5 (Coastal zones and low laying areas) of the IPCC Fifth Report. Founding member and president of the Moroccan Association of Regional Sciences (AMSR).

**Said Lahssini** holds Forest Engineer Diploma from National School of Forestry Engineering (Morocco) and PhD from UCL (Belgium). After working for Moroccan Forestry Department for more than 12 years, he become Professor at National School of Forest Engineering. He occupied many responsibilities and lead many projects and was actively involved in numerous forest inventory and management studies.

**Hayat Lionboui** holds an AgroEconomist Engineer's degree from the Agronomic and veterinarian Institute, Morocco. He joined the National Institute of Agricultural Research (INRA) in January 2009. She is experienced in monitoring and management of water resources using modeling. The author has published many papers in refereed journals.

**Adad Moulay Abdeslam** is a researcher from the Graduate School, Institute of Agronomy and Veterinary Medicine Hassan II, Rabat, Morocco Head of The Department of the Production in Land Registration in the National Agency of Land Registry, Cadastre and Cartography. ANCFCC. Rabat Surveying Engineer at the ANCFCC.

**Hassan Rhinane**, Professor at the University Hassan II, Casablanca, is an expert in Geographic Information Systems (GIS), remote sensing, and in environmental sciences. During the last decade, he supervised many PhD researches in his laboratory. Since 2016 he has held the position of Director of GIS and Land Management Research Center affiliated to the Hassan II University of Casablanca. He is also author of many papers published in indexed journals.

**Nabil Rifai** has a PhD in Biological Sciences from the University of Hassan II (Casablanca, Morocco) and Diploma of Forest Engineer from the National School of Forestry Engineering – in Forest Management Planning - (Morocco). Forest engineer at the High Commission on Water, Forests and Combating against Desertification (Morocco) since 2008. Contribute in forest management planning and forest-rangeland studies in different ecosystems. Coordinator of the national programs for regeneration and sylviculture in forests and rang-lands. Leading the development of a national strategy on sylvo-pastoralisme.

**Bernard Tychon** has a PhD in Environmental Sciences from the Luxembourg University Foundation (1993) and Agricultural Engineer graduated in 1987 from the Catholic University of Louvain, Bernard Tychon leads a research group in agrometeorology. The group has developed a series of forecasting tools for farmers. In addition to implementing these tools in several European, African and Asian countries, Bernard Tychon. He is a full lecturer in agrometeorology, data processing and modeling, remote sensing and geographic information systems, diffuse pollution. He has published more than a hundred scientific papers. Finally, he was awarded the Outstanding Young Person Award of the Belgian Junior Economical Chamber in 2004 and since 2004 has been a Belgian representative at the European Society of Agronomy.

# Index

## T



## U



## W

www.ingramcontent.com/pod-product-compliance
Lightning Source LLC
LaVergne TN
LVHW080536300926
849162LV00007B/1621
* 9 7 8 1 5 2 2 5 5 9 3 9 9 *